SOLUTIONS MANUAL

for the

MECHANICAL ENGINEERING REVIEW MANUAL

Michael R. Lindeburg, P.E.

PROFESSIONAL PUBLICATIONS, INC.
San Carlos, CA 94070

In the *ENGINEERING REVIEW MANUAL SERIES*

Engineer-In-Training Review Manual
Quick Reference Cards for the E-I-T Exam
Mini-Exams for the E-I-T Exam
Civil Engineering Review Manual
Seismic Design for the Civil P.E. Exam
Timber Design for the Civil P.E. Exam
Mechanical Engineering Review Manual
Electrical Engineering Review Manual
Chemical Engineering Review Manual
Chemical Engineering Practice Exam Set
Expanded Interest Tables
Engineering Law, Design Liability, and Professional Ethics

Distributed by: Professional Publications, Inc.
Post Office Box 199
Department 77
San Carlos, CA 94070
(415) 593-9119

SOLUTIONS MANUAL
for the
MECHANICAL ENGINEERING REVIEW MANUAL

Printed in the United States of America

ISBN 0-932276-44-X

Professional Engineering Registration Program
Post Office Box 911, San Carlos, CA 94070

TABLES OF CONTENTS

PROFESSIONAL ENGINEERING REGISTRATION PROGRAM • P.O. Box 911, San Carlos, CA 94070

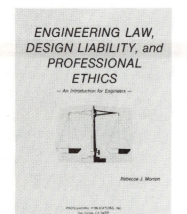

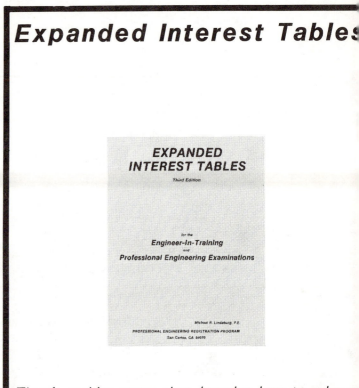

Mathematics

WARM-UPS

1 $\sum_{J=1}^{5}(J+1)^2-1 = (1+1)^2-1 + (2+1)^2-1 + (3+1)^2-1 + (4+1)^2-1 + (5+1)^2-1$

$= 2^2+3^2+4^2+5^2+6^2-5 = 85$

2 THE ACTUAL VALUE IS

$y(2.7) = 3(2.7)^{.93} + 4.2 = 11.756$

$y(2) = 3(2)^{.93} + 4.2 = 9.916$

$y(3) = 3(3)^{.93} + 4.2 = 12.534$

THE ESTIMATED VALUE IS

$9.916 + .7(12.534 - 9.916) = 11.749$

THE ERROR IS

$\frac{11.756 - 11.749}{11.756} = .0006$ OR $.06\%$

3 LET d BE THE DIAMETER

$V_{SPHERE} = \frac{4}{3}\pi r^3 = \frac{4}{3}\pi\left(\frac{d}{2}\right)^3 = .524 d^3$

$V_{CONE} = \frac{\pi}{3} r^2 h = \frac{\pi}{3}\left(\frac{d}{2}\right)^2 h = .262 d^2 h$

BUT $.524 d^3 = .262 d^2 h$

$h = 2.00 d$

4

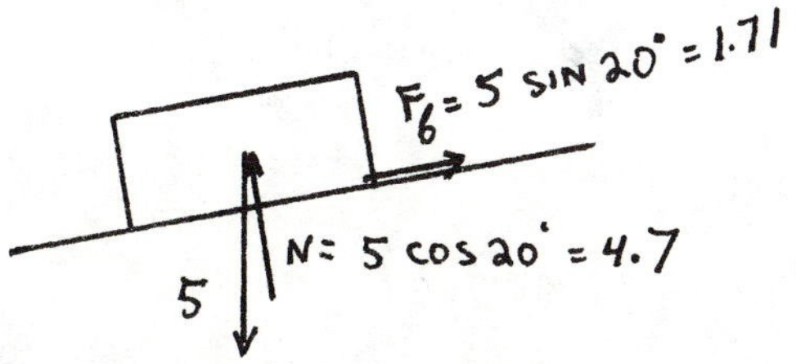

5 EXPAND BY 2ND COLUMN

$-2\begin{vmatrix} 4 & 3 \\ 9 & 5 \end{vmatrix} = -2(20-27) = 14$

6 FROM EQN 6.3

$250° + 460° = 710°R$

$\frac{5}{9}(250-32) = 121.1°C$

7 FROM PAGE 1-42,

$K = 1.71 \text{ EE-}9 \frac{BTU}{FT^2\text{-}HR\text{-}°R^4}$

$$\frac{(1.71 \text{ EE-}9)\frac{BTU}{FT^2\text{-}HR\text{-}°R^4}(17.57)\frac{WATT\text{-}MIN}{BTU}\left(\frac{1}{60}\right)\frac{HR}{MIN}}{(.3048)^2 \frac{M^2}{FT^2}\left(\frac{5}{9}\right)^4 K^4/°R^4}$$

$= 5.66 \text{ EE-}8 \frac{WATTS}{M^2\text{-}°K^4}$

8 $y = 6 + .75(2-6) = 3.0$

9 THE SLOPE IS $\frac{9.5-3.4}{8.3-1.7} = .924$

USING THE FIRST POINT,

$(y - 3.4) = .924(x - 1.7)$

10 LET x BE THE NUMBER OF ELAPSED PERIODS OF .1 SECOND. LET Y_x BE THE AMOUNT PRESENT AFTER x PERIODS

$Y_1 = 1.001\, Y_0$

$Y_2 = (1.001)^2 Y_0$

$Y_N = (1.001)^N Y_0$

NOW $\frac{Y_x}{Y_0} = 2 = (1.001)^N$

$\log(2) = N \log(1.001)$

$N = 693.5$ PERIODS

$t = 69.35$ SECONDS

CONCENTRATES

1 FIRST, REARRANGE

$x + y = -4$

$x + z = 1$

$3x - y + 2z = 4$

NOW, USE CRAMER'S RULE (PAGE 1-6)

$\begin{vmatrix} 1 & 1 & 0 \\ 1 & 0 & 1 \\ 3 & -1 & 2 \end{vmatrix} = 1\begin{vmatrix} 0 & 1 \\ -1 & 2 \end{vmatrix} - 1\begin{vmatrix} 1 & 1 \\ 3 & 2 \end{vmatrix} = (0+1) - (2-3)$

$= 1 + 1 = 2$

$\begin{vmatrix} -4 & 1 & 0 \\ 1 & 0 & 1 \\ 4 & -1 & 2 \end{vmatrix} = -2 \quad \begin{vmatrix} 1 & -4 & 0 \\ 1 & 1 & 1 \\ 3 & 4 & 2 \end{vmatrix} = -6 \quad \begin{vmatrix} 1 & 1 & -4 \\ 1 & 0 & 1 \\ 3 & -1 & 4 \end{vmatrix} = 4$

$x^* = \frac{-2}{2} = -1 \quad y^* = \frac{-6}{2} = -3 \quad z^* = \frac{4}{2} = 2$

2 ALWAYS GRAPH THE DATA FIRST TO SEE IF IT IS A STRAIGHT LINE. IN THIS CASE, IT IS.

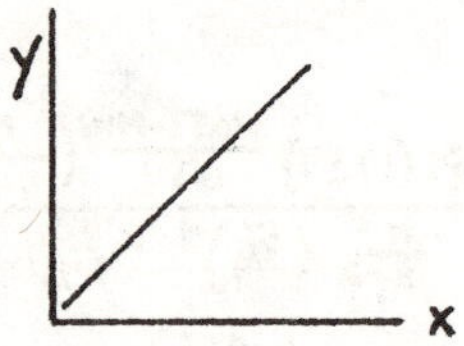

USE LINEAR REGRESSION (PAGE 1-13)

$N = 7$

$\Sigma x = 12550$ $\quad \Sigma y = 12300$

$\bar{x} = 1792.9$ $\quad \bar{y} = 1757.1$

$\Sigma x^2 = 3.117 \text{ EE7}$ $\quad \Sigma y^2 = 3.017 \text{ EE7}$

$(\Sigma x)^2 = 1.575 \text{ EE8}$ $\quad (\Sigma y)^2 = 1.513 \text{ EE8}$

$\Sigma xy = 3.067 \text{ EE7}$

$$m = \frac{7(3.067 \text{ EE7}) - (12550)(12300)}{7(3.117 \text{ EE7}) - (12550)^2} = .994$$

$$b = 1757.1 - .994(1792.9) = -25.0$$

so $y = .994x - 25.0$

THE CORRELATION COEFFICIENT IS

$$r = \frac{7(3.067 \text{ EE7}) - (12550)(12300)}{\sqrt{\left[(7)(3.117 \text{ EE7}) - (12550)^2\right]\left[(7)(3.017 \text{ EE7}) - (12300)^2\right]}}$$

$$= \sim 1.00$$

3 UPON GRAPHING THE DATA, WE SEE THAT IT IS NOT A STRAIGHT LINE

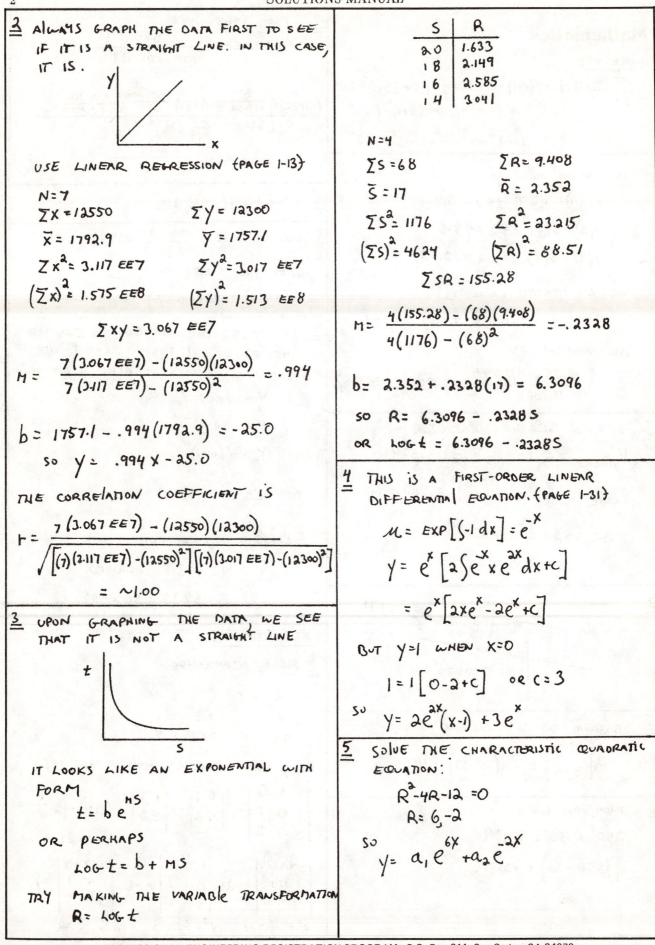

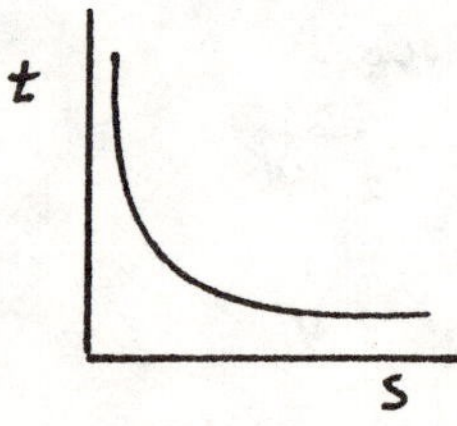

IT LOOKS LIKE AN EXPONENTIAL WITH FORM

$$t = be^{ms}$$

OR PERHAPS

$$\log t = b + ms$$

TRY MAKING THE VARIABLE TRANSFORMATION $R = \log t$

S	R
20	1.633
18	2.149
16	2.585
14	3.041

$N = 4$

$\Sigma S = 68$ $\quad \Sigma R = 9.408$

$\bar{S} = 17$ $\quad \bar{R} = 2.352$

$\Sigma S^2 = 1176$ $\quad \Sigma R^2 = 23.215$

$(\Sigma S)^2 = 4624$ $\quad (\Sigma R)^2 = 88.51$

$\Sigma SR = 155.28$

$$m = \frac{4(155.28) - (68)(9.408)}{4(1176) - (68)^2} = -.2328$$

$$b = 2.352 + .2328(17) = 6.3096$$

so $R = 6.3096 - .2328S$

or $\log t = 6.3096 - .2328S$

4 THIS IS A FIRST-ORDER LINEAR DIFFERENTIAL EQUATION. (PAGE 1-31)

$$\mu = \text{EXP}\left[\int -1\,dx\right] = e^{-x}$$

$$y = e^{x}\left[2\int e^{-x} x e^{2x}\,dx + c\right]$$

$$= e^{x}\left[2xe^{x} - 2e^{x} + c\right]$$

BUT $y = 1$ WHEN $x = 0$

$1 = 1\left[0 - 2 + c\right]$ or $c = 3$

so

$$y = 2e^{2x}(x-1) + 3e^{x}$$

5 SOLVE THE CHARACTERISTIC QUADRATIC EQUATION:

$$R^2 - 4R - 12 = 0$$

$$R = 6, -2$$

so

$$y = a_1 e^{6x} + a_2 e^{-2x}$$

6 LET X_t = POUNDS OF SALT IN TANK AT TIME t

$X_0 = 60$

X' = RATE AT WHICH SALT CONTENT CHANGES

2 = POUNDS OF SALT ENTERING EACH MINUTE

3 = GALLONS LEAVING EACH MINUTE.

THE SALT LEAVING EACH MINUTE IS

$$3\left(\begin{matrix}\text{CONCENTRATION}\\ \text{IN LB/GAL}\end{matrix}\right) = 3\left(\frac{\text{SALT CONTENT}}{\text{VOLUME}}\right) = 3\left(\frac{X}{100-t}\right)$$

$$X' = 2 - 3\left(\frac{X}{100-t}\right)$$

OR

$$X' + \frac{3X}{100-t} = 2$$

THIS IS FIRST ORDER LINEAR (PAGE 1-31)

$$\mu = \text{EXP}\left[3\int\frac{dt}{100-t}\right] = (100-t)^{-3}$$

$$X = (100-t)^3\left[2\int\frac{dt}{(100-t)^3} + K\right]$$

$$= 100 - t + K(100-t)^3$$

BUT $X = 60$ AT $t = 0$

SO $K = -.00004$

$$X = 100 - t - .00004(100-t)^3$$

$X_{60} = 37.44$ POUNDS

7 IF C IS POSITIVE, THEN $N(\infty) = \infty$, WHICH IS CONTRARY TO THE GIVEN DATA. SO $C \le 0$.

IF $C = 0$, THEN $N(\infty) = \dfrac{a}{1+b} = 100$

WHICH IS POSSIBLE DEPENDING ON a, b

IF $C = 0$, THEN $N(0) = \dfrac{a}{1+b} = 10$, WHICH CONFLICTS WITH THE PREVIOUS STEP.

SO $C < 0$. THEN $N(\infty) = a$, SO $\underline{\underline{a = 100}}$

NOW $N(0) = \dfrac{a}{1+b} = \dfrac{100}{1+b} = 10$, SO $\underline{\underline{b = 9}}$

$$\frac{dN}{dt} = -100(1+9e^{ct})^{-2}(9)e^{ct}(c)$$

IF $t = 0$, THEN $C = \underline{\underline{-.0556}}$

8

$$\frac{dy}{dx} = 3X^2 - 18X$$

$3X^2 - 18X = 0$ AT ALL EXTREME POINTS

$X^2 - 6X = 0$ AT $X = 0$, $X = 6$

$$\frac{d^2Y}{dX^2} = 6X - 18$$

$6X - 18 = 0$ AT INFLECTION POINTS

$X = 3$ IS AN INFLECTION POINT

$6(0) - 18 = -18$, SO $X = 0$ IS A MAXIMUM

$6(6) - 18 = 18$, SO $X = 6$ IS A MINIMUM

9 THE ENERGY CONTAINED IN ONE GRAM OF ANY SUBSTANCE IS

$$E = MC^2 = (.001)\text{ KG }(3\text{ EE8})^2(\text{M/S})^2$$

$= 9$ EE13 JOULES

$$(9\text{ EE13})\text{ J}\left(\frac{1}{1000}\right)\frac{\text{KJ}}{\text{J}}(.9478)\frac{\text{BTU}}{\text{KJ}}$$

$= 8.53$ EE10 BTU

$$\#\text{ TONS} = \frac{8.53\text{ EE10 BTU}}{(13{,}000)\frac{\text{BTU}}{\text{LB}}(2000)\frac{\text{LB}}{\text{TON}}}$$

$= 3281$ TONS

10

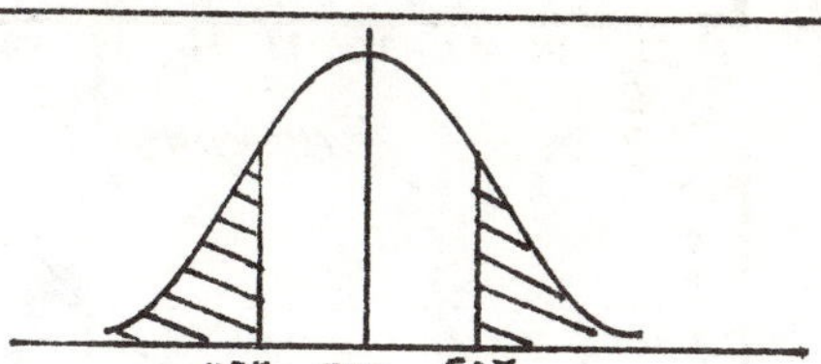

THE STANDARD NORMAL VARIABLES ARE

$$z_1 = \frac{.502 - .497}{.005} = 1$$

$$z_2 = \frac{.507 - .502}{.005} = 1$$

a) $p\{\text{DEFECTIVE}\} = 2[.5 - .3413] = .3174$

b) $p\{3,2\} = \dfrac{3!}{(3-2)!\,2!}(.3174)^2(1-.3174)^1 = .2063$

c) $(8)(200)(.3174) = 507.8$

11

THE RANGE OF SPEEDS IS $(48-20)=28$. SINCE THERE ARE NOT A LOT OF OBSERVATIONS, 10 CELLS WOULD BE BEST. CHOOSE THE CELL WIDTH AS $\left(\frac{28}{10}\right) \approx 3$

INTERVAL	MID-POINT	FREQ.	CUM FREQ.	CUM %
20-22	21	1	1	.03
23-25	24	3	4	.10
26-28	27	5	9	.23
29-31	30	8	17	.43
32-34	33	3	20	.50
35-37	36	4	24	.60
38-40	39	3	27	.68
41-43	42	8	35	.88
44-46	45	3	38	.95
47-49	48	2	40	1.00

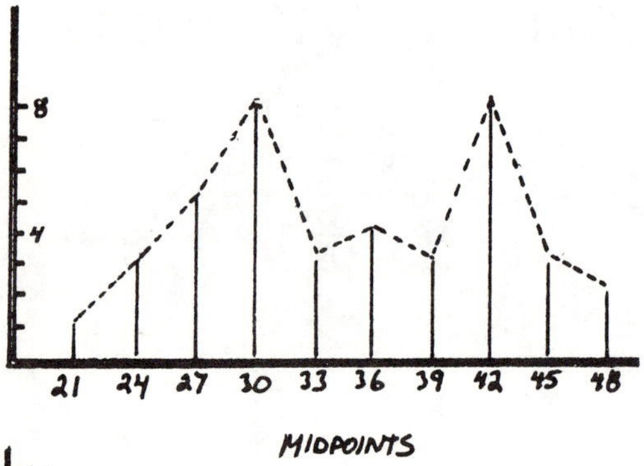

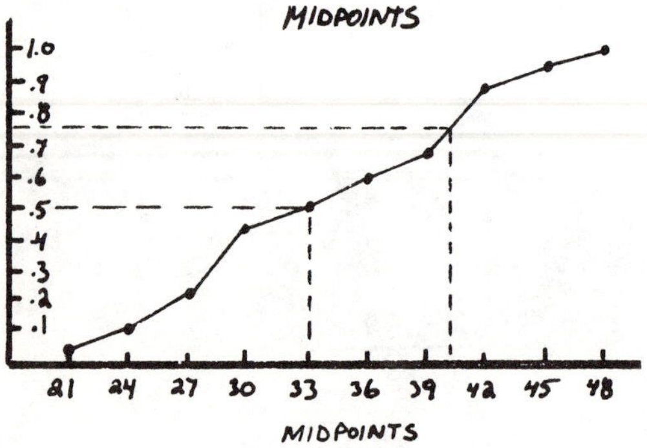

f) USE THE CUMULATIVE DISTRIBUTION GRAPH. FOR 75% (.75), THE CELL MID-POINT IS APPROXIMATELY 40

g) USE THE CUMULATIVE GRAPH TO FIND THE MID-POINT FOR 50%. THIS OCCURS AT APPROXIMATELY 33.

$\Sigma x_i = 1390$, SO MEAN $= \frac{1390}{40} = 34.75$

h) $\Sigma x^2 = 50496$

$$\sigma = \sqrt{\left(\frac{50496}{40}\right) - \left(\frac{1390}{40}\right)^2} = 7.405$$

i) $s = \sqrt{\frac{N}{N-1}}(\sigma) = \sqrt{\frac{40}{39}}(7.405) = 7.500$

j) $s^2 = 56.24$

12 NO CONTRACT DEADLINE WAS GIVEN, SO ASSUME 36 AS A SCHEDULED TIME. LOOK FOR A PATH WHERE $(LS-ES)=0$ EVERYWHERE

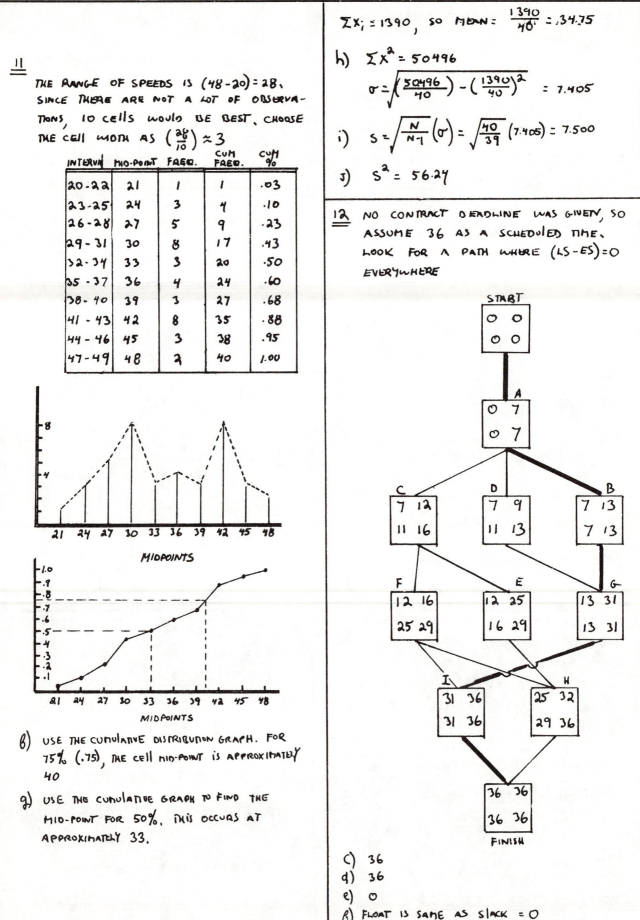

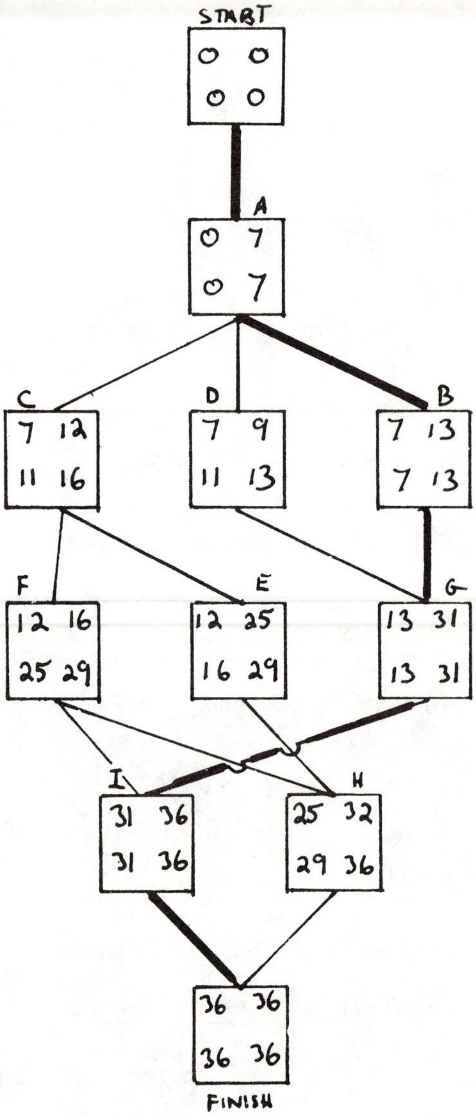

c) 36

d) 36

e) 0

f) FLOAT IS SAME AS SLACK = 0

13 TO SOLVE THIS AS A REGULAR CPM PROBLEM, IT IS NECESSARY TO CALCULATE t_{MEAN} AND σ FOR EACH ACTIVITY. FOR ACTIVITY A,

$$t_{MEAN} = \frac{1}{6}\left[1 + (4)(2) + 5\right] = 2.33$$

$$\sigma_A = \frac{1}{6}(5-1) = .67$$

THE FOLLOWING TABLE IS GENERATED IN THE SAME MANNER.

ACTIVITY	t_{MEAN}	σ
START	0	0
A	2.33	.67
B	10.5	2.17
C	11.83	2.17
D	4.17	.83
FINISH	0	0
	28.83	

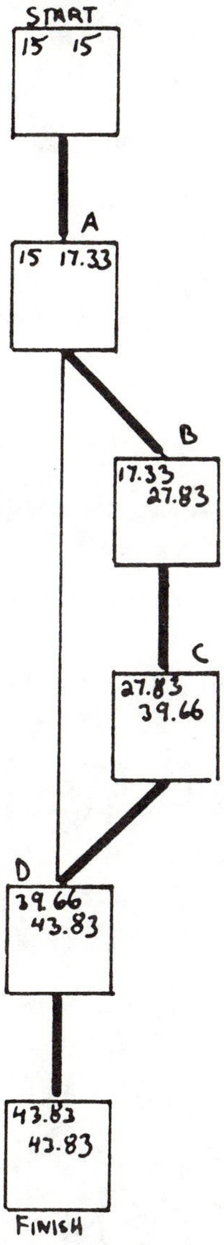

BY OBSERVATION, THE CRITICAL PATH IS START-A-B-C-D-FINISH, THE PROJECT VARIANCE IS

$$\sigma^2 = (.67)^2 + (2.17)^2 + (2.17)^2 + (.83)^2 = 10.56$$

THE PROJECT STANDARD DEVIATION IS

$$\sigma = \sqrt{10.56} = 3.25$$

SO, WE ASSUME THE COMPLETION TIMES ARE NORMALLY DISTRIBUTED WITH A MEAN OF 28.83 AND A STANDARD DEVIATION OF 3.25

THE STANDARD NORMAL VARIABLE IS

$$z = \frac{28.83 + 15 - 42}{3.25} = .56$$

AREA UNDER TAIL FOR $z = .56$ IS .2123
SO - $.5 - .2123 = .2877$ (28.77%)

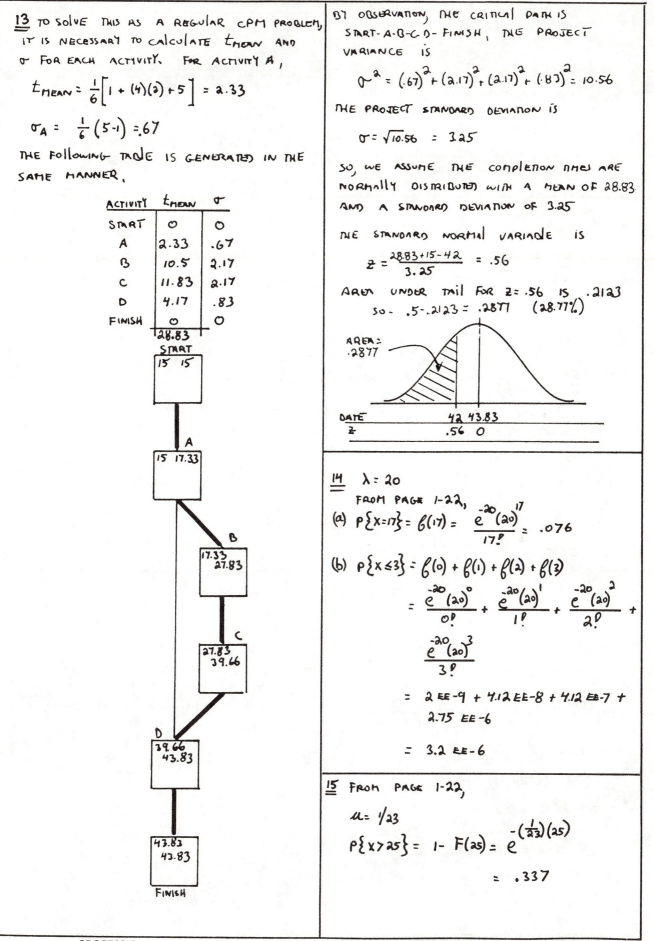

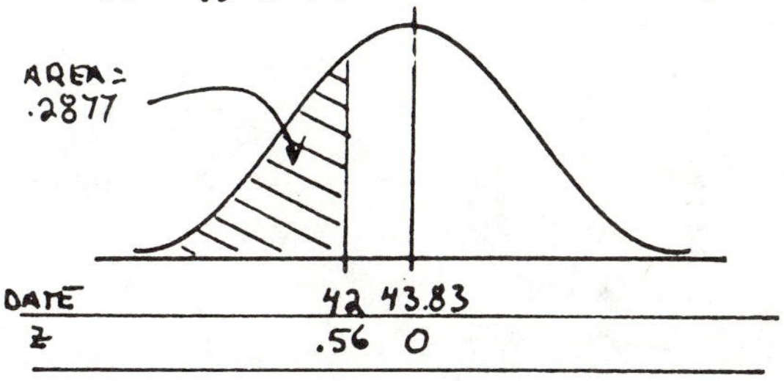

14 $\lambda = 20$

FROM PAGE 1-22,

(a) $p\{x=17\} = f(17) = \dfrac{e^{-20}(20)^{17}}{17!} = .076$

(b) $p\{x \le 3\} = f(0) + f(1) + f(2) + f(3)$

$$= \frac{e^{-20}(20)^0}{0!} + \frac{e^{-20}(20)^1}{1!} + \frac{e^{-20}(20)^2}{2!} + \frac{e^{-20}(20)^3}{3!}$$

$= 2\text{ EE-9} + 4.12\text{ EE-8} + 4.12\text{ EE-7} + 2.75\text{ EE-6}$

$= 3.2\text{ EE-6}$

15 FROM PAGE 1-22,

$\mu = 1/23$

$$p\{x > 25\} = 1 - F(25) = e^{-\left(\frac{1}{23}\right)(25)}$$

$= .337$

16 FIRST, PLOT THE DATA TO SEE IF IT IS LINEAR

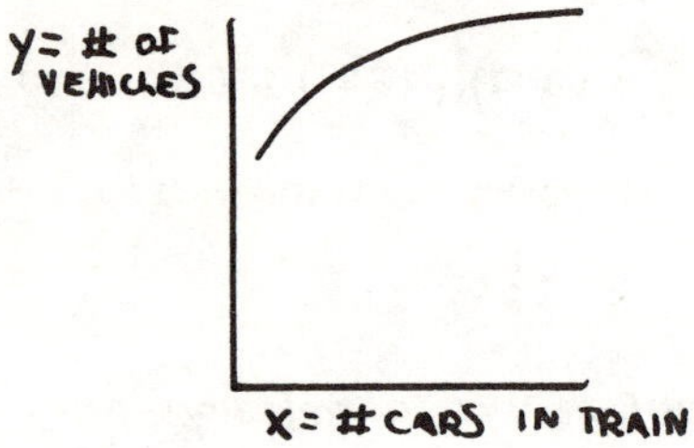

IT DOESN'T LOOK LINEAR, SO TRY THE FORM

$y = a + bz$

WHERE $Z = \log_{10} X$

Z	Y
.3	14.8
.7	18.0
.9	20.4
1.08	23.0
1.43	29.9

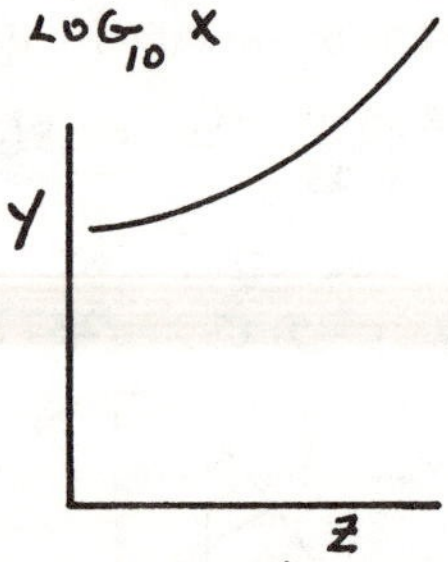

THAT ISN'T LINEAR EITHER. WE'VE 'OVERCOMPENSATED' THE CURVE. TRY THE FORM

$y = a + bw$

WHERE $w = z^2 = (\log_{10} x)^2$

W	Y
.09	14.8
.49	18.0
.81	20.4
1.17	23.0
2.04	29.9

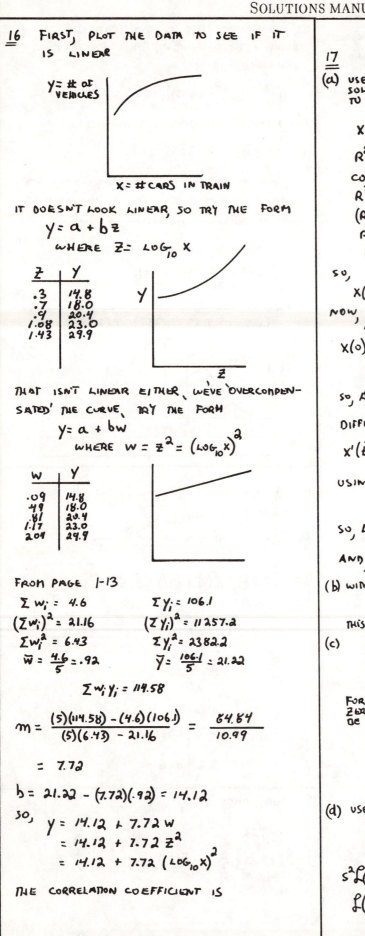

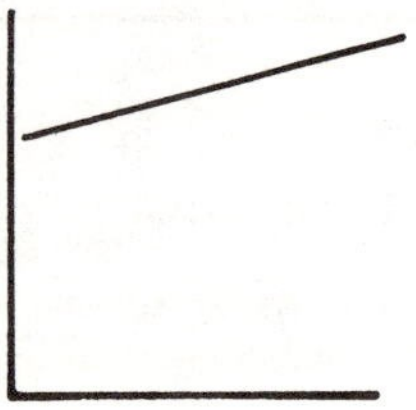

FROM PAGE 1-13

$\Sigma w_i = 4.6$ $\qquad \Sigma y_i = 106.1$

$(\Sigma w_i)^2 = 21.16$ $\qquad (\Sigma y_i)^2 = 11257.2$

$\Sigma w_i^2 = 6.43$ $\qquad \Sigma y_i^2 = 2382.2$

$\bar{w} = \frac{4.6}{5} = .92$ $\qquad \bar{y} = \frac{106.1}{5} = 21.22$

$$\Sigma w_i y_i = 114.58$$

$$m = \frac{(5)(114.58) - (4.6)(106.1)}{(5)(6.43) - 21.16} = \frac{64.84}{10.99}$$

$$= 7.72$$

$$b = 21.22 - (7.72)(.92) = 14.12$$

SO,

$$y = 14.12 + 7.72\,w$$
$$= 14.12 + 7.72\,z^2$$
$$= 14.12 + 7.72\,(\log_{10} x)^2$$

THE CORRELATION COEFFICIENT IS

17

(a) USE THE 'CHARACTERISTIC EQUATION' METHOD TO SOLVE THE HOMOGENEOUS CASE. (IT IS MUCH QUICKER TO USE LAPLACE TRANSFORMS, HOWEVER)

$x'' + 2x' + 2x = 0$ DIFF. EQ.

$R^2 + 2R + 2 = 0$ CHARACTERISTIC EQ.

COMPLETE THE SQUARE TO FIND R

$R^2 + 2R = -2$

$(R+1)^2 = -2 + 1$

$R + 1 = \pm\sqrt{-1}$

$R = -1 \pm i$

SO,

$$x(t) = A_1 e^{-t} \cos t + A_2 e^{-t} \sin t$$

NOW, USE THE INITIAL CONDITIONS TO FIND A_1 AND A_2.

$x(0) = 0$

$$0 = A_1(1)(1) + A_2(1)(0)$$

SO, $A_1 = 0$

DIFFERENTIATING THE SOLUTION,

$$x'(t) = A_2\left[e^{-x}\cos x - \sin x\, e^{-x}\right]$$

USING $x'(0) = 1$

$$1 = A_2\left[(1)(1) - (0)(1)\right]$$

SO, $A_2 = 1$

AND, THE SOLUTION IS $x(t) = e^{-t} \sin t$

(b) WITH NO DAMPING, THE DIFFERENTIAL EQUATION WOULD BE

$$x'' + 2x = 0$$

THIS HAS A SOLUTION OF $x = \sin\sqrt{2}\,t$, SO $\omega_{nat} = \sqrt{2}$

(c) $x(t) = e^{-t} \sin t$

$x'(t) = e^{-t}\cos t - \sin t\, e^{-t}$

$= e^{-t}(\cos t - \sin t)$

FOR x TO BE MAXIMUM, $x'(t) = 0$. SINCE e^{-t} IS NOT ZERO UNLESS t IS VERY LARGE, $\cos t - \sin t$ MUST BE ZERO. THIS OCCURS AT $t = .785$ RADIANS, SO

$$x(.785) = e^{-.785} \sin(.785)$$
$$= .322$$

(d) USE THE LAPLACE TRANSFORM METHOD

$$x'' + 2x' + 2x = \sin(t)$$

$$\mathcal{L}(x'') + 2\mathcal{L}(x') + 2\mathcal{L}(x) = \mathcal{L}(\sin(t))$$

$$s^2\mathcal{L}(x) - 1 + 2s\mathcal{L}(x) + 2\mathcal{L}(x) = \frac{1}{s^2+1}$$

$$\mathcal{L}(x)\left[s^2 + 2s + 2\right] - 1 = \frac{1}{s^2+1}$$

$$\mathcal{L}(x) = \frac{1}{s^2+2s+2} + \frac{1}{(s^2+1)(s^2+2s+2)}$$

$$= \frac{1}{(s+1)^2+1} + \frac{1}{(s^2+1)(s^2+2s+2)}$$

USE PARTIAL FRACTIONS TO EXPAND THE SECOND TERM:

$$\frac{1}{(s^2+1)(s^2+2s+2)} = \frac{A_1+B_1 s}{s^2+1} + \frac{A_2+B_2 s}{s^2+2s+2}$$

CROSS MULTIPLYING,

$$= \frac{A_1 s^2+2A_1 s+2A_1+B_1 s^3+2B_1 s^2+2B_1 s+A_2 s^2+A_2+B_2 s^3+B_2 s}{(s^2+1)(s^2+2s+2)}$$

$$= \frac{s^3[B_1+B_2]+s^2[A_1+A_2+2B_1]+s[2A_1+2B_1+B_2]+2A_1+A_2}{(s^2+1)(s^2+2s+2)}$$

COMPARING NUMERATORS, WE OBTAIN THE FOLLOWING 4 SIMULTANEOUS EQUATIONS:

$$\begin{aligned} B_1 + B_2 &= 0 \\ A_1 + A_2 + 2B_1 &= 0 \\ 2A_1 + 2B_1 + B_2 &= 0 \\ 2A_1 + A_2 &= 1 \end{aligned}$$

USE CRAMER'S RULE TO FIND A_1

$$A_1 = \frac{\begin{vmatrix} 0 & 0 & 1 & 1 \\ 0 & 1 & 2 & 0 \\ 0 & 0 & 2 & 1 \\ 1 & 1 & 0 & 0 \end{vmatrix}}{\begin{vmatrix} 0 & 0 & 1 & 1 \\ 1 & 1 & 2 & 0 \\ 2 & 0 & 2 & 1 \\ 2 & 1 & 0 & 0 \end{vmatrix}} = \frac{-1}{-5} = 1/5$$

SO, THE REST OF THE COEFFICIENTS FOLLOW EASILY FROM THE EQUATIONS:

$$A_1 = 1/5$$
$$A_2 = 3/5$$
$$B_1 = -2/5$$
$$B_2 = 2/5$$

THEN,

$$\mathcal{L}(x) = \frac{1}{(s+1)^2+1} + \frac{1/5}{s^2+1} + \frac{-2/5\, s}{s^2+1} + \frac{3/5}{s^2+2s+2} + \frac{2/5\, s}{s^2+2s+2}$$

TAKING THE INVERSE TRANSFORM,

$$x = \mathcal{L}^{-1}\{\mathcal{L}(x)\} = e^{-t}\sin t + \tfrac{1}{5}\sin t - \tfrac{2}{5}\cos t + 3/5\, e^{-t}\sin t - \tfrac{2}{5}\left[e^{-t}\cos t - e^{-t}\sin t\right]$$

$$= 2e^{-t}\sin t - \tfrac{2}{5}e^{-t}\cos t + \tfrac{1}{5}\sin t - \tfrac{2}{5}\cos t$$

TIMED

1. This is a typical hypothesis test of 2 population means. The two populations are the original population from which the manufacturer got his 1600 hour average life value and the new population from which the sample was taken. We know the mean (x = 1520 hours) of the sample and its standard deviation (s = 120 hours), but we do not know the mean and standard deviation of a population of average lifetimes. Therefore, we assume that

a) the average lifetime population mean and the sample mean are identical ($x = u_{\bar{x}}$ = 1520 hours.)

b) the standard deviation of the average lifetime population is

$$\sigma_{\bar{x}} = \frac{s}{\sqrt{n}} = \frac{120}{\sqrt{100}} = 12 \qquad \{\text{EQN } 1.220\}$$

The manufacturer can be reasonably sure that his claim of a 1600 hour average life is justified if the average test life is near 1600 hours. 'Reasonably sure' must be evaluated based on an acceptable probability of being incorrect. If he is willing to be wrong with a 5% probability, then a 95% confidence level is required.

Since the direction of bias is known, a one-tailed test is required. We want to know if the mean has shifted downward. We test this by seeing if 1600 hours is within the 95% limits of a distribution with a mean of 1520 hours and a standard deviation of 12 hours. From page 1-29, 5% of a standard normal population is outside of z = 1.645. The 95% confidence limit is, therefore,

$$1520 + 1.645(12) = 1540$$

The manufacturer can be 95% certain that the average lifetime of his bearings is less than 1600 since 1600 is not between 1520 and 1540.

If the manufacturer is willing to be wrong with a probability of only 1%, then a 99% confidence limit is required. From page 1-24, z = 2.33 and the 99% confidence limit is

$$1520 + 2.33(12) = 1548$$

The manufacturer can be 99% certain that the average bearing life is less than 1600 hours.

2

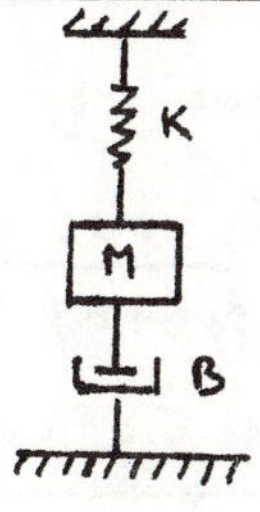

$$M = \frac{8 \text{ LBM}}{32.2 \frac{\text{FT}}{\text{SEC}^2}} = .25 \text{ SLUGS}$$

$$K = \frac{8 \text{ LBF}}{6 \text{ IN/FT}} = 16 \text{ LBF/FT}$$

$$B = .25 \ \frac{\text{LBF-SEC}}{\text{FT}}$$

THE DIFFERENTIAL EQUATION OF MOTION IS

$$Ma + BV + KX = F(t)$$

$$.25x'' + .25x' + 16X = 4\cos(2t)$$

$$x'' + x' + 64X = 16\cos(2t)$$

TIMED #2 CONTINUED

THE INITIAL CONDITIONS ARE $x(0) = x'(0) = 0$

TAKING THE LAPLACE TRANSFORM OF BOTH SIDES,

$$\mathcal{L}(x'') + \mathcal{L}(x') + \mathcal{L}(64x) = \mathcal{L}(16\cos(2t))$$

$$s^2\mathcal{L}(x) - s(x(0)) - x'(0) + s\mathcal{L}(x) - x(0) + 64\mathcal{L}(x)$$
$$= 16\left(\frac{s}{s^2+4}\right)$$

$$\mathcal{L}(x) = \frac{.266371\,s}{s^2+4} + \frac{.01774}{s^2+4}$$

$$- .266371\left[\frac{s}{s^2+s+64}\right] - .28411\left[\frac{1}{s^2+s+64}\right]$$

$$= \frac{.266371\,s}{s^2+4} + \frac{.01774}{s^2+4}$$

$$- .266371\left[\frac{s+\frac{1}{2}}{(s+\frac{1}{2})^2+63.75}\right]$$

$$- .28411\left[\frac{1 - .46878}{(s+\frac{1}{2})^2+63.75}\right]$$

$$x = \mathcal{L}^{-1}\left[\mathcal{L}(x)\right]$$

$$= .266371\cos(2t) + .00887\sin(2t)$$
$$- .266371\,e^{-.5t}\cos(7.9844t)$$
$$- .01890\,e^{-.5t}\sin(7.9844t)$$

$$\mathcal{L}(x)\left[s^2+s+64\right] = \frac{16s}{s^2+4}$$

$$\mathcal{L}(x) = \frac{16s}{(s^2+4)(s^2+s+64)}$$

USE PARTIAL FRACTIONS (PAGE 1-8)

$$\mathcal{L}(x) = \frac{A_1s+B_1}{s^2+4} + \frac{A_2s+B_2}{s^2+s+64}$$

$$= \frac{A_1s^3+A_1s^2+64A_1s+B_1s^2+B_1s+64B_1+A_2s^3+4A_2s+B_2s^2+4B_2}{(s^2+4)(s^2+s+64)}$$

$$= \frac{s^3(A_1+A_2) + s^2(A_1+B_1+B_2) + s(64A_1+B_1+4A_2) + (64B_1+4B_2)}{(s^2+4)(s^2+s+64)}$$

SO $A_1 + A_2 = 0$
$A_1 \quad\quad + B_1 + B_2 = 0$
$64A_1 + 4A_2 + B_1 \quad = 16$
$\quad\quad\quad 64B_1 + 4B_2 = 0$

USE CRAMER'S RULE TO FIND A_1

$$\begin{vmatrix} 1 & 1 & 0 & 0 \\ 1 & 0 & 1 & 1 \\ 64 & 4 & 1 & 0 \\ 0 & 0 & 64 & 4 \end{vmatrix} = -3604$$

$$\begin{vmatrix} 0 & 1 & 0 & 0 \\ 0 & 0 & 1 & 1 \\ 16 & 4 & 1 & 0 \\ 0 & 0 & 64 & 4 \end{vmatrix} = -960$$

$$A_1^* = \frac{-960}{-3604} = .266371$$

SUBSTITUTING IN GIVES
$$A_2^* = -.266371$$
$$B_1^* = .01774$$
$$B_2^* = -.28411$$

SO $\mathcal{L}(x) = \dfrac{.266371\,s + .01774}{s^2+4} - \dfrac{.266371\,s + .28411}{s^2+s+64}$

THIS CAN BE REARRANGED TO FIT THE FORMS FOUND IN LAPLACE TABLES

RESERVED FOR FUTURE USE

Engineering Economics

WARM-UPS

1 $F = 1000\,(F/P, 6\%, 10)$

$= 1000\,(1.7908) = 1790.80$

2 $P = 2000\,(P/F, 6\%, 4)$

$= 2000\,(.7921) = 1584.20$

3 $P = 2000\,(P/F, 6\%, 20)$

$= 2000\,(.3118) = 623.60$

4 $500 = A\,(P/A, 6\%, 7)$

$= A\,(5.5824)$

$A = \frac{500}{5.5824} = 89.57$

5 $F = 50\,(F/A, 6\%, 10)$

$= 50\,(13.1808) = 659.04$

6 EACH YEAR IS INDEPENDENT

$\frac{200}{1.06} = 188.68$

7 $2000 = A\,(F/A, 6\%, 5)$

$= A\,(5.6371)$

$A = \frac{2000}{5.6371} = 354.79$

8 $F = 100\,[(F/P, 6\%, 10) + (F/P, 6\%, 8) + (F/P, 6\%, 6)]$

$= 100\,(1.7908 + 1.5938 + 1.4185)$

$= 480.31$

9 $r = .06$

$\phi = \frac{.06}{12} = .005$

$N = 5(12) = 60$

$F = 500\,(1.005)^{60} = 674.43$

10 $120 = 80\,(F/P, ?, 7)$

$(F/P, ?, 7) = \frac{120}{80} = 1.5$

SEARCHING THE TABLES, $? = 6\%$

CONCENTRATES

1 $EUAC = (17000 + 5000)(A/P, 6\%, 5)$

$- (14000 + 2500)(A/F, 6\%, 5) + 200$

$= (22000)(.2374) - (16500)(.1774)$

$+ 200$

$= 2495.70$

2 ASSUME THE BRIDGE WILL BE THERE FOREVER.

KEEP OLD BRIDGE

THE GENERALLY ACCEPTED METHOD IS TO CONSIDER THE SALVAGE VALUE AS A BENEFIT LOST (COST).

$EUAC = (9000 + 13000)(A/P, 8\%, 20)$

$- 10,000\,(A/F, 8\%, 20) + 500$

$= (22000)(.1019) - (10,000)(.0219)$

$+ 500$

$= 2522.80$

REPLACE

$EUAC = 40,000\,(A/P, 8\%, 25)$

$- 15000\,(A/F, 8\%, 25) + 100$

$= (40,000)(.0937) - 15000(.0137) + 100$

$= 3642.50$

KEEP OLD BRIDGE

3 $D = \frac{150,000}{15} = 10,000$

$0 = -150,000 + (32000)(1 - .48)(P/A, ?, 15)$

$- 7530\,(1 - .48)(P/A, ?, 15)$

$+ 10000\,(.48)(P/A, ?, 15)$

$150,000 = [16640 - 3915.60 + 4800]\,(P/A, ?, 15)$

$(P/A, ?, 15) = \frac{150,000}{17524.40} = 8.5595$

SEARCHING THE TABLES, $i = 8\%$

4 a) $\frac{1,500,000 - 300,000}{1,000,000} = 1.2$

b) $1,500,000 - 300,000 - 1,000,000$

$= 200,000$

5 ANNUAL RENT IS $(12)(75) = 900$

$$F = (14000 + 1000)(F/P, 10\%, 10) + (150 + 250 - 900)(F/A, 10\%, 10)$$
$$= 15000(2.5937) - 500(15.9374)$$
$$= 30936.80$$

6 $2000 = 89.30\,(P/A, ?, 30)$

$$(P/A, ?, 30) = \frac{2000}{89.30} = 22.396$$

$? = 2\%$ PER MONTH

$$i = (1.02)^{12} - 1 = .2682 \text{ OR } 26.82\%$$

7 SL $D = \dfrac{500{,}000 - 100{,}000}{25} = 16000$

SOYD $T = \frac{1}{2}(25)26 = 325$

$$D_1 = \frac{25}{325}(500{,}000 - 100{,}000) = 30769$$
$$D_2 = \frac{24}{325}(400{,}000) = 29538$$
$$D_3 = \frac{23}{325}(400{,}000) = 28308$$

DDB $D_1 = \frac{2}{25}(500{,}000) = 40{,}000$

$$D_2 = \frac{2}{25}(500{,}000 - 40{,}000) = 36{,}800$$
$$D_3 = \frac{2}{25}(500{,}000 - 40{,}000 - 36{,}800) = 33{,}856$$

8 $P = -12000 + 2000(P/F, 10\%, 10) - 1000(P/A, 10\%, 10) - 200(P/G, 10\%, 10)$

$$= -12000 + 2000(.3855) - 1000(6.1446) - 200(22.8913)$$
$$= -21951.86$$

$$EUAC = 21951.86\,(A/P, 10\%, 10)$$
$$= 21951.86(.1627) = 3571.56$$

9 ASSUME THAT THE PROBABILITY OF FAILURE IN ANY OF THE N YEARS IS $1/N$

$$EUAC(9) = 1500(A/P, 6\%, 20) + \tfrac{1}{9}(.35)(1500) + (.04)(1500)$$
$$= 1500\left[.0872 + (.35)\left(\tfrac{1}{9}\right) + .04\right] = 249.13$$

$$EUAC(14) = 1600\left[.1272 + (.35)\left(\tfrac{1}{14}\right)\right] = 243.52$$
$$EUAC(30) = 1750\left[.1272 + (.35)\left(\tfrac{1}{30}\right)\right] = 243.01$$
$$EUAC(52) = 1900\left[.1272 + (.35)\left(\tfrac{1}{52}\right)\right] = 254.47$$
$$EUAC(86) = 2100\left[.1272 + (35)\left(\tfrac{1}{86}\right)\right] = 275.67$$

CHOOSE THE 30 YEAR PIPE

10 $EUAC(7) = .15(25000) = 3750$

$$EUAC(8) = 15000(A/P, 10\%, 20) + .10(25000)$$
$$= 15000(.1175) + .10(25000)$$
$$= 4262.50$$

$$EUAC(9) = 20000(.1175) + .07(25000)$$
$$= 4100.00$$

$$EUAC(10) = 30{,}000(.1175) + .03(25000)$$
$$= 4275$$

CHEAPEST TO DO NOTHING

TIMED

1 $EUAC(1) = 10{,}000(A/P, 20, 1) + 2000 - 8000(A/F, 20\%, 1)$

$$= 10{,}000(1.2000) + 2000 - 8000(1.0000) = 6000$$

$$EUAC(2) = 10{,}000(A/P, 20, 2) + 2000 + 1000(A/G, 20\%, 2) - 7000(A/F, 20\%, 2)$$
$$= 10{,}000(.6545) + 2000 + 1000(.4545) - 7000(.4545) = 5818.00$$

$$EUAC(3) = 10{,}000(A/P, 20\%, 3) + 2000 + 1000(A/G, 20\%, 3) - 6000(A/F, 20\%, 3)$$
$$= 10000(.4747) + 2000 + 1000(.8791) - 6000(.2747) = 5977.90$$

$$EUAC(4) = 10{,}000(A/P, 20, 4) + 2000 + 1000(A/G, 20\%, 4) - 5000(A/F, 20\%, 4)$$
$$= 10{,}000(.3863) + 2000 + 1000(1.2742) - 5000(.1863) = 6205.7$$

TIMED #1, CONTINUED

$$EUAC(5) = 10{,}000\,(A/P,20\%,5) + 2000 + 1000\,(A/G,20\%,5) - 4000\,(A/F,20\%,5)$$
$$= 10{,}000\,(.3344) + 2000 + 1000\,(1.6405) - 4000\,(.1344) = 6446.9$$

SELL AT END OF 2ND YEAR

b) MAINTENANCE AND DROP IN SALVAGE:

$$6000 + (5000 - 4000) = 7000$$

2 THE MAN SHOULD CHARGE HIS COMPANY ONLY FOR THE COSTS DUE TO THE BUSINESS TRAVEL:

INSURANCE $300 - 200 = 100$

MAINTENANCE $200 - 150 = 50$

SALVAGE $(1000 - 500)$ IN 5 YEARS

$$500\,(A/F,10\%,5) = 500(.1638) = 81.90$$

GASOLINE $\dfrac{5000\,(.60)}{15} = 200$

$$\text{EUAC PER MILE} = \frac{100 + 50 + 81.9 + 200}{5000} = \$.0864$$

a) YES. $\$.10 > \$.0864$ SO IT IS ADEQUATE

b) $$(.10)X = 5000\,(A/P,10\%,5) + 250 + 200 - 800\,(A/F,10\%,5) + \frac{X}{15}(.60)$$
$$= 5000\,(.2638) + 250 + 200 - 800(.1638) + .04X$$

$$.10X = 1637.96 + .04X$$
$$.06X = 1637.96$$
$$X = 27299 \text{ miles}$$

3 USE EUAC SINCE LIVES ARE DIFFERENT

$$P\{A\} = -80{,}000 - 5000(P/F,10\%,10) + 7000(P/F,10\%,20) - 2000\,(P/A,10\%,20) + 500\,(P/A,10\%,10) + 500\,(P/A,10\%,5)$$
$$= -80{,}000 - 5000(.3855) + 7000(.1486) - 2000\,(8.5136) + 500\,(6.1446 + 3.7908)$$
$$= -92947$$

$$EUAC\{A\} = 92947\,(A/P,10\%,20) = 92947\,(.1175) = 10921$$

$$P\{B\} = -35000 - 4000\,(P/A,10\%,10) + 1000\,(P/A,10\%,5)$$
$$= -35000 - 4000(6.1446) + 1000(3.7908)$$
$$= -55788$$

$$EUAC\{B\} = 55788\,(A/P,10\%,10) = 55788\,(.1627) = 9077$$

B HAS THE LOWEST COST

4 LET X BE THE NUMBER OF MILES DRIVEN PER YEAR. THE EUAC FOR BOTH ALTERNATIVES ARE

$$EUAC(A) = .15X$$
$$EUAC(B) = .04X + 500 + 5000\,(A/P,10\%,3) - 1200\,(A/F,10\%,3)$$
$$= .04X + 500 + 5000(.4021) - 1200\,(.3021)$$
$$= .04X + 2148$$

SETTING THESE EQUAL,

$$.15X = .04X + 2148$$

X = 19527 MILES

5 a) THE OPERATING COSTS PER YEAR ARE

A: $(10.50)(365)(24) = 91980$

B: $(8)(365)(24) = 70080$

$$EUAC\{A\} = (13000 + 10000)(A/P,7\%,10) + 91980 - 5000\,(A/F,7\%,10)$$
$$= (23000)(.1424) + 91980 - 5000(.0724)$$
$$= 94893$$

THE COST PER TON-YEAR IS

$$\text{COST}(A) = \frac{94893}{50} = \boxed{1898}$$

NOW WORK WITH B. EXCLUSIVE OF OPERATING COSTS, THE PRESENT WORTH IS

$$P\{B\} = -8000 - (7000 + 2200 - 2000)\,(P/F,7\%,5) + 2000\,(P/F,7\%,10)$$
$$= -8000 - (7200)(.7130) + 2000\,(.5083)$$
$$= -12117$$

(MORE)

TIMED #5 CONTINUED

$$EUAC\{B\} = 70080 + (12117)(A/P, 7\%, 10)$$
$$= 70080 + (12117)(.1424)$$
$$= 71805$$

THE COST PER TON-YEAR FOR B IS

$$\text{COST B} = \frac{71805}{20} = \boxed{3590}$$

(b)

	COST OF USING A's	COST OF USING B's	CHEAPEST
0-20	94893 (1X)	71805 (1X)	B
20-40	94893 (1X)	143610 (2X)	A
40-50	94893 (1X)	215415 (3X)	A
50-60	189,786 (2X)	215415 (3X)	A
60-80	189,786 (2X)	287220 (4X)	A

6 FIRST YEAR COST = (1)(37,440) = 37440

GRADIENT = (.1)(37440) = 3744

(a)
$$EUAC = 60{,}000\,(A/P, 7\%, 20) + 37{,}440 + 3744\,(A/G, 7\%, 20) - 10{,}000\,(A/F, 7\%, 20)$$
$$= 60{,}000\,(.0944) + 37440 + 3744\,(7.3163) - 10{,}000\,(.0244)$$
$$= 70{,}252$$

THE REQUIRED FARE IS

$$\frac{70252}{80{,}000} = \boxed{\$.878}$$

(b)
$$.878 = .35 + (\text{INCREASE})(A/G, 7\%, 20)$$
$$.878 = .35 + (\text{INCREASE})(7.3163)$$
$$\text{INCREASE} = \boxed{.072 \text{ PER YEAR}}$$

(c)
$$(80{,}000)(.35) = 28{,}000$$
$$(80{,}000)(.05) = 4000$$

$$PW = -60{,}000 + (28{,}000 - 37440)(P/A, 7\%, 20) + (4000 - 3744)(P/G, 7\%, 20) + (10{,}000)(P/F, 7\%, 20)$$
$$= -60{,}000 - (9440)(10.594) + (256)(77.5091) + 10{,}000(.2584)$$
$$= \boxed{-137{,}581}$$

RESERVED FOR FUTURE USE

Fluid Statics

WARM-UPS

1 FROM PAGE 3-36 FOR AIR AT 14.7 PSIA AND 80°F,

$\mu = 3.85\text{EE-}7 \frac{\text{LB-SEC}}{\text{FT}^2}$ {INDEPENDENT OF PRESSURE}

$$\rho = P/RT = \frac{(70)(144)}{(53.3)(80+460)} = .350$$

FROM EQN 3.12,

$$\nu = \frac{\mu g}{\rho} = \frac{(3.85\text{EE-}7)\frac{\text{LB-SEC}}{\text{FT}^2}(32.2)\frac{\text{FT}}{\text{SEC}^2}}{(.35)\ \text{LB/FT}^3}$$

$$= 3.54\text{EE-}5\ \text{FT}^2/\text{SEC}$$

2 $P = 14.7 - 8.7 = 6.0$ PSIA

3 FROM PAGE 3-25,

$$P_1 - P_2 = 7(.491) - 7(.0361) = 3.184 \text{ PSI}$$

4 THE PERIMETER IS

$$p = \pi d = \pi(10) = 31.42'$$

ASSUME AN ELLIPTICAL CROSS SECTION.

FROM PAGE 1-2,

$$b = \tfrac{1}{2}(7.2) = 3.6$$

THEN

$$31.42 = 2\pi\sqrt{\tfrac{1}{2}(a^2 + (3.6)^2)}$$

SO $a = 6.09''$

THE AREA IN FLOW IS

$$A = \tfrac{1}{2}(\pi ab) = \tfrac{1}{2}(\pi)(6.09)(3.6) = 34.44 \text{ IN}^2$$

THEN FROM EQN 3.65

$$r_H = \frac{34.44}{\frac{1}{2}(31.42)} = 2.19''$$

5 ASSUME SCHEDULE 40 PIPE AND 70°F WATER

$D_i = 6.065'' = .5054$ FT

$A_i = .2006\ \text{FT}^2$

$$v = \frac{1.5\ \text{FT}^3/\text{SEC}}{.2006\ \text{FT}^2} = 7.478\ \text{FT/SEC}$$

FROM PAGE 3-37 AT 70°F

$\nu = 1.059\text{EE-}5\ \text{FT}^2/\text{SEC}$

FROM EQN 3.62

$$N_{Re} = \frac{Dv}{\nu} = \frac{(.5054)\text{FT}(7.478)\text{FT/SEC}}{(1.059\text{EE-}5)\ \text{FT}^2/\text{SEC}} = 3.57\text{EE}5$$

6 FROM PAGE 3-20, $\epsilon = .0008$

$$\epsilon/D = \frac{.0008}{.5054} = .00158 \text{ \{SAY, .0016\}}$$

FROM PAGE 3-20, $f \approx .022$

FROM EQN 3.71,

$$h_f = \frac{(.022)(1200)\text{FT}(7.478)^2\ \text{FT}^2/\text{SEC}^2}{(.5054)\text{FT}(2)(32.2)\text{FT/SEC}^2}$$

$$= 45.4 \text{ FT}$$

7 ASSUME SCHEDULE 40 PIPE

AT A
$D_A = .5054$ FT
$A_A = .2006\ \text{FT}^2$
$$V_A = \frac{5\ \text{FT}^3/\text{SEC}}{.2006\ \text{FT}^2} = 24.925\ \text{FT/SEC}$$
$P_A = 10$ PSIA $= 1440$ PSF
$z_A = 0$

AT B
$D_B = 1.4063$ FT
$A_B = 1.5533\ \text{FT}^2$
$V_B = 5/1.5533 = 3.219$ FT/SEC
$P_B = 7$ PSIA $= 1008$ PSF
$z_B = 15$

IGNORE THE MINOR LOSSES AND ASSUME $\rho = 62.3$ AT 70°F. THEN, THE TOTAL ENERGY AT A AND B FROM EQN 5.21 IS

$$(TH)_A = \frac{1440}{62.3} + \frac{(24.925)^2}{(2)(32.2)} + 0 = 32.76$$

$$(TH)_B = \frac{1008}{62.3} + \frac{(3.219)^2}{(2)(32.2)} + 15 = 31.3$$

FLOW IS FROM HIGH ENERGY TO LOW.

$A \to B$

8 FROM EQUATION 16.38,

$$r = \frac{V_0^2 \sin 2\phi}{g}$$

$$= \frac{(50)^2 \sin(2(45))}{32.2} = 77.64 \text{ FT}$$

9 THE WEIGHT OF WATER AT 70°F IS

$$(100)\frac{\text{FT}^3}{\text{SEC}}(62.4)\frac{\text{LBM}}{\text{FT}^3} = 6240\ \frac{\text{LBM}}{\text{SEC}}$$

THE ENERGY LOSS IN FT OF WATER IS

$$\frac{(30+5)\frac{\text{LBF}}{\text{IN}^2}(144)\frac{\text{IN}^2}{\text{FT}^2}}{62.4\ \text{LBM/FT}^3} = 80.77 \text{ FT}$$

THE HORSEPOWER IS

$$\frac{6240\frac{\text{LBM}}{\text{SEC}}(80.77)\text{ FT}}{550\frac{\text{FT-LBF}}{\text{HP-SEC}}} = 916.37 \text{ HP}$$

10 ASSUME SCHEDULE 40 PIPE AND 70°F WATER

$D_1 = 7.981''$

$D_2 = 6.065''$

(MORE)

WARM-UP #10 CONTINUED

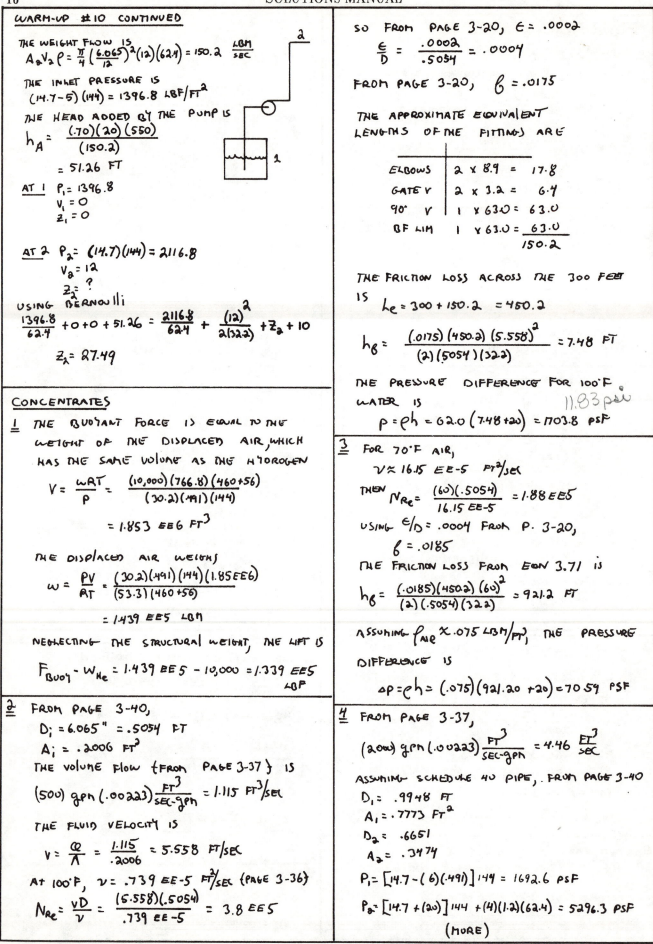

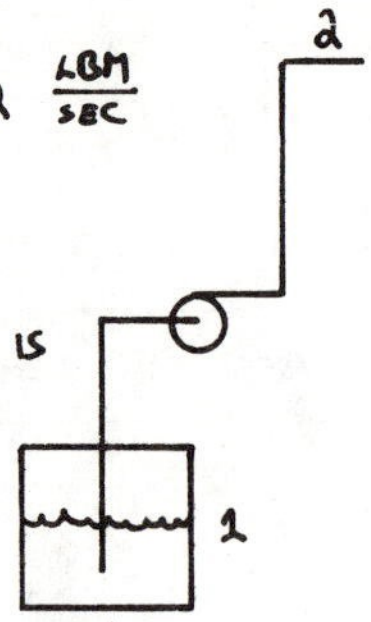

THE WEIGHT FLOW IS

$$A_2 V_2 \rho = \frac{\pi}{4}\left(\frac{6.065}{12}\right)^2(12)(62.4) = 150.2 \ \frac{LBM}{SEC}$$

THE INLET PRESSURE IS

$$(14.7-5)(144) = 1396.8 \ LBF/FT^2$$

THE HEAD ADDED BY THE PUMP IS

$$h_A = \frac{(.70)(20)(550)}{(150.2)} = 51.26 \ FT$$

AT 1 $P_1 = 1396.8$
$V_1 = 0$
$Z_1 = 0$

AT 2 $P_2 = (14.7)(144) = 2116.8$
$V_2 = 12$
$Z_2 = ?$

USING BERNOULLI

$$\frac{1396.8}{62.4} + 0 + 0 + 51.26 = \frac{2116.8}{62.4} + \frac{(12)^2}{2(32.2)} + Z_2 + 10$$

$$Z_2 = 27.49$$

CONCENTRATES

1 THE BUOYANT FORCE IS EQUAL TO THE WEIGHT OF THE DISPLACED AIR, WHICH HAS THE SAME VOLUME AS THE HYDROGEN

$$V = \frac{wRT}{P} = \frac{(10,000)(766.8)(460+56)}{(30.2)(.491)(144)} = 1.853 \ EE6 \ FT^3$$

THE DISPLACED AIR WEIGHS

$$w = \frac{PV}{RT} = \frac{(30.2)(.491)(144)(1.85 EE6)}{(53.3)(460+56)} = 1.439 \ EE5 \ LBM$$

NEGLECTING THE STRUCTURAL WEIGHT, THE LIFT IS

$$F_{BUOY} - W_{He} = 1.439 \ EE5 - 10,000 = 1.339 \ EE5 \ LBF$$

2 FROM PAGE 3-40,

$D_i = 6.065'' = .5054 \ FT$
$A_i = .2006 \ FT^2$

THE VOLUME FLOW (FROM PAGE 3-37) IS

$$(500) \ gpm \ (.00223)\frac{FT^3}{SEC\text{-}gpm} = 1.115 \ FT^3/SEC$$

THE FLUID VELOCITY IS

$$v = \frac{Q}{A} = \frac{1.115}{.2006} = 5.558 \ FT/SEC$$

AT 100°F, $\nu = .739 \ EE\text{-}5 \ FT^2/SEC$ (PAGE 3-36)

$$N_{Re} = \frac{vD}{\nu} = \frac{(5.558)(.5054)}{.739 \ EE\text{-}5} = 3.8 \ EE5$$

SO FROM PAGE 3-20, $\epsilon = .0002$

$$\frac{\epsilon}{D} = \frac{.0002}{.5054} = .0004$$

FROM PAGE 3-20, $f = .0175$

THE APPROXIMATE EQUIVALENT LENGTHS OF THE FITTINGS ARE

ELBOWS	2 x 8.9 =	17.8
GATE V	2 x 3.2 =	6.4
90° V	1 x 63.0 =	63.0
BF LIM	1 x 63.0 =	63.0
		150.2

THE FRICTION LOSS ACROSS THE 300 FEET IS

$$L_e = 300 + 150.2 = 450.2$$

$$h_f = \frac{(.0175)(450.2)(5.558)^2}{(2)(.5054)(32.2)} = 7.48 \ FT$$

THE PRESSURE DIFFERENCE FOR 100°F WATER IS

$$p = \rho h = 62.0(7.48 + 20) = 1703.8 \ PSF$$

11.83 psi

3 FOR 70°F AIR,

$\nu \approx 16.15 \ EE\text{-}5 \ FT^2/SEC$

THEN

$$N_{Re} = \frac{(60)(.5054)}{16.15 \ EE\text{-}5} = 1.88 \ EE5$$

USING $\epsilon/D = .0004$ FROM P. 3-20,

$f = .0185$

THE FRICTION LOSS FROM EQN 3.71 IS

$$h_f = \frac{(.0185)(450.2)(60)^2}{(2)(.5054)(32.2)} = 921.2 \ FT$$

ASSUMING $\rho_{AIR} \approx .075 \ LBM/FT^3$, THE PRESSURE DIFFERENCE IS

$$\Delta p = \rho h = (.075)(921.20 + 20) = 70.59 \ PSF$$

4 FROM PAGE 3-37,

$$(200) \ gpm \ (.00223)\frac{FT^3}{SEC\text{-}gpm} = 4.46 \ \frac{FT^3}{SEC}$$

ASSUMING SCHEDULE 40 PIPE, FROM PAGE 3-40

$D_1 = .9948 \ FT$
$A_1 = .7773 \ FT^2$
$D_2 = .6651$
$A_2 = .3474$

$$P_1 = [14.7 - (6)(.491)]144 = 1692.6 \ PSF$$

$$P_2 = [14.7 + (20)]144 + (4)(1.2)(62.4) = 5296.3 \ PSF$$

(MORE)

CONCENTRATE #4 CONTINUED

$$v_1 = \frac{4.46}{.7773} = 5.74 \text{ FPS}$$

$$v_2 = \frac{4.46}{.3474} = 12.84 \text{ FPS}$$

THE TOTAL HEADS AT 1 AND 2 {EQN 3.56} ARE

$$(TH)_1 = \frac{1692.6}{(62.4)(1.2)} + \frac{(5.7)^2}{(2)(32.2)} = 23.11$$

$$(TH)_2 = \frac{5296.3}{(62.4)(1.2)} + \frac{(12.84)^2}{(2)(32.2)} = 73.29$$

THE PUMP MUST ADD $(73.29 - 23.11) = 50.18$ FT OF HEAD. THE POWER REQUIRED IS

$$P = \Delta h \dot{m}$$

$$= \frac{(50.18)\text{FT}(1.2)(62.4)\frac{\text{LBM}}{\text{FT}^3}(4.46)\text{FT}^3/\text{SEC}}{550 \frac{\text{FT-LBF}}{\text{HP-SEC}}}$$

$$= 30.47 \text{ HP}$$

THE INPUT HORSEPOWER IS

$$\frac{30.47}{.85} = 35.85 \text{ HP}$$

5 THIS CANNOT BE SOLVED CORRECTLY WITH THE ENGINEERING REVIEW MANUAL. REFER TO PAGE 2-13 OF CRANE'S TECHNICAL PAPER 410.

$$K_B = K_1 + (N-1)\left(0.25 f \pi \left(\frac{r}{D}\right) + .5K_1\right)$$

ASSUME TYPE K COPPER TUBING. THEN

$$D_i = 2.125 - 2(.083) = 1.959'' = .1633 \text{ FT}$$

$$\frac{r}{D} = \frac{1}{.1633} = 6.126$$

FROM PAGE A-29 OF CRANE FOR A 90° BEND

$$K_1 \approx 17.4 f$$

$$N = \#90° \text{ BENDS} = (4)(5) = 20$$

ASSUMING SMOOTH COPPER PIPE {PAGE 3-20}

$$\epsilon = .000005$$

$$\frac{\epsilon}{D} = \frac{.000005}{.1633} = .000031$$

FOR 200°F WATER, $\nu = .341$ EE-5

$$N_{Re} = \frac{vD}{\nu} = \frac{(10)(.1633)}{.341 \text{ EE-5}} = 4.79 \text{ EE5}$$

FROM PAGE 3-20, $f = .0135$

THEN $K_1 = (17.4)(.0135) = .235$

$$K_B = .235 + (20-1)\left[(.25)(.0135)(\pi)(6.126) + (.5)(.235)\right]$$

$$= 3.70$$

$$h_f = K_B \frac{v^2}{2g} = (3.70)\frac{(10)^2}{(2)(32.2)} = 5.75 \text{ FT}$$

6 $\dot{Q} = \frac{\pi}{4}\left(\frac{2}{12}\right)^2 \text{FT}^2 (40)\text{FT/SEC} = .8727 \text{ FT}^3/\text{SEC}$

FROM EQN 3.168

$$\dot{Q}' = \frac{40-15}{40}(.8727) = .5454 \text{ FT}^3/\text{SEC}$$

FROM EQNS 3.171 AND 3.172

$$F_x = \frac{-(.5454)(62.4)}{32.2}(40-15)(1-\cos 60°)$$

$$= -13.2 \text{ LBF}$$ {FORCE OF FLUID IS TO THE LEFT}

$$F_y = \frac{(.5454)(62.4)}{32.2}(40-15)(\sin 60°)$$

$$= 22.9$$ {FORCE ON FLUID IS UPWARDS}

$$F = \sqrt{(13.2)^2 + (22.9)^2} = 26.4 \text{ LBF}$$

7 FOR DYNAMIC SIMILITUDE, USE EQN 3.190

$$N_{Re,\text{MODEL}} = N_{Re,\text{TRUE}}$$

$$\left(\frac{vL}{\nu}\right)_M = \left(\frac{vL}{\nu}\right)_T$$

$$\left(\frac{vL\rho}{\mu g}\right)_M = \left(\frac{vL\rho}{\mu g}\right)_T$$

BUT $\rho = P/RT$ FOR IDEAL GASES

$$\left(\frac{vLp}{\mu g RT}\right)_M = \left(\frac{vLp}{\mu g RT}\right)_T$$

$V_M = V_t$

$L_M = L_t/20$

$T_M = T_t$

$g = g$

$R = R$

$\mu_M = \mu_t$ {INDEPENDENT OF PRESSURE}

SO $P_M = 20 P_t$

8 FROM EQN 3.190

$$(N_{Re})_M = (N_{Re})_t$$

$$\left(\frac{vD}{\nu}\right)_M = \left(\frac{vD}{\nu}\right)_t$$

V IS A LINEAR VELOCITY, AND

$V \propto (\text{RPM})(\text{DIAMETER})$

SO $V_M \propto 2(\text{RPM})_M$

$V_t \propto (1)(1000)$

$D_M = 2D_t$

$\nu_{\text{AIR},68°} = 16.0$ EE-5 FT²/SEC {FROM P. 3-38} (MORE)

CONCENTRATE # 8 CONTINUED

$\nu_{\text{CASTOR OIL}} \approx 1110\text{ EE-5 FT}^2/\text{SEC}$

THEN

$$\frac{(2)(\text{RPM})_M(2)D_t}{16\text{ EE-5}} = \frac{(1000)D_t}{1110\text{ EE-5}}$$

$$(\text{RPM})_M = 3.6$$

9 THE SPECIFIC GRAVITY OF BENZENE AT 60°F IS ≈ .885 {PAGE 4-37 OF CAMERON}

THE DENSITY IS $(.885)(62.4) = 55.2\ \frac{\text{LBM}}{\text{FT}^3}$

$= .0319\ \frac{\text{LBM}}{\text{IN}^3}$

THE PRESSURE DIFFERENCE IS

$$\Delta P = 4(.491 - .0319) = 1.836\text{ PSI}$$
$$= 264.4\text{ PSF}$$

FROM EQN 3.132,

$$F_{VA} = \frac{1}{\sqrt{1-\left(\frac{3.5}{8}\right)^4}} = 1.019$$

$$A_2 = \frac{\pi}{4}\left(\frac{3.5}{12}\right)^2 = .0668\text{ FT}^2$$

FROM EQN 3.134,

$$Q = (1.019)(.99)(.0668)\sqrt{\frac{(2)(32.2)(264.4)}{55.2}}$$

$$= 1.184\text{ FT}^3/\text{SEC}$$

10 ASSUME SCHED 40 PIPE, THEN

$D_i = .9948$

$A_i = .7773\text{ FT}^2$

$V_1 = \frac{Q}{A} = \frac{10}{.7773} = 12.87\text{ FT/SEC}$

FOR 70°F WATER, $\nu = 1.059\text{ EE-5}$

$$N_{Re} = \frac{VD}{\nu} = \frac{(12.87)(.9948)}{(1.059\text{ EE-5})} = 1.21\text{ EE6}$$

USE EQN 3.142

$$Q = C_f A_o\sqrt{\frac{2g(P_1-P_2)}{\rho}}$$

$$10 = C_f A_o\sqrt{(2)(32.2)(25)}$$

$$C_f A_o = .249$$

BOTH C_f AND A_o DEPEND ON D_o

ASSUME $D_o = 6"$

$$A_o = \frac{\pi}{4}\left(\frac{6}{12}\right)^2 = .1963\text{ FT}^2$$

FROM FIGURE 3.25,

$$\frac{A_o}{A_1} = \frac{.1963}{.7773} = .25$$

AND $N_{Re} = 1.21\text{ EE6}$,

$C_f = .63$

$C_f A_o = (.63)(.1963) = .124$ TOO SMALL

ASSUME $D_o = 9"$

$$A_o = \frac{\pi}{4}\left(\frac{9}{12}\right)^2 = .442$$

$$\frac{A_o}{A_1} = \frac{.442}{.7773} = .57$$

FROM FIGURE 3.25, $C_f \approx .73$

$C_f A_o = (.73)(.442) = .322$ TOO HIGH

INTERPOLATING

$$D_o = 6" + (9-6)\frac{.249 - .124}{.322 - .124} = 7.9$$

FURTHER ITERATIONS YIELD

$D_o \approx 8.1$

$C_f A_o \approx .243$

TIMED

1

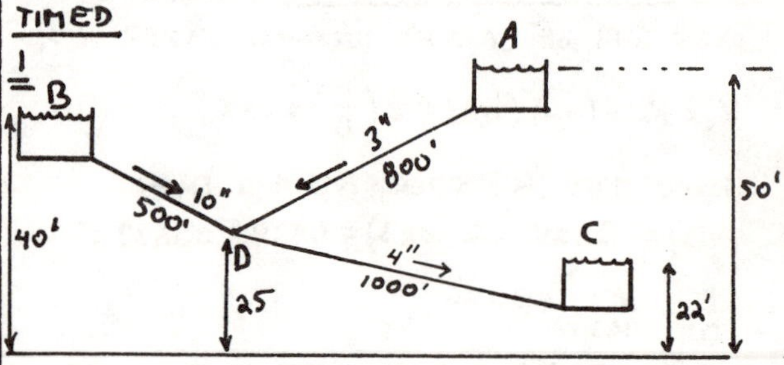

WE ASSUME THE RESERVOIRS A, B, FLOW TOWARDS D AND THE FLOW IS TOWARDS C, THEN

$$Q_{A-D} + Q_{B-D} - Q_{D-C} = 0$$

$$A_A V_{A-D} + A_B V_{B-D} - A_C V_{D-C} = 0$$

ASSUME SCHEDULE 40 PIPE SO THAT

$A_A = .05134\text{ FT}^2$ $D_A = .2557\text{ FT}$

$A_B = .5476\text{ FT}^2$ $D_B = .8350\text{ FT}$

$A_C = .08841\text{ FT}^2$ $D_C = .3355\text{ FT}$

SO

$$.05134V_{A-D} + .5476V_{B-D} - .08841V_{D-C} = 0 \quad (1)$$

IGNORING THE VELOCITY HEADS, THE CONSERVATION OF ENERGY EQN 3.70 BETWEEN POINTS A AND D IS

$$Z_A = \frac{P_D}{\rho} + Z_D + h_{f,A-D}$$

TIMED #1 CONTINUED

$$50 = \frac{P_D}{62.4} + 25 + \frac{(.02)(800)(V_{A-D})^2}{(2)(.2557)(32.2)}$$

or

$$V_{A-D} = \sqrt{25.73 - .0165 P_D} \quad (2)$$

SIMILARLY FOR B-D

$$40 = \frac{P_D}{62.4} + 25 + \frac{(.02)(500)(V_{B-D})^2}{(2)(.8350)(32.2)}$$

OR $$V_{B-D} = \sqrt{80.66 - .0862 P_D} \quad (3)$$

AND FOR D-C

$$22 = \frac{P_D}{62.4} + 25 - \frac{(.02)(1000)(V_{D-C})^2}{(2)(.3355)(32.2)}$$

OR $$V_{D-C} = \sqrt{3.24 + .0173 P_D} \quad (4)$$

ASSUME $P_D = 935$ PSF, WHICH IS THE LARGEST IT CAN BE TO KEEP EQUATIONS 2 AND 3 REAL. THEN

$V_{A-D} = 3.21$
$V_{B-D} = .25$
$V_{D-C} = 4.41$

FROM EQN 1

$$(.05134)(3.21) + (.5476)(.25) - (.08841)(4.41) = -.088$$

THIS IS REPEATED USING INTERPOLATION.

TRIAL	P_D	RIGHT-HAND SIDE OF EQN 1	HOW WAS P_D CHOSEN?
1	935	−.088	TO KEEP 2,3 REAL
2	900	.745	ARBITRARILY
3	931.3	.116	INTERPOLATION WITH TRIALS 1,2
4	933.4	.02	INTERPOLATION, 1+3
5	933.7	.005	INTERPOLATION, 1+4

IF $P_D = 933.7$

$V_{AD} = 3.21$ FT/SEC

$V_{BD} = .418$

$V_{DC} = 4.40$

FLOW IS OUT OF B

2 ASSUME 70°F WATER. FROM PAGE 3.36

$E = 320$ EE3 PSI

$\rho = 62.3$

ALTHOUGH E VARIES CONSIDERABLY WITH THE CLASS OF THE CAST IRON, USE $E = 20$ EE6 {P. 14-3}

IGNORING THE CONTRIBUTION OF THE PIPE WALL, THE WATER HAMMER VELOCITY IS

$$c = \sqrt{\frac{(320\text{ EE }3)}{62.3}(144)(32.2)}$$

$= 4880$ FT/SEC

FROM EQN 3.177

$$\Delta p = \frac{(62.3)(4880)(6)}{32.2} = 56{,}475 \text{ PSF}$$

FROM EQN 3.176, THE MAXIMUM CLOSURE TIME ALLOWED WITHOUT REDUCING Δp IS

$$t = \frac{(2)(500)}{4880} = .205 \text{ SEC}$$

RESERVED FOR FUTURE USE

Hydraulic Machines

WARM-UPS

1 FROM PAGE 1-40,

$$(72)\ \text{gpm}\ (2.228\ EE{-3})\ \frac{\text{CFS}}{\text{gpm}} = .1604\ \text{CFS}$$

2 FROM EQN 3.37,

$$HP = \frac{Q\,H\,(SG)_{oil}}{3960}$$

$$Q = 37\ \text{gpm}$$

$$H = \frac{\Delta P}{\rho} = \frac{(40)\frac{\text{LBF}}{\text{IN}^2}(144)\frac{\text{IN}^2}{\text{FT}^2}}{(SG)_{oil}\,(62.4)\frac{\text{LBM}}{\text{FT}^3}} = \frac{92.308}{(SG)_{oil}}$$

THEN

$$HP = \frac{(37)\left(\frac{92.308}{SG}\right)(SG)}{3960} = .862\ HP$$

3 FROM EQN 4.33,

$$(HP)_2 = (.5)\left(\frac{2000}{1750}\right)^3 = .746$$

4 USE EQN 4.27

$$N = 900\ \text{RPM}$$

$$Q = \left(\tfrac{1}{2}\right)(300)\frac{\text{GAL}}{\text{SEC}}(60)\frac{\text{SEC}}{\text{MIN}} = 9000$$

↖ THIS TERM BECAUSE THE PUMP IS DOUBLE SUCTION

$$H = 20\ \text{FT}$$

$$N_s = \frac{(900)\sqrt{9000}}{(20)^{.75}} = 9028$$

5 ASSUME EACH STAGE ADDS 150 FEET OF HEAD. THE SUCTION LIFT IS 10 FEET. FROM PAGE 4-19 FOR SINGLE SUCTION,

$$N_s \approx 2050$$

CONCENTRATES

1 FROM PAGE 3-34

$$D_i = .3355$$

$$A_i = .08841$$

SO $V = \frac{Q}{A} = \frac{1.25}{.08841} = 14.139$

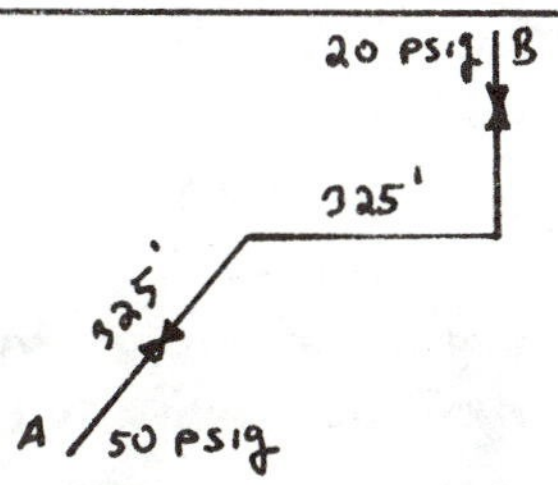

ASSUME REGULAR, SCREWED, STEEL FITTINGS. THE APPROXIMATE EQUIVALENT LENGTHS ARE:

ELBOWS	2 x 13 = 26
GATE V.	1 x 2.5 = 2.5
CHECK V.	1 x 38. = 38.0
	66.5

ASSUME 70°F, SO $\nu = 1.059\ EE{-5}$

$$N_{Re} = \frac{Dv}{\nu} = \frac{(.3355)(14.139)}{1.059\ EE{-5}} = 4.479\ EE5$$

FROM PAGE 3-20, $\epsilon = .0002$ FT, SO

$$\frac{\epsilon}{D} = \frac{.0002}{.3355} \approx .0006$$

FROM PAGE 3-20, $f = .0185$. FROM EQN 3.71,

$$h_f = \frac{(.0185)(700+66.5)(14.139)^2}{(2)(.3355)(32.2)} = 131.20$$

FROM EQN 3.70, THE HEAD ADDED IS

$$h_A = \frac{p_2}{\rho} + \frac{v_2^2}{2g} + z_2 + h_f - \frac{p_1}{\rho} - \frac{v_1^2}{2g} - z_1$$

BUT $v_1 = 0$ AND $z_1 = 0$

$$h_A = \frac{(20+14.7)144}{62.3} + \frac{(14.139)^2}{(2)(32.2)} + 50 + 131.20 - \frac{(50+14.7)(144)}{62.3} = 114.96$$

ASSUME 70°F WATER. $\rho = 62.3$ LBM/FT³

$$\dot{m} = (62.3)\frac{\text{LBM}}{\text{FT}^3}(1.25)\frac{\text{FT}^3}{\text{sec}} = 77.875\ \text{LBM/sec}$$

$$HP = \frac{(77.875)\frac{\text{LBM}}{\text{SEC}}(114.96)\ \text{FT}}{550\ \frac{\text{FT-LBF}}{\text{HP-SEC}}} = 16.28$$

2

THE PROPELLER IS LIMITED BY CAVITATION. THIS will OCCUR WHEN

$$h_{atmos} - h_{velocity} < h_{vapor\ pressure}$$

ASSUME THE DENSITY OF SEA WATER IS 64.0 LBM/FT³

$$h_{atmos} = \frac{(14.7)(144)}{64.0} = 33.075'$$

$$h_{depth} = 8$$

$$h_{velocity} = \frac{v_{prop}^2}{2g} = \frac{(4.2\,V_{boat})^2}{(2)(32.2)} = .2739\,V_{boat}^2$$

THE VAPOR PRESSURE OF 70°F FRESH WATER IS

$$h_v = \frac{(.3631)\frac{\text{LBF}}{\text{IN}^2}(144)}{62.4} = .837\ \text{ft}$$

RAOULT'S LAW PREDICTS THE ACTUAL VAPOR PRESSURE OF THE SOLUTION

$$P_{vapor,\ solution} = P_{vapor,\ solvent}\left(\text{mole fraction of solvent}\right)$$

ASSUME 2½% SALT BY WEIGHT. OUT OF 100 POUNDS OF SEA WATER, WE'LL HAVE 2.5 LBM SALT AND 97.5 LBM WATER. THE MOLECULAR WEIGHT OF SALT IS (23.0 + 35.5) = 58.5. SO, THE # OF MOLES OF SALT IN 100 POUNDS OF SEA WATER IS

$$N_s = \frac{2.5}{58.5} = .043$$

SIMILARLY, THE MOLECULAR WEIGHT OF WATER IS 2(1) + 16 = 18. THE NUMBER OF MOLES OF WATER IS

$$N_w = \frac{97.5}{18} = 5.417$$

THE MOLE FRACTION OF WATER IS

$$\frac{5.417}{5.417 + .043} = .992$$

SO, $p_{\text{VAPOR SEA WATER}} = (.992)(.837) = .83$ ft

then

$$8 + 33.075 - .2739\,v^2 = .83$$

or $v = 12.12$ ft/sec

3

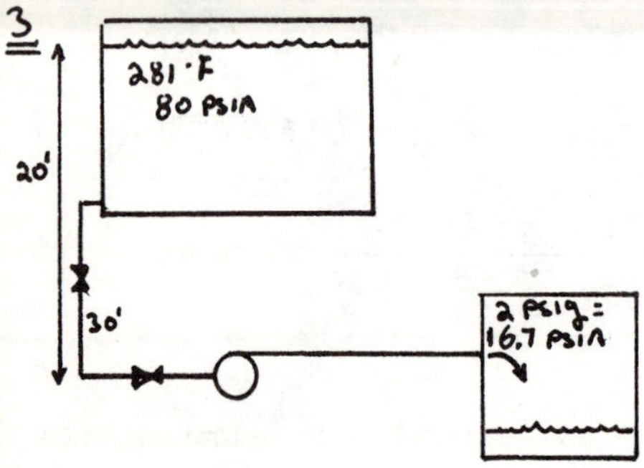

ASSUME SCHEDULE 40 PIPE. FROM PAGE 3-39,

$D_i = .1342$

$A_i = .01414$

FROM PAGE 3-37

$(100)\text{ gpm} = (100)(2.228\text{ EE-3}) = .2228\ \text{ft}^3/\text{sec}$

$$v = \frac{Q}{A} = \frac{.2228}{.01414} = 15.76\ \text{ft/sec}$$

ASSUME REGULAR, SCREWED, STEEL FITTINGS. THE APPROXIMATE FITTING EQUIVALENT LENGTHS ARE

ELBOWS	2 × 3.4	= 6.8
GATE V	2 × 1.2	= 2.4
		9.2

SO, THE TOTAL EQUIVALENT LENGTH IS

$$30 + 9.2 = 39.2$$

FROM PAGE 3-20

$\epsilon = .0002$

$$\frac{\epsilon}{D} = \frac{.0002}{.1342} = .0015$$

AT 281°F, $\nu = .239$ EE-5 (INTERPOLATION)

$$N_{Re} = \frac{vD}{\nu} = \frac{(15.76)(.1342)}{.239\text{ EE-5}} = 8.85\text{ EE5}$$

FROM PAGE 3-20 $f = .022$

$$h_f = \frac{(.022)(39.2)(15.76)^2}{(2)(.1342)(32.2)} = 24.78$$

AT 281°, $\rho = 1/.01727 = 57.9\ \frac{\text{lbm}}{\text{ft}^3}$ (PAGE 6-29)

ALSO FROM PAGE 6-29, $p_{\text{VAPOR}} = 50.04$ psi

FROM EQN. 4.12

$$\text{NPSHA} = \frac{(80)\text{psia}\,(144)\,\frac{\text{in}^2}{\text{ft}^2}}{57.9\ \frac{\text{lbm}}{\text{ft}^3}} + 20 - 24.78 - \frac{(50.04)(144)}{57.9}$$

$= 69.73$

SINCE NPSHR = 10, THE PUMP WILL NOT CAVITATE AT ANY TIME AS LONG AS THE PRESSURE REMAINS AT 80 PSI. NOTICE THAT THE DISCHARGE LINE DOES NOT AFFECT NPSHA.

4

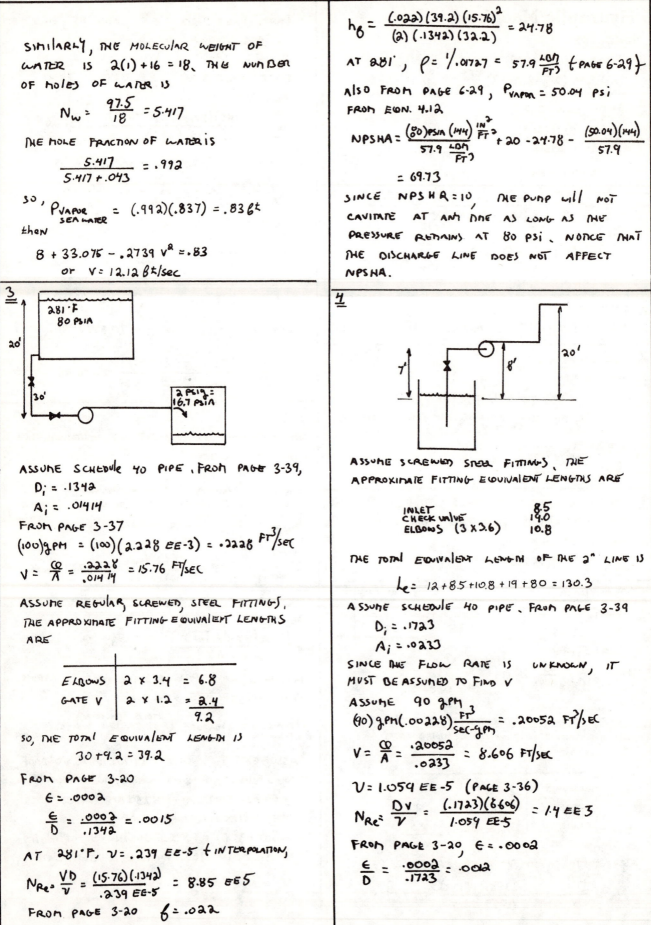

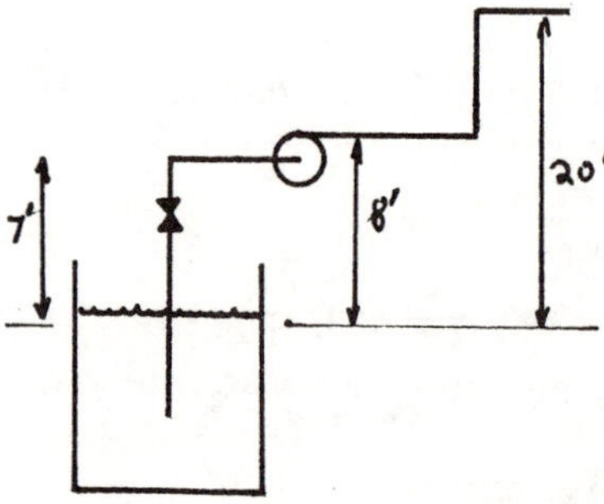

ASSUME SCREWED STEEL FITTINGS. THE APPROXIMATE FITTING EQUIVALENT LENGTHS ARE

INLET	8.5
CHECK VALVE	19.0
ELBOWS (3 × 3.6)	10.8

THE TOTAL EQUIVALENT LENGTH OF THE 2" LINE IS

$$L_e = 12 + 8.5 + 10.8 + 19 + 80 = 130.3$$

ASSUME SCHEDULE 40 PIPE. FROM PAGE 3-39

$D_i = .1723$

$A_i = .0233$

SINCE THE FLOW RATE IS UNKNOWN, IT MUST BE ASSUMED TO FIND V

ASSUME 90 gpm

$$(90)\text{ gpm}\,(.00228)\frac{\text{ft}^3}{\text{sec-gpm}} = .20052\ \text{ft}^3/\text{sec}$$

$$v = \frac{Q}{A} = \frac{.20052}{.0233} = 8.606\ \text{ft/sec}$$

$\nu = 1.059$ EE-5 (PAGE 3-36)

$$N_{Re} = \frac{Dv}{\nu} = \frac{(.1723)(8.606)}{1.059\text{ EE-5}} = 1.4\text{ EE3}$$

FROM PAGE 3-20, $\epsilon = .0002$

$$\frac{\epsilon}{D} = \frac{.0002}{.1723} = .0012$$

FROM PAGE 3-20, $f = .022$

THE FRICTION LOSS IN THE LINE AT 90 gpm IS

$$h_f = \frac{(.022)(130.3)(8.606)^2}{(2)(.1723)(32.2)} = 19.1$$

AT 90 gpm, THE VELOCITY HEAD IS

$$h_v = \frac{v^2}{2g} = \frac{(8.606)^2}{(2)(32.2)} = 1.2$$

IN GENERAL, THE VELOCITY HEAD IS $1.2\left(\frac{Q_2}{90}\right)^2$

THE TOTAL SYSTEM HEAD IS

$$H = (\Delta z) + h_v + h_f$$

$$= 20 + (1.2 + 19.1)\left(\frac{Q_2}{90}\right)^2$$

Q_2	H	Q_2	H
0	20.0	60	29.0
10	20.3	70	32.3
20	21.0	80	36.0
30	22.3	90	40.3
40	24.0	100	45.0
50	26.3	110	50.3

THE INTERSECTION OF THE SYSTEM AND PUMP CURVES IS AT 94 gpm ($H = 40$ FT)

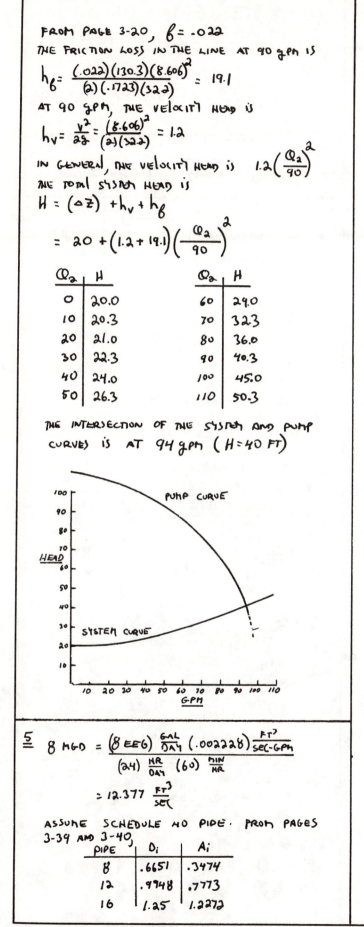

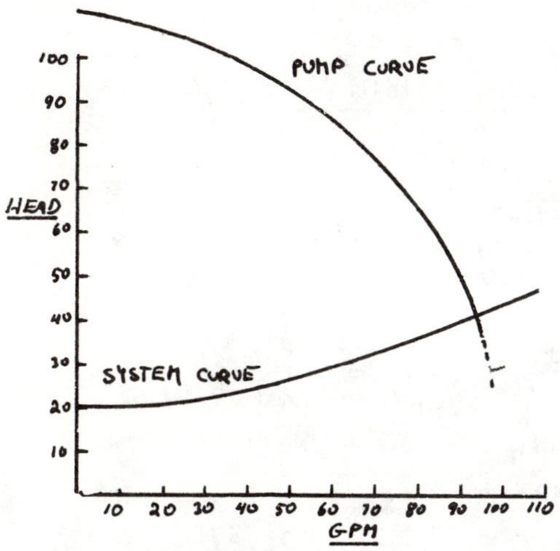

5 $8 \text{ MGD} = \dfrac{(8 \text{ EE6}) \frac{\text{GAL}}{\text{DAY}} (.002228) \frac{\text{FT}^3}{\text{SEC-GPM}}}{(24) \frac{\text{HR}}{\text{DAY}} (60) \frac{\text{MIN}}{\text{HR}}}$

$$= 12.377 \ \frac{\text{FT}^3}{\text{SEC}}$$

ASSUME SCHEDULE 40 PIPE. FROM PAGES 3-39 AND 3-40,

PIPE	D_i	A_i
8	.6651	.3474
12	.9948	.7773
16	1.25	1.2272

THE VELOCITY IN THE INLET PIPE IS

$$v = \frac{Q}{A} = \frac{12.377}{.3474} = 35.63 \text{ FT/SEC}$$

ASSUME 70°F, SO $\nu = 1.59$ EE-5 {PAGE 3-36}

THEN

$$N_{Re} = \frac{Dv}{\nu} = \frac{(.6651)(35.63)}{1.059 \text{ EE-5}} = 2.24 \text{ EE6}$$

FROM PAGE 3-20 FOR UNCOATED CAST IRON, $\epsilon = .0008$. THEN $\frac{\epsilon}{D} = \frac{.0008}{.6651} = .0012$

FROM PAGE 3-20 $f = .021$ SO

$$h_{f,1} = \frac{(.021)(1000)(35.63)^2}{(2)(.6651)(32.2)} = 622.4 \text{ FT}$$

FOR THE OUTLET PIPE,

$$v = \frac{Q}{A} = \frac{12.377}{.7773} = 15.92 \text{ FPS}$$

$$N_{Re} = \frac{(.9948)(15.92)}{1.059 \text{ EE-5}} = 1.5 \text{ EE6}$$

USING $\epsilon = .0008$, $\frac{\epsilon}{D} = \frac{.0008}{.9948} = .0008$

SO $f = .0185$

$$h_{f,2} = \frac{(.0185)(1500)(15.92)^2}{(2)(.9948)(32.2)} = 109.8$$

NOW, ASSUME A 50% SPLIT THROUGH THE 2 BRANCHES. IN THE UPPER BRANCH,

$$v = \frac{Q}{A} = \frac{\frac{1}{2}(12.377)}{.3474} = 17.81 \text{ FPS}$$

$$N_{Re} = \frac{(.6651)(17.81)}{(1.059 \text{ EE-5})} = 1.1 \text{ EE6}$$

USING $\frac{\epsilon}{D} = .012$, $f = .0205$

FOR THE 16" PIPE,

$$v = \frac{Q}{A} = \frac{\frac{1}{2}(12.377)}{1.2272} = \frac{6.189}{1.2272} = 5.04$$

$\epsilon = .0008$ SO $\frac{\epsilon}{D} = \frac{.0008}{1.25} = .0006$

$$N_{Re} = \frac{(1.25)(5.04)}{1.059 \text{ EE-5}} = 5.9 \text{ EE5}$$

FROM PAGE 3-20, $f = .018$

THESE VALUES OF f FOR THE 2 BRANCHES ARE FAIRLY INSENSITIVE TO CHANGES IN Q, SO THEY WILL BE USED FOR THE REST OF THE PROBLEM.

IN THE UPPER BRANCH, ASSUMING 4 MGD,

$$h_{f,\text{UPPER}} = \frac{(.0205)(500)(17.81)^2}{(2)(.6651)(32.2)} = 75.91$$

EQN 4.32 PREDICTS THE LOSS FOR ANY OTHER FLOW.

$$H_2 = H_1\left(\frac{Q_2}{Q_1}\right)^2 = 75.91\left(\frac{Q}{6.189}\right)^2 = 1.98\ Q^2$$

SIMILARLY FOR THE LOWER BRANCH,

IN THE 8" SECTION

$$h_{f,\,lower,\,8'} = \frac{(.0205)(250)(17.81)^2}{(2)(.6651)(32.2)}$$

$$= 37.95$$

IN THE 16" SECTION

$$h_{f,\,lower,\,16"} = \frac{(.018)(1000)(5.04)^2}{(2)(1.25)(32.2)}$$

$$= 5.68'$$

THE TOTAL LOSS IN THE LOWER BRANCH IS

$$h_{f,\,lower} = 37.95 + 5.68 = 43.63$$

FOR ANY OTHER FLOW, THE LOSS will be

$$H_2 = 43.63\left(\frac{Q}{6.189}\right)^2 = 1.14\,(Q)^2$$

LET X BE THE FRACTION FLOWING IN THE UPPER BRANCH. THEN, BECAUSE THE FRICTION LOSSES ARE EQUAL,

$$1.98\,[(x)\,12.377]^2 = 1.14\,[(1-x)\,12.377]^2$$

OR $x = .432$

THEN

$$Q_{upper} = .432\,(12.377) = 5.347$$

$$Q_{lower} = (1-.432)(12.377) = 7.03$$

$$H_2 = 1.98\,(5.347)^2 = 56.61$$

$$h_{f,\,total} = 622.4 + 56.61 + 109.8 = 788.81$$

6 a) AT 70°F, THE HEAD DROPPED IS

$$H = \frac{(500-30)\,\frac{LBF}{IN^2}\,(144)\,\frac{IN^2}{FT}}{62.4\,\frac{LBM}{FT^3}} = 1084.6\ FT$$

FROM EQN 4.52, THE SPECIFIC SPEED OF A TURBINE IS

$$N_s = \frac{N\sqrt{bhp}}{(H)^{1.25}} = \frac{1750\sqrt{250}}{(1084.6)^{1.25}} = 4.445$$

SINCE THE LOWEST SUGGESTED VALUE OF N_s FOR A REACTION TURBINE IS 10 {PAGE 4-16} WE RECOMMEND AN IMPULSE {PELTON} WHEEL.

b) FROM PAGE 3-30 {APPLICATION E}

$$Q = Av = \frac{\pi}{4}\left(\frac{4}{12}\right)^2(35) = 3.054\ CFS$$

$$Q' = \frac{(35-10)(3.054)}{35} = 2.181\ CFS \quad \{EQN\ 3.168\}$$

ASSUMING 62.4 LBM/FT³

$$F_x = -\frac{(2.181)(62.4)}{32.2}(35-10)(1-\cos 80°) = -87.32\ LBF$$

$$F_y = \frac{(2.181)(62.4)}{32.2}(35-10)(\sin 80°) = 104.06\ LBF$$

$$R = \sqrt{(87.32)^2 + (104.06)^2} = 135.84\ LBF$$

7

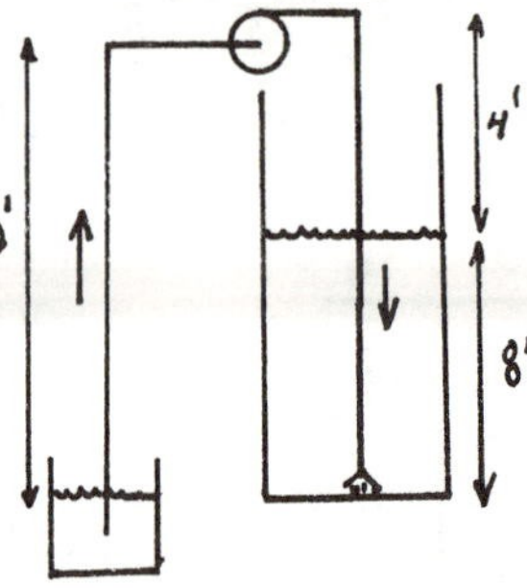

$$80\ gpm = (80)(.002228) = .1782\ CFS$$

$$A = \frac{\pi}{4}\left(\frac{2}{12}\right)^2 = .0218$$

$$V = \frac{Q}{A} = \frac{.1782}{.0218} = 8.174\ FPS$$

USE 80° WATER DATA.

$\nu = .93\ EE{-5}$ {PAGE 3-36}

$$N_{Re} = \frac{VD}{\nu} = \frac{(8.174)(2/12)}{(.93\ EE{-5})} = 1.46\ EE5$$

ASSUME THE RUBBER HOSE IS SMOOTH. THEN, FROM PAGE 3-20, $f = .016$

$$h_f = \frac{(.016)(50)(8.174)^2}{(2)\left(\frac{2}{12}\right)(32.2)} = 5'$$

$$h_v = \frac{(8.174)^2}{2(32.2)} = 1.0$$

$$h_A = 5 + 1 + 12 - 4 = 14\ FT$$

THIS NEGLECTS ENTRANCE LOSSES

TIMED

1 THIS IS SIMILAR TO EXAMPLE 4.10.

a) FROM EQN 3.56, THE TOTAL HEAD ENTERING IS

$$H = 92.5 + \frac{(12)^2}{(2)(32.2)} + 5.26 = 100'$$

b) THE WATER HORSEPOWER IS

$$\frac{\dot{m}H}{550} = \frac{(25)\ FT^3/sec\ (62.4)\ LBM/FT^3\ (100)\ FT}{550}$$

$$= 283.6\ HP$$

$$\eta = \frac{250}{283.6} = .881$$

c) FROM EQN 6.33

$$N_2 = N_1\sqrt{H_2/H_1} = 610\sqrt{\frac{225}{100}} = 915\ RPM$$

d) FROM EQN 4.40,

$$HP_2 = HP_1\left(\frac{H_2}{H_1}\right)^{1.5} = 250\left(\frac{225}{100}\right)^{1.5} = 843.75 \text{ HP}$$

e) FROM EQN 4.39,

$$Q_2 = Q_1\sqrt{H_2/H_1} = 25\sqrt{\frac{225}{100}} = 37.5 \text{ CFS}$$

2 (a) 1,000,000 BTUH SYSTEM

THE ENTHALPY DROP ACROSS THE RADIATOR IS

$$(167.99 - 147.92) = 20.07 \text{ BTU/LBM}$$

THE MASS FLOW RATE THROUGH THE SYSTEM IS

$$\frac{(1{,}000{,}000)\ \frac{\text{BTU}}{\text{HR}}}{(60)\ \frac{\text{MIN}}{\text{HR}}\ (20.07)\ \frac{\text{BTU}}{\text{LBM}}} = 830.4\ \frac{\text{LBM}}{\text{MIN}}$$

AT THE BULK TEMPERATURE OF 190°F,

$$\rho = \frac{1}{.01657} = 60.35\ \frac{\text{LBM}}{\text{FT}^3}$$

THE VOLUME FLOW RATE IS

$$Q = \frac{(830.4)\ \frac{\text{LBM}}{\text{MIN}}\ (7.48)\ \frac{\text{GAL}}{\text{FT}^3}}{(60.35)\ \frac{\text{LBM}}{\text{FT}^3}} = 102.9 \text{ gpm}$$

NEXT, PLOT THE PUMP CURVES AND GET THE HEAD LOSSES AT 102.9 gpm

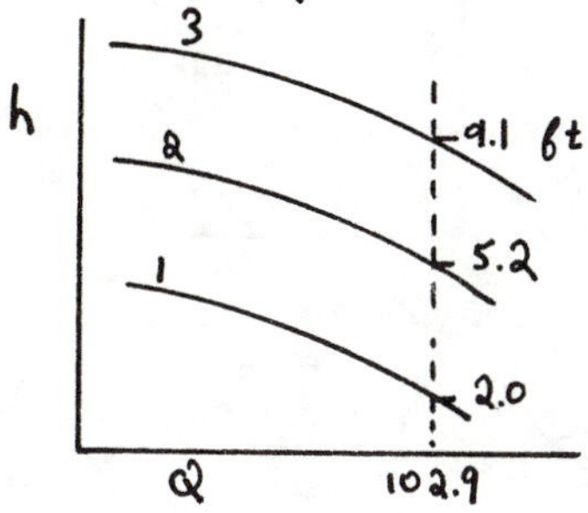

THE LOSS PER FOOT FOR THESE 3 PUMPS IS

PUMP 1 $(2.0)(12)/420 = .057$ IN/FT - TOO LOW

PUMP 2 $(5.2)(12)/420 = .12$ IN/FT - TOO LOW

PUMP 3 $(9.1)(12)/420 = .26$ IN/FT - OK

SO, CHOOSE **PUMP #3**

SINCE THE PIPE MATERIAL WAS NOT SPECIFIED, ASSUME $f = .020$ (GOOD FOR TURBULENT FLOW IN STEEL PIPE 1" TO 3" DIAMETER).

KEEP VELOCITY BETWEEN 2 AND 4 FPS FOR NOISE

$$h_f = \frac{fLv^2}{2Dg} \quad \text{SO}$$

$$D = \frac{fLv^2}{2h_f g} = \frac{(.02)(420)(4)^2}{(2)(9.1)(32.2)} = .229 \text{ FT}$$

SAY D = 3"

TRY A DIFFERENT APPROACH

$$A = \frac{Q}{V} = \frac{(102.9)\ \frac{\text{GAL}}{\text{MIN}}}{(60)\ \frac{\text{SEC}}{\text{MIN}}\ (7.48)\ \frac{\text{GAL}}{\text{FT}^3}\ (4)\ \frac{\text{FT}}{\text{SEC}}}$$

$$= .057 \text{ FT}^2$$

$$D = \sqrt{\frac{4A}{\pi}} = \sqrt{\frac{(4)(.057)}{\pi}} = .27 \text{ FT}$$

SAY **D = 3½"**

THE ACTUAL PIPE INSIDE DIAMETER WILL DEPEND ON WHETHER COPPER, STEEL, OR CAST IRON PIPE IS USED.

(b) 300,000 BTUH

PROCEED SIMILARLY.

$$Q = \frac{(300{,}000)(7.48)}{(60)(20.07)(60.35)} = 30.88 \text{ gpm}$$

THE HEAD LOSS AT 30.88 gpm FOR EACH PUMP IS

PUMP 1 $\frac{(5.5)(12)}{420} = .157"$ TOO LOW

PUMP 2 $\frac{(9.0)(12)}{420} = .257$ OKAY

PUMP 3 $\frac{(13.4)(12)}{420} = .38$ OKAY

PUMPS 2 AND 3 MEET THE SPECIFICATIONS

$$A = \frac{30.88}{(60)(7.48)(4)} = .0172$$

$$D = \sqrt{\frac{(4)(.0172)}{\pi}} = .148'$$

SAY, **1¾" PIPE**

3 REDRAW AS A TOPOGRAPHIC MAP

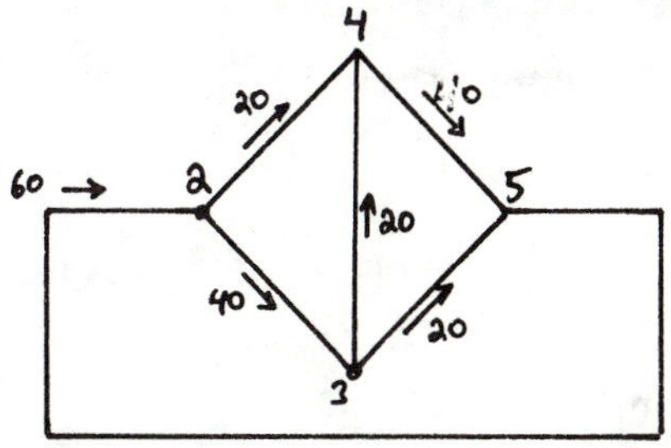

THIS CANNOT BE SIMPLIFIED WITHOUT USING WYE-DELTA TRANSFORMATIONS.

HOWEVER, THE PRESSURE DROP FROM 2-5 IS INDEPENDENT OF PATH 3-4.

THE DROP 2-4-5 IS THE SAME AS THE DROP 2-3-5.

ALTHOUGH THE DARCY EQUATION COULD BE USED TO CALCULATE FRICTION LOSSES, IT IS MORE CONVENIENT TO USE STANDARD TABLES FOR TYPE L TUBING.

FROM THE ASHRAE GUIDE AND DATA BOOK (APPLICATIONS), PAGE 157 1966-1967

SECTION	LOSS PER FT	LENGTH	TOTAL LOSS
5-2	.3" W.G.	40	12.0
2-4	.47	70	32.9
4-5	.4	50	20.0

THE TOTAL LOSS ALONG PATH 2-4-5-2 IS

$$\frac{12 + 32.9 + 20}{12} = \boxed{5.4 \text{ FT}}$$

{IF PATH 2-3-5-2 HAD BEEN USED, THE LOSS WOULD HAVE BEEN 5.2 FT - CLOSE ENOUGH FOR THIS ROUNDED DATA}

ASSUME THE WATER IS APPROXIMATELY 200°F. THEN $\rho = 60.1$ LBM/FT^3

$$\dot{m} = (60)\,\text{GPM}\,(.1337)\,\frac{\text{FT}^3}{\text{GAL}}\,(60.1)\,\frac{\text{LBM}}{\text{FT}^3}$$

$$= 482\ \frac{\text{LBM}}{\text{MIN}}$$

$$\text{WHP} = \frac{(482)(5.4)\ \text{FT}}{(33{,}000)\ \frac{\text{FT-LBF}}{\text{HP-MIN}}} = .079\ \text{HP}$$

$$\text{MOTOR HP} = \frac{.079}{.45} = .175 \quad \boxed{\text{SAY } \tfrac{1}{4} \text{ HP}}$$

$$q = \dot{m} C_p \Delta T$$

$$= (482)\frac{\text{LBM}}{\text{MIN}}(60)\frac{\text{MIN}}{\text{HR}}(1)\frac{\text{BTU}}{\text{LBM-°F}}(20)°\text{F}$$

$$= \boxed{578{,}400 \text{ BTUH}}$$

4 ASSUME REGULAR 90° ELBOWS

THE EQUIVALENT LENGTH OF PIPE AND FITTINGS IS

$$500 + 6(4.4) + 2(2.8) = 532 \text{ FT}$$

FOR 3" STEEL SCHEDULE 40 PIPE,

$$A_i = .05134 \text{ FT}^2$$

$$D_i = .2557 \text{ FT}$$

ASSUME 100 GPM. (THE ACTUAL VALUE YOU CHOOSE FOR THIS FIRST ESTIMATE IS NOT IMPORTANT)

$$v = \frac{Q}{A} = \frac{(100)\,\text{GPM}\,(.1337)\frac{\text{FT}^3}{\text{GAL}}}{(.05134)\,\text{FT}^2\,(60)\frac{\text{SEC}}{\text{MIN}}} = 4.34\ \text{FT/SEC}$$

$$N_{Re} = \frac{Dv}{\nu} = \frac{(.2557)(4.34)}{6\text{EE-}6} = 1.85\ \text{EE}5$$

USE $\epsilon = .0002$

$$\text{SO}\quad \frac{\epsilon}{D} = \frac{.0002}{.2557} = .0008$$

FROM THE MOODY CHART,

$f \approx .02$ (GOES TO .018 AT FULL TURBULENCE) WHICH THIS WILL BE AT HIGHER FLOWS. SO, USE $f = .018$.

$$h_f = \frac{(.018)(532)(4.34)^2}{(2)(.2557)(32.2)} = 10.95 \text{ FT OF GASOLINE}$$

USING $\frac{h_1}{h_2} = \left(\frac{Q_1}{Q_2}\right)^2$ THE OTHER SYSTEM CURVE POINTS CAN BE FOUND

$$h_2 = \frac{10.95}{(100)^2} Q_2^2 = .001095\, Q_2^2$$

Q	h	h+60
100	10.95	71.0
200	43.8	103.8
300	98.6	158.6
400	175.2	235.2
500	273.8	333.8
600	394.2	454.2

PLOTTING THE SYSTEM AND PUMP CURVES GIVES

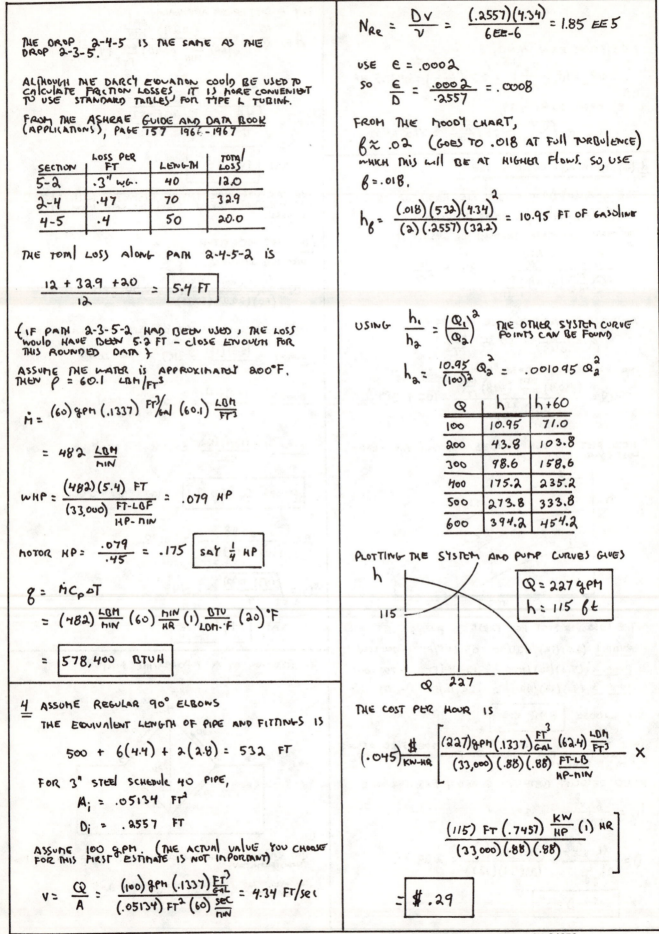

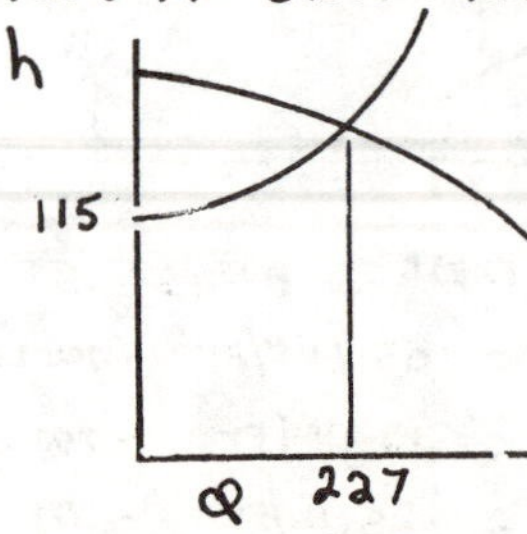

$$\boxed{\begin{aligned} Q &= 227 \text{ GPM} \\ h &= 115 \text{ ft} \end{aligned}}$$

THE COST PER HOUR IS

$$(.045)\frac{\$}{\text{KW-HR}}\left[\frac{(227)\,\text{GPM}\,(.1337)\frac{\text{FT}^3}{\text{GAL}}\,(62.4)\frac{\text{LBM}}{\text{FT}^3}}{(33{,}000)(.88)(.88)\ \frac{\text{FT-LB}}{\text{HP-MIN}}} \times \frac{(115)\ \text{FT}\ (.7457)\frac{\text{KW}}{\text{HP}}\ (1)\ \text{HR}}{(33{,}000)(.88)(.88)}\right]$$

$$= \boxed{\$.29}$$

RESERVED FOR FUTURE USE

Fans and Ductwork

1 FROM EQN 5.30 WITH R=4, THE SHORT SIDE IS

$$a = \frac{18''(4+1)^{.25}}{1.3(4)^{.625}} = 8.705''$$

THE LONG SIDE IS

$$b = (R)(a) = (4)(8.705) = 34.82$$

2 FROM EQN 5.11, OTHER POINTS ON THE SYSTEM CURVE CAN BE FOUND

$$\frac{P_2}{P_1} = \left(\frac{Q_2}{Q_1}\right)^2 \quad \text{OR} \quad Q_2 = Q_1\sqrt{P_2/P_1}$$

$$Q_2 = \frac{(10{,}000)\sqrt{P_2}}{\sqrt{4}} = 5000\sqrt{P_2}$$

P_2	Q
1"	5000 CFM
2"	7070
3"	8660
4"	10,000
5"	11,180
6"	12,250

3 AT SEA LEVEL, $\rho = .07651$
AT 5000 FT, $\rho = .06592$

DENSITY CAN BE CALCULATED FROM DATA ON P. 8-20

FROM EQN 5.7,

$$(AHP) = \frac{(40)\text{ CFM }(.5)''\text{ wg}}{6356 \frac{\text{CFM-wg}}{\text{HP}}} = 3.147\text{ EE-3}$$

FROM EQN 5.17,

$$(AHP)_2 = (3.147\text{ EE-3})\left(\frac{1}{\frac{1}{8}}\right)^5\left(\frac{\frac{1}{2}(300)}{300}\right)^3\left(\frac{.06592}{.07651}\right)$$

$$= 11.1$$

4 FROM PAGE 5-7, THE FRICTION LOSS IS .45" WATER PER 100 FT AND V=2700 FPM. THE FRICTION LOSS DUE TO THE DUCT IS

$$h_{f,1} = (.45)\text{ IN wg}\left(\frac{750}{100}\right) = 3.375''\text{ wg}$$

FROM TABLE 5.3, THE EQUIVALENT LENGTH OF EACH ELBOW IS 12 D. FOR THE TWO,

$$L_e = 4\left[(12)\left(\frac{20\text{ IN}}{12\text{ IN/FT}}\right)\right] = 80'$$

THIS CREATES A FRICTION LOSS OF

$$h_{f,2} = (.45)\left(\frac{80}{100}\right) = .36''\text{ wg}$$

ALSO, FROM TABLE 5.3, FOR $\frac{E}{D} = \frac{2}{20} = .10$, $C_1 = .2$. USING EQN 5.33,

$$h_{f,3} = 2\left[(.2)\left(\frac{2700}{4005}\right)^2\right] = .182''\text{ wg}$$

THE TOTAL LOSS IS

$$\Sigma h_f = 3.375 + .36 + .182 = 3.917''\text{ wg}$$

5 IN THE 18" DUCT,

$$A = \frac{\frac{\pi}{4}(18)^2}{144} = 1.767\text{ FT}^2$$

$$\text{SO } V = \frac{Q}{A} = \frac{1500}{1.767} = 848.9\text{ FPM}$$

IN THE 14" DUCT, $A = \frac{\frac{\pi}{4}(14)^2}{144} = 1.069\text{ FT}^2$

$$\text{SO } V = \frac{(1500-400)}{1.069} = 1029$$

FROM EQN 5.34

$$\Delta P_s = .75\left(\frac{(848.9)^2 - (1029)^2}{(4005)^2}\right) = -.0158''\text{ wg LOSS}$$

CONCENTRATES

1

FROM TABLE 5.5, SELECT 1600 FPM AS THE MAIN DUCT VELOCITY. FROM PAGE 24-9 WITH 1600 FPM AND 1500 CFM

$$h_f = .27''\text{ wg PER }100'$$

$$D_{FAN\text{-}A} = 13''$$

FROM EQN 5.30,

$$a = \frac{D(1.5+1)^{.25}}{1.3(1.5)^{.625}} = .75\,D$$

$$b = Ra = (1.5)(.75\,D) = 1.125\,D$$

$$\text{SO, } a_{FAN\text{-}A} = (.75)(13) = 9.75$$

AND

$$b_{FAN\text{-}A} = (1.125)(13) = 14.63$$

PROCEEDING SIMILARLY FOR ALL SECTIONS, THE FOLLOWING TABLE IS OBTAINED:

SECTION	Q	D	a x b
FAN–A	1500	13	9.8 x 14.6
A–B	1200	11.8	8.9 x 13.3
B–C	900	11.0	8.3 x 12.4
C–D	600	9.1	6.8 x 10.2
D–E	400	8.0	6 x 9
E–F	200	6.0	4.5 x 6.8

ASSUME $R/D = 1.5$ FOR THE BENDS. FROM TABLE 5.3, $L_e = 12D$. THEN, FOR THE 2 ELBOWS,

$$L_e = 12\left[\frac{13}{12} + \frac{9.1}{12}\right] = 22.1\text{ FT}$$

THE DISTANCE FROM FAN TO F IS

15 + 45 + 30 + 30 + 20 + 10 + 20 + 20 = 190

THEN

$$h_f = \left(\frac{190+22.1}{100}\right)(.27''\ \text{wg}/100\ \text{FT}) = .57''\ \text{wg}$$

THE FAN MUST SUPPLY THE TERMINAL PRESSURE ALSO

$$.57 + .25 = .82''\ \text{wg}$$

THE EQUAL FRICTION METHOD IGNORES ANY VELOCITY HEAD CONTRIBUTION TO STATIC PRESSURE.

2 CHOOSE 1600 FPM AS THE MAIN DUCT VELOCITY.

$$Q_{FAN} = (12)(300) = 3600\ \text{CFM}$$

FROM PAGE 5-7 {WITH Q = 3600 AND V = 1600 FPM}

$$h = .16''\ \text{wg PER 100 FT}$$

$$D_{FAN-1ST} = 20''$$

PROCEEDING ACCORDING TO THE EQUAL FRICTION METHOD RESULTS IN THE FOLLOWING TABLE:

SECTION	Q	D	SECTION	Q	D
FAN-1ST	3600	20	2ND-E	1200	13.2
1ST-2ND	2400	17.2	E-F	900	12
2ND-3RD	1200	13.2	F-G	600	10.1
1ST-A	1200	13.2	G-H	300	7.9
A-B	900	12	I-J	900	12
B-C	600	10.1	J-K	600	10.1
C-D	300	7.9	K-L	300	7.9

THE LONGEST RUN IS (FAN-L). ITS LENGTH IS

$$L_{LONGEST} = 25 + 35 + 20 + 20 + 10 + 20 + 20 + 20 = 170$$

FROM TABLE 5.3, THE ELBOWS HAVE $L_e = 14.5\ D$ {INTERPOLATED}. SO

$$L_e = 170 + 14.5\left(\frac{20'' + 13.2''}{12\ \frac{\text{IN}}{\text{FT}}}\right) = 210.1$$

THE FAN SUPPLIES

$$\left(\frac{210.1}{100}\right)(.16''\ \text{wg PER } 100') + .15 = .486''$$

3 CHOOSE 1600 FPM AS THE MAIN DUCT VELOCITY. THEN Q = 3600 CFM, AND FROM PAGE 5-7

$$h_f = .16''\ \text{wg PER 100 FEET}$$

$$D = 20''$$

AS IN PROBLEM 2, THE BENDS HAVE $L_e = 14.5\ D$. THEN, THE LOSS UP TO THE FIRST TAKE-OFF IS

$$h_{f,\text{FAN-1ST}} = \frac{.16\left[25 + 35 + 14.5\left(\frac{20}{12}\right)\right]}{100} = .135''\ \text{wg}$$

ASSUME R = .75

AFTER THE FIRST TAKE OFF,

$$Q = 3600 - 1200 = 2400$$

$$L_e = 20$$

$$L_e/Q^{.61} = \frac{20}{(2400)^{.61}} = .173$$

FROM FIGURE 5.6 OR EQN 5.40 $V_2 = 1390$. PROCEEDING SIMILARLY, THE FOLLOWING TABLE IS DEVELOPED:

SECTION	Q	L_e	$L_e/Q^{.61}$	V **	D
FAN-1ST BR	3600	80.8	.547	1600	20
1ST BR-2ND BR	2400	20	.173	1390	17.8
2ND - I *	1200	49.3	.65	960	15.1
I-J	900	20	.31	800	14.4
J-K	600	20	.40	650	13.0
K-L	300	20	.62	500	10.5

* L_e FOUND ASSUMING D = 16

$$L_e = 20 + 14.5\left(\frac{16}{12}\right) + 10 = 49.3$$

** SOLVED GRAPHICALLY.

SINCE THE SYSTEM IS SYMMETRICAL, THE OTHER BRANCHES ARE THE SAME. THE FAN MUST SUPPLY

$$.135'' + .15'' = .285''\ \text{wg}$$

TIMED

1 NO DAMPER IS NEEDED IN DUCT A.

THE AREA OF SECTION A IS

$$\frac{\pi}{4}\left(\frac{12}{12}\right)^2 = .7854\ \text{FT}^2$$

THE VELOCITY IN SECTION A IS

$$V_A = \frac{Q}{A} = \frac{3000\ \text{FT}^3/\text{MIN}}{(60)\frac{\text{SEC}}{\text{MIN}}(.7854)\ \text{FT}^2} = 63.66\ \text{FT/SEC}$$

$$= 3819.7\ \text{FT/MIN}$$

FROM PAGE 32, FIGURE 8, 1969 ASHRAE <u>EQUIPMENT GUIDE + DATA BOOK</u>, FOR 4-PIECE ELLS, WITH $r/d = 1.5$

$$L_e \approx 14\ D$$

ASSUME D = 10".

THE TOTAL EQUIVALENT LENGTH OF RUN C IS

$$L_e = 50+10+10+2\left[14\left(\frac{10''}{12\,\text{in/ft}}\right)\right] = 93.3 \text{ FT}$$

FOR ANY DIAMETER, D, IN INCHES OF SECTION C, THE VELOCITY WILL BE

$$V_C = \frac{Q}{A} = \frac{2000\ \text{FT}^3/\text{MIN}}{\left(\frac{\pi}{4}\right)\left(\frac{D}{12}\right)^2 \text{FT}^2\ 60\ \frac{\text{SEC}}{\text{MIN}}} = \frac{6111.5}{D^2}$$

THE FRICTION LOSS IN SECTION C WILL BE (EQN 3.71)

$$h_{f,C} = \frac{(.02)(93.3)(6111.5/D^2)^2}{(2)\left(\frac{D}{12}\right)(32.2)} = \frac{1.3\text{ EE7}}{D^5} \text{ FT. OF AIR}$$

THE REGAIN BETWEEN A AND C WILL BE

$$h_{REGAIN} = .65\left[\frac{V_A^2 - V_C^2}{2g}\right]$$

$$h_{REGAIN} = .65\left[\frac{(63.66)^2 - \left(\frac{6111.5}{D^2}\right)^2}{(2)(32.2)}\right]$$

$$= 40.9 - \frac{3.77\text{ EE5}}{D^4} \text{ FT OF AIR}$$

THE PRINCIPLE OF STATIC REGAIN IS THAT

$h_f = h_{REGAIN}$ SO

$$\frac{1.3\text{ EE7}}{D^5} = 40.9 - \frac{3.77\text{ EE5}}{D^4}$$

BY TRIAL AND ERROR, $D = 13.5$

SINCE $D = 10''$ WAS ASSUMED TO FIND THE EQUIVALENT LENGTH OF THE ELLS, THIS PROCESS SHOULD BE REPEATED.

$$L_e = 70 + 2\left[(14)\left(\frac{13.5}{12}\right)\right] = 101.5$$

$$h_f = \frac{(.02)(101.5)(6111.5/D^2)^2}{(2)\left(\frac{D}{12}\right)(32.2)} = \frac{1.41\text{ EE7}}{D^5}$$

THEN

$$\frac{1.41\text{ EE7}}{D^5} = 40.9 - \frac{3.77\text{ EE5}}{D^4}$$

BY TRIAL AND ERROR, $D = 13.63''$

THIS RESULTS IN A FRICTION LOSS OF

$$h_f = \frac{1.41\text{ EE7}}{(13.63)^5} = 30 \text{ FT OF AIR.}$$

HOWEVER, THE REGAIN CANCELS THIS, SO THE PRESSURE LOSS FROM A TO C IS ZERO, NO DAMPERS ARE NEEDED IN DUCT C.

FOR ANY DIAMETER, D, IN INCHES IN SECTION B, THE VELOCITY WILL BE

$$V_B = \frac{Q}{A} = \frac{1000}{\left(\frac{\pi}{4}\right)\left(\frac{D}{12}\right)^2(60)} = \frac{3055.8}{D^2}$$

THE FRICTION LOSS IN SECTION B WILL BE

$$h_f = \frac{(.02)(10)\left(\frac{3055.8}{D^2}\right)^2}{(2)\left(\frac{D}{12}\right)(32.2)} = \frac{3.48\text{ EE5}}{D^5} \text{ FT OF AIR}$$

THE 90° TAKE-OFF WILL ALSO CREATE A LOSS. FROM EQN 5.33,

$$h_f = C_b\left(\frac{V_A}{4005}\right)^2 \text{ INCHES OF WATER}$$

AT THIS POINT, ASSUME $\frac{V_B}{V_A} = 1.00$. THEN, FROM TABLE 5.4 FOR A 45° FITTING, $C_b = .5$

SO

$$h_f = .5\left(\frac{3819.7}{4005}\right)^2 = .455 \text{ INCHES WATER} = 31.5 \text{ FT OF AIR}$$

THE REGAIN BETWEEN A AND B WILL BE

$$h_{REGAIN} = .65\left[\frac{V_A^2 - V_B^2}{2g}\right]$$

$$h_{REGAIN} = .65\left[\frac{(63.66)^2 - (3055.8/D^2)^2}{(2)(32.2)}\right]$$

$$= 40.9 - \frac{9.42\text{ EE4}}{D^4}$$

THEN SETTING REGAIN EQUAL TO LOSS

$$40.9 - \frac{9.42\text{ EE4}}{D^4} = \frac{3.48\text{ EE5}}{D^5} + 31.5$$

BY TRIAL AND ERROR,

$$D_B = 10.77''$$

THIS MAKES $V_B = \frac{3055.8}{(10.77)^2} = 26.34$ FPS

AND SINCE $\frac{V_B}{V_A} = \frac{26.34}{63.6} \approx .4$, THE VALUE OF C_b IS STILL ABOUT .5

FOR SECTION B, THE FRICTION THAT IS CANCELLED BY THE REGAIN IS

$$h_f = 31.5 + \frac{3.48\text{ EE5}}{(10.77)^2} = 33.9 \text{ FT}$$

A DAMPER IS REQUIRED IN B TO ENSURE THAT ONLY 1000 CFM PASS THROUGH B.

RESERVED FOR FUTURE USE

Thermodynamics

WARM-UPS

1 FROM PAGE 6-29,

$h = 218.48 + .92(945.5) = 1088.34$ BTU/LBM

THE MOLECULAR WEIGHT OF STEAM (WATER) IS 18, SO

$H = 18(1088.34) = 19590$ BTU/PMOLE

2 FROM PAGE 6-30,

$h_1 = 393.84 + .95(809) = 1162.4$

FROM THE MOLLIER DIAGRAM FOR AN ISENTROPIC PROCESS,

$h_2 = 1031$ BTU/LBM

THE MAXIMUM WORK OUTPUT IS

$h_1 - h_2 = 1162.4 - 1031 = 131.4$ BTU/LBM

3 THE AVE. VALUE OF C_p FOR IRON IS

$\bar{C}_p = .10$ (TABLE 6.1)

$$Q = (.10)\frac{\text{BTU}}{\text{LBM-°F}}(780-80)\text{°F} = 70.0 \text{ BTU/LBM}$$

4 FROM PAGE 10-33

$C_p = .250$ BTU/LBM-°F

FROM TABLE 6.4, $\longrightarrow R = 53.3 \frac{\text{FT-LBF}}{\text{LBM-°R}}$

$$C_v = C_p - \frac{R}{J} = .250\frac{\text{BTU}}{\text{LBM-°F}} - \frac{53.3 \frac{\text{FT-LBF}}{\text{LBM-°R}}}{778 \frac{\text{FT-LBF}}{\text{BTU}}}$$

$= .1815$

SO $K = \frac{C_p}{C_v} = \frac{.250}{.1815} = 1.377$

5 a) FROM EQN 3.21,

$$C_{AIR} = \sqrt{kgRT}$$

$$= \sqrt{(1.4)(32.2)(53.3)(460+70)} = 1128.5 \text{ FT/sec}$$

b) FROM TABLE 14.1

$E_{STEEL} = 30$ EE6 PSI

$\rho_{STEEL} = .283$ LBM/IN^3

FROM EQN 3.20,

$$C_{STEEL} = \sqrt{\frac{(30 \text{ EE6})\frac{\text{LBF}}{\text{IN}^2}(32.2)\frac{\text{FT}}{\text{SEC}^2}}{(.283)\frac{\text{LBM}}{\text{IN}^3}(12)\frac{\text{IN}}{\text{FT}}}} = 16866 \text{ FT/sec}$$

c) FROM PAGE 3-36

$E = 320$ EE3 LBF/IN^2

$\rho = 62.3$ LBM/FT^3

FROM EQN 3.20,

$$C_{WATER} = \sqrt{\frac{(320 \text{ EE3})(144)\,32.2}{62.3}} = 4880 \text{ FT/sec}$$

6 FROM TABLE 6.4

$R_{He} = 386.3$

FROM EQN 6.45

$$\rho = \frac{P}{RT} = \frac{(14.7)\frac{\text{LBF}}{\text{IN}^2}(144)\frac{\text{IN}^2}{\text{FT}^2}}{(386.3)\frac{\text{FT-LBF}}{\text{LBM-°R}}(460+600)\text{°R}}$$

$= .00517$ LBM/FT^3

7 FROM PAGE 6-30

$h_1 = 1187.2$ BTU/LBM

FROM THE MOLLIER DIAGRAM (PAGE 6-34)

$h_2 = 953$ BTU/LBM (AT 3 PSIA, $S_2 = S_1$)

$h_2' = 1022$ BTU/LBM

FROM EQN 7.11

$$\eta_s = \frac{1187.2 - 1022}{1187.2 - 953} = .705$$

8 FROM PAGE 7-33

$$(470)\text{ BTU}(2.93 \text{ EE-4})\frac{\text{KW-HRS}}{\text{BTU}} = 1.377 \text{ EE-1 KW-HRS}$$

9 THE AIR WEIGHT (FROM EQN

$$M = \frac{P_1V_1}{RT_1} = \frac{(14.7)\text{ LBF/IN}^2(144)\text{ IN}^2/\text{FT}^2(8)\text{ FT}^3}{(53.3)\frac{\text{FT-LBF}}{\text{LBM-°R}}(460+180)\text{°R}}$$

$= .4964$ LBM

THEN, ALSO FROM EQN 6.41

$$V_2 = \frac{wRT_2}{P_2} = \frac{(.4964)\text{ LBM}(53.3)\frac{\text{FT-LBF}}{\text{LBM-°R}}(460+100)\text{°R}}{(14.7)\frac{\text{LBF}}{\text{IN}^2}(144)\text{ IN}^2/\text{FT}^2}$$

$= 7.00$ FT^3

FROM PAGE 6-24 FOR A CONSTANT PRESSURE PROCESS

$$W = P(V_2 - V_1) = (14.7)(144)\frac{\text{LBF}}{\text{FT}^2}(7.00-8)\text{ FT}^3$$

$= -2116.8$ FT-LBF

(NEGATIVE BECAUSE WORK IS DONE ON SYSTEM)

10 ASSUME THE BUILDING NEEDS 3 EE5 CFH OF 75°F AIR. THEN

$$\dot{M}_{AIR} = \frac{PV}{RT} = \frac{(14.7)(144)(3 \text{ EE5})}{(53.3)(460+75)} = 2.227 \text{ EE4 LBM/HR}$$

THIS IS A CONSTANT PRESSURE PROCESS, SO

$$Q = \dot{M}C_p\Delta T = \frac{(2.227 \text{ EE4})\frac{\text{LBM}}{\text{HR}}(.241)\frac{\text{BTU}}{\text{LBM-°R}}(75-35)\text{°F}}{3600\frac{\text{SEC}}{\text{HR}}}$$

$= 59.63$ BTU/SEC

FOR THE WATER

$$Q = \dot{M}_W C_p \Delta T = V\rho C_p \Delta T$$

(MORE)

WARM-UP #10 CONTINUED

DENSITY OF WATER AT 165°F ≈ 61 LBM/FT³

SO $59.63 = (\text{gpm})(.002228)\frac{\text{FT}^3}{\text{SEC-gpm}}(61)\frac{\text{LBM}}{\text{FT}^3} \times$

$\times (1)\frac{\text{BTU}}{\text{LBM-°F}}(180-150)\,°\text{F}$

SO gpm = 14.63

CONCENTRATES

1 a) FROM PAGE 7-33

$(5000)\,\text{KW}\,(3412.9)\frac{\text{BTU}}{\text{HR-KW}} = 1.706\text{ EE7 BTU/HR}$

T_{SAT} FOR 200 PSIA STEAM IS 381.8°F

SO THIS STEAM IS AT 381.8 + 100 = 481.8°F

$h_1 \approx 1258$ BTU/LBM

$h_2 \approx 868$ {FROM MOLLIER, ASSUMING ISENTROPIC EXPANSION}

SO $\Delta h = 1258 - 868 = 390$ BTU/LBM

THE STEAM FLOW RATE IS

$$\dot{m} = \frac{(1.706\text{ EE7})\text{ BTU/HR}}{(390)\text{ BTU/LBM}} = 4.374\text{ EE4 LBM/HR}$$

FROM PAGE 7-2, THE WATER RATE IS

$$WR = \frac{\dot{M}}{\text{KW}} = \frac{4.374\text{ EE4 }\frac{\text{LBM}}{\text{HR}}}{5000\text{ KW}} = 8.749\frac{\text{LBM}}{\text{KW-HR}}$$

b) IF, WHEN THE LOAD DECREASES, THE STEAM FLOW IS REDUCED ACCORDINGLY, THERE WILL BE NO LOSS OF ENERGY. IF, HOWEVER, THE STEAM IS THROTTLED TO REDUCE THE AVAILABILITY,

$$\text{LOSS} = \tfrac{1}{2}(h_1 - h_2) = \tfrac{1}{2}(390) = 195\text{ BTU/LBM}$$

2 $h_1 \approx 1390$ BTU/LBM {FROM MOLLIER}

$h_2 \approx 935$ {IF EXPANSION IS ISENTROPIC}

THE 'ADIABATIC HEAT DROP' IS

$h_1 - h_2 = 1390 - 935 = 455$ BTU/LBM

3 THE WATER RATE IS

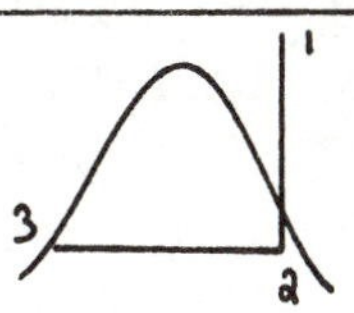

$$WR = \frac{(\dot{M})\text{ LBM/HR}}{(P)\text{ KW}}$$

OR $P = \frac{\dot{M}}{WR}$

ALSO, $\dot{M} = \frac{(P)\text{ KW }(3412.9)\frac{\text{BTU}}{\text{HR-KW}}}{(\Delta h)\text{ BTU/LBM}}$

SO, $\Delta h = \frac{3412.9}{WR} = \frac{3412.9}{20} = 170.65$

THE ACTUAL STEAM FLOW IS

$\dot{M} = (P)\text{ KW }(WR)\frac{\text{LBM}}{\text{HR-KW}} = (750)(20) = 15{,}000\text{ LBM/HR}$

$P_1 = 150$ PSIG = 164.7 PSIA {SAY 165}

$T_{SAT} = 366°$F FOR 165 PSIA STEAM

$T_1 = T_{SAT} + T_{SUPERHEAT} = 366 + 50 = 416°$F

$h_1 \approx 1226$ FROM MOLLIER DIAGRAM

$P_2 = (26)\text{ IN}_{Hg}(.491)\frac{\text{LBF}}{\text{IN}^3} = 12.77$ PSIA

$h_2 = 1226 - 170.65 = 1055.4$ BTU/LBM

$h_{f,3} \approx 171.6$ BTU/LBM {INTERPOLATED FROM PAGE 6-30},

SO, THE HEAT REMOVAL IS

$(1055.4 - 171.6)\text{ BTU/LBM }(15000)\text{ LBM/HR} = 1.326\text{ EE7 BTU/HR}$

THE SATURATION TEMPERATURE CORRESPONDING TO P_2 = 12.77 PSI IS $T_2 = T_3 \approx 204°$F

ASSUMING THE WATER AND STEAM LEAVE IN THERMAL EQUILIBRIUM, THE COOLING WATER BALANCE IS

$C_p = .999$ BTU/LBM-°F {PAGE 3-28}

$q_{IN} = \dot{m} C_p \Delta T$

$(1.326\text{EE7}) = (\dot{m}_w)(.999)(204-65)$

$\dot{M}_W = 9.55\text{ EE4 LBM/HR}$

4 T_{SAT} FOR 4.45 PSIA STEAM ≈ 157°F

ASSUME COUNTER FLOW OPERATION TO CALCULATE ΔT_M

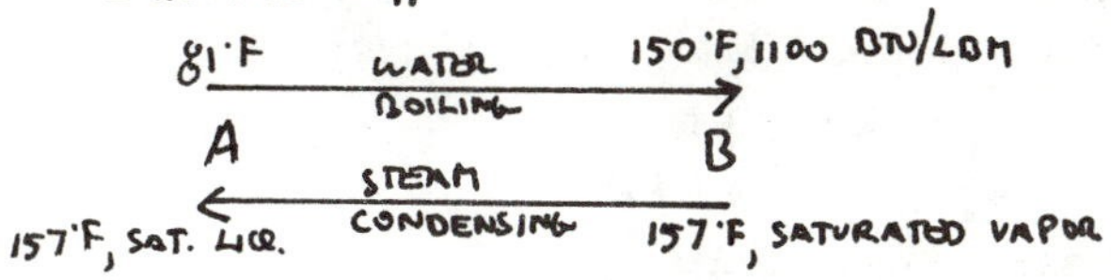

THE AVERAGE WATER TEMPERATURE IS

$\tfrac{1}{2}(81 + 150) = 115.5$, SO

$C_p \approx .999$ BTU/LBM-°F {PAGE 3-28}

ASSUMING 81°F SATURATED WATER,

$h_{w,1} \approx 49$ BTU/LBM

THE HEAT TRANSFERRED TO THE WATER IS

$Q = \dot{m}\Delta h = (332{,}000)\frac{\text{LBM}}{\text{HR}}(1100-49)\frac{\text{BTU}}{\text{LBM}}$

$= 3.489\text{ EE8 BTU/HR}$

FROM EQN 10.66

$$\Delta T_M = \frac{(157-81) - (157-150)}{\ln\left(\frac{157-81}{157-150}\right)} = 28.93$$

SINCE $T_{STEAM,A} = T_{STEAM,B}$ THE CORRECTION FACTOR FOR ΔT_M {F_c IN EQN 10.68} IS ONE.

{MORE}

CONCENTRATE #4 CONTINUED

FROM EQN 10.67,

$$U = \frac{Q}{A\Delta T_M} = \frac{(3.489\ \text{EE}8)\ \text{BTU/HR}}{(1850)\ \text{FT}^2 (28.93)\ ^\circ\text{F}} = 6519\ \frac{\text{BTU}}{\text{HR-FT}^2\text{-}^\circ\text{F}}$$

b) THE ENTHALPY OF SATURATED 4.45 PSIA STEAM IS

$h_1 \approx 1128$ BTU/LBM

THE ENTHALPY OF SATURATED 4.45 PSIA WATER

$h_2 \approx 125$ BTU/LBM

THE ENTHALPY CHANGE IS

$h_1 - h_2 = 1128 - 125 = 1003$ BTU/LBM

FROM AN ENERGY BALANCE WITH THE WATER,

$(3.489\ \text{EE}8) = \dot{m}(1003)$

$\dot{m} = 3.479\ \text{EE}5$ LBM/HR STEAM

THEN, FROM PAGE 7-2, THE EXTRACTION RATE IS

$$ER = \dot{m}h_1 = (3.479\ \text{EE}5)\ \frac{\text{LBM}}{\text{HR}} (1128)\ \frac{\text{BTU}}{\text{LBM}} = 3.92\ \text{EE}8\ \text{BTU/HR}$$

5

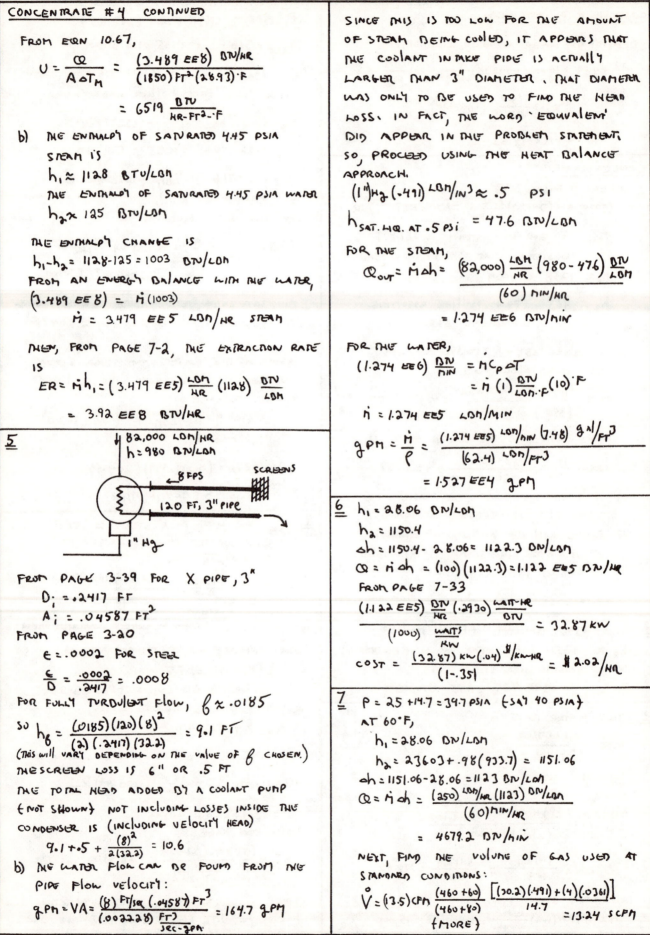

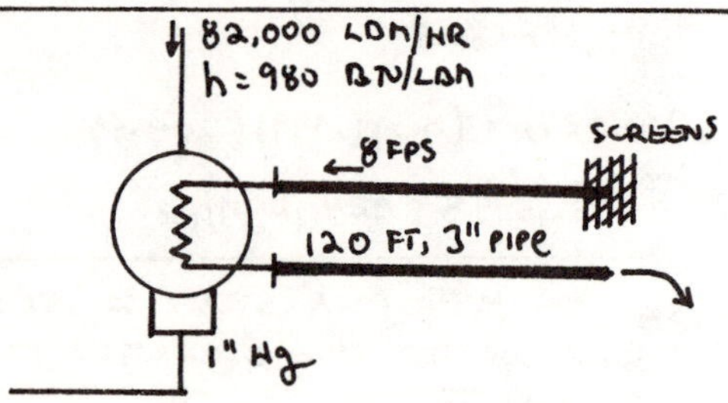

FROM PAGE 3-39 FOR X PIPE, 3"

$D_i = .2417$ FT

$A_i = .04587$ FT^2

FROM PAGE 3-20

$\epsilon = .0002$ FOR STEEL

$$\frac{\epsilon}{D} = \frac{.0002}{.2417} = .0008$$

FOR FULLY TURBULENT FLOW, $f \approx .0185$

SO $$h_f = \frac{(.0185)(120)(8)^2}{(2)(.2417)(32.2)} = 9.1\ \text{FT}$$

(THIS WILL VARY DEPENDING ON THE VALUE OF f CHOSEN.)

THE SCREEN LOSS IS 6" OR .5 FT

THE TOTAL HEAD ADDED BY A COOLANT PUMP (NOT SHOWN) NOT INCLUDING LOSSES INSIDE THE CONDENSER IS (INCLUDING VELOCITY HEAD)

$$9.1 + .5 + \frac{(8)^2}{2(32.2)} = 10.6$$

b) THE WATER FLOW CAN BE FOUND FROM THE PIPE FLOW VELOCITY:

$$\text{gpm} = VA = \frac{(8)\ \text{FT/SEC}\ (.04587)\ \text{FT}^3}{(.002228)\ \frac{\text{FT}^3}{\text{SEC-GPM}}} = 164.7\ \text{GPM}$$

SINCE THIS IS TOO LOW FOR THE AMOUNT OF STEAM BEING COOLED, IT APPEARS THAT THE COOLANT INTAKE PIPE IS ACTUALLY LARGER THAN 3" DIAMETER. THAT DIAMETER WAS ONLY TO BE USED TO FIND THE HEAD LOSS. IN FACT, THE WORD 'EQUIVALENT' DID APPEAR IN THE PROBLEM STATEMENT. SO, PROCEED USING THE HEAT BALANCE APPROACH.

$(1")\text{Hg}\ (.491)\ \text{LBM/IN}^3 \approx .5$ PSI

$h_{\text{SAT. LIQ. AT .5 PSI}} = 47.6$ BTU/LBM

FOR THE STEAM,

$$Q_{out} = \dot{m}\Delta h = \frac{(82{,}000)\ \frac{\text{LBM}}{\text{HR}} (980 - 476)\ \frac{\text{BTU}}{\text{LBM}}}{(60)\ \text{MIN/HR}} = 1.274\ \text{EE}6\ \text{BTU/MIN}$$

FOR THE WATER,

$$(1.274\ \text{EE}6)\ \frac{\text{BTU}}{\text{MIN}} = \dot{m}c_p\Delta T = \dot{m}\ (1)\ \frac{\text{BTU}}{\text{LBM-}^\circ\text{F}} (10)\ ^\circ\text{F}$$

$\dot{m} = 1.274\ \text{EE}5$ LBM/MIN

$$\text{gpm} = \frac{\dot{m}}{\rho} = \frac{(1.274\ \text{EE}5)\ \text{LBM/MIN}\ (7.48)\ \text{GAL/FT}^3}{(62.4)\ \text{LBM/FT}^3} = 1.527\ \text{EE}4\ \text{GPM}$$

6 $h_1 = 28.06$ BTU/LBM

$h_2 = 1150.4$

$\Delta h = 1150.4 - 28.06 = 1122.3$ BTU/LBM

$Q = \dot{m}\Delta h = (100)(1122.3) = 1.122\ \text{EE}5$ BTU/HR

FROM PAGE 7-33

$$\frac{(1.122\ \text{EE}5)\ \frac{\text{BTU}}{\text{HR}} (.2930)\ \frac{\text{WATT-HR}}{\text{BTU}}}{(1000)\ \frac{\text{WATTS}}{\text{KW}}} = 32.87\ \text{KW}$$

$$\text{COST} = \frac{(32.87)\ \text{KW}\ (.04)\ \$/\text{KW-HR}}{(1 - .35)} = \$2.02/\text{HR}$$

7 $P = 25 + 14.7 = 39.7$ PSIA (SAY 40 PSIA)

AT 60°F,

$h_1 = 28.06$ BTU/LBM

$h_2 = 236.03 + .98(933.7) = 1151.06$

$\Delta h = 1151.06 - 28.06 = 1123$ BTU/LBM

$$Q = \dot{m}\Delta h = \frac{(250)\ \text{LBM/HR}\ (1123)\ \text{BTU/LBM}}{(60)\ \text{MIN/HR}} = 4679.2\ \text{BTU/MIN}$$

NEXT, FIND THE VOLUME OF GAS USED AT STANDARD CONDITIONS:

$$\dot{V} = (13.5)\ \text{CFM}\ \frac{(460 + 60)}{(460 + 80)} \left[\frac{(30.2)(.491) + (4)(.0361)}{14.7}\right] = 13.24\ \text{SCFM}$$

(MORE)

CONCENTRATE #7 CONTINUED

$$\eta = \frac{4679.2 \text{ BTU/MIN}}{(1324)\text{ FT}^3\text{/MIN } (550) \text{ BTU/FT}^3} = .643$$

8

THE TANGENTIAL BLADE SPEED IS

$$V_b = \frac{(\pi)\left(\frac{18}{12}\right)\text{ FT } (12000) \text{ RPM}}{(60) \text{ SEC/MIN}} = 942.5 \text{ FT/SEC}$$

THE JET SPEED IS

$$V = \frac{942.5}{.4} = 2356.3 \text{ FT/SEC}$$

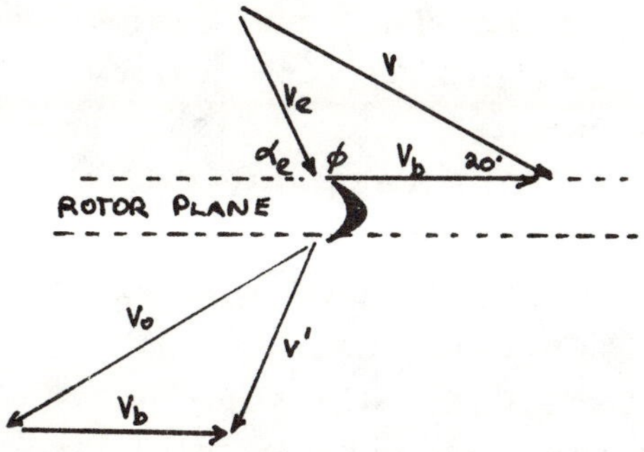

THE RELATIVE JET SPEED IS

$$V_e = \sqrt{(2356.3)^2 + (942.5)^2 - 2(2356.3)(942.5)\cos 20°}$$
$$= 1505.6 \text{ FT/SEC}$$

THE ANGLE ϕ CAN BE FOUND FROM THE LAW OF SINES

$$\frac{1505.5}{\sin 20°} = \frac{2356.2}{\sin \phi} \quad \text{OR} \quad \phi = 147.6°$$

SO $\alpha_e = 180 - \phi = 32.4$

THE ACTUAL STEAM EXIT VELOCITY IS

$$V' = \sqrt{(942.5)^2 + (1505.6)^2 - 2(942.5)(1505.6)\cos 32.4°}$$
$$= 871.1 \text{ FT/SEC}$$

THE REDUCTION IN KINETIC ENERGY IS

$$\Delta h = \Delta E_K = \frac{(2356.3)^2 - (871.1)^2}{(2)(32.2)(778)\frac{\text{FT-LBF}}{\text{BTU}}} = 95.67 \text{ BTU/LBM}$$

9 STEP 1 DETERMINE THE ACTUAL GRAVIMETRIC ANALYSIS OF THE COAL AS FIRED. USE THE 'SUCCESSIVE DELETION' METHOD ON A PER-POUND BASIS. ONE POUND OF COAL CONTAINS .02 LBM MOISTURE, LEAVING .98 LBM COAL. OF THIS, 5% IS ASH, SO THE WEIGHT OF ASH IS $(.05)(.98) = .049$ LBM. THE REMAINDER $(.98 - .049) = .931$ IS ASSUMED TO BE CARBON

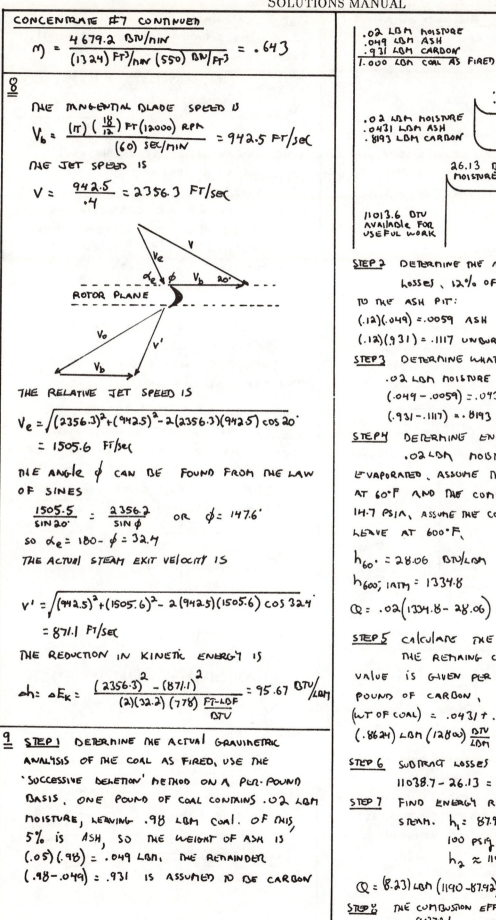

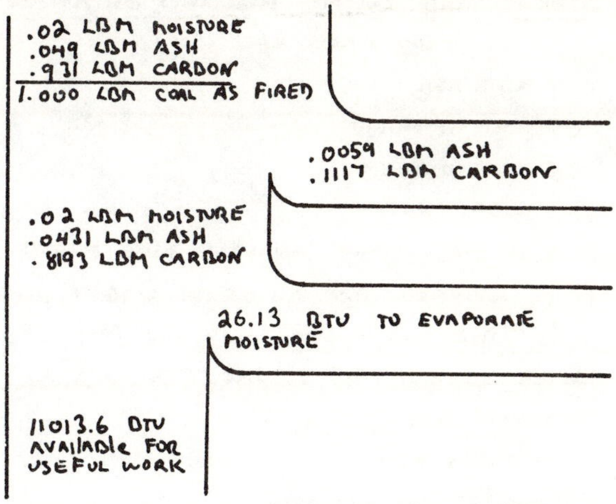

STEP 2 DETERMINE THE ASH PIT MATERIAL LOSSES. 12% OF THE DRY COAL GOES TO THE ASH PIT:

$(.12)(.049) = .0059$ ASH

$(.12)(.931) = .1117$ UNBURNED CARBON

STEP 3 DETERMINE WHAT REMAINS

.02 LBM MOISTURE

$(.049 - .0059) = .0431$ LBM FLY ASH

$(.931 - .1117) = .8193$ LBM CARBON

STEP 4 DETERMINE ENERGY LOSSES. THE .02 LBM MOISTURE HAS TO BE EVAPORATED. ASSUME THE COAL IS INITIALLY AT 60°F AND THE COMBUSTION OCCURS AT 14.7 PSIA. ASSUME THE COMBUSTION PRODUCTS LEAVE AT 600°F.

$h_{60°} = 28.06$ BTU/LBM

$h_{600;\,1ATM} = 1334.8$

$Q = .02(1334.8 - 28.06) = 26.13$ BTU

STEP 5 CALCULATE THE HEATING VALUE OF THE REMAINING COAL. THE HEATING VALUE IS GIVEN PER POUND OF COAL, NOT POUND OF CARBON.

(WT OF COAL) $= .0431 + .8193 = .8624$ LBM

$(.8624) \text{ LBM } (12800) \frac{\text{BTU}}{\text{LBM}} = 11038.72$ BTU

STEP 6 SUBTRACT LOSSES

$11038.7 - 26.13 = 11012.6$ BTU

STEP 7 FIND ENERGY REQUIRED TO VAPORIZE STEAM. $h_1 = 87.92$ BTU/LBM

100 PSIG $\approx$ 115 PSIA

$h_2 \approx 1190$ BTU/LBM

$$Q = (8.23) \text{ LBM } (1190 - 87.92) \text{ BTU/LBM} = 9070.1 \text{ BTU}$$

STEP 8 THE COMBUSTION EFFICIENCY IS

$$\eta = \frac{9070.1}{11012.6} = .824$$

10 STEP 1 THE INCOMING REACTANTS ON A PER-POUND BASIS ARE

.07 LBM ASH
.05 LBM HYDROGEN
.05 LBM OXYGEN
.83 LBM CARBON

THIS IS AN ULTIMATE ANALYSIS (SEE P. 9-2). IT IS ASSUMED THAT ALL OF THE OXYGEN AND 1/8 OF THE HYDROGEN IS IN THE FORM OF WATER (SEE P. 9-7). THE REACTANTS AS COMPOUNDS ARE

.07 LBM ASH
.05625 LBM MOISTURE
.04375 LBM HYDROGEN
.83 LBM CARBON

THE AIR IS 23.15% OXYGEN BY WEIGHT (SEE TABLE 9.9.) SO OTHER REACTANTS ARE

$(.2315)(26) = 6.019$ LBM OXYGEN
$(.7685)(26) = 19.981$ LBM NITROGEN

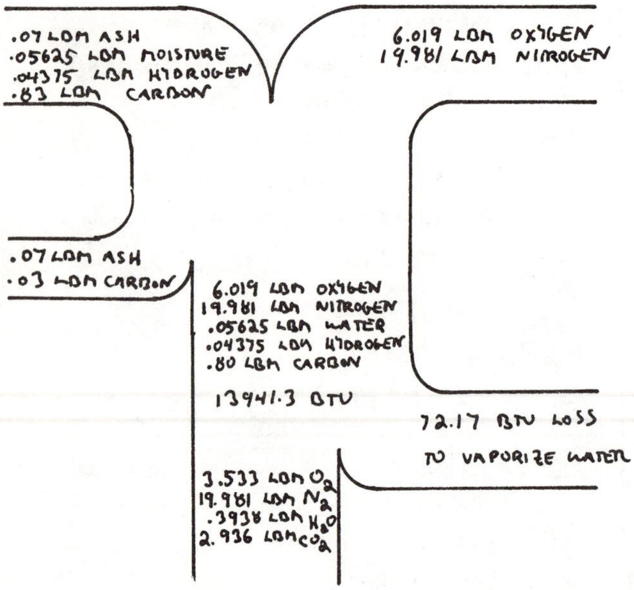

STEP 2 ASSUME ASH PIT MATERIAL LOSSES. ASSUME A .1 LBM LOSS, WHICH INCLUDES ALL OF THE ASH.
.07 LBM ASH
.03 LBM UNBURNED COAL

STEP 3 DETERMINE WHAT REMAINS

STEP 4 DETERMINE THE ENERGY LOSS IN VAPORIZING THE MOISTURE. ASSUME THE COAL IS INITIALLY AT 60°F AND THAT COMBUSTION OCCURS AT 14.7 PSIA.

$h_{60°} = 28.06$ BTU/LBM
$h_{550} = 1311$ BTU/LBM

$Q = .05625(1311 - 28.06) = 72.17$ BTU

STEP 5 CALCULATE THE HEATING VALUE OF THE REMAINING FUEL COMPONENTS. FROM PAGE 9-22

$$Q_{CARBON} = (.80)\text{LBM}(14093)\frac{\text{BTU}}{\text{LBM}} = 11274.4 \text{ BTU}$$
$$Q_{HYDROGEN} = (.04375)(60958) = 2666.9$$
$$13941.3 \text{ BTU}$$

THE HEATING VALUE AFTER THE COAL MOISTURE IS EVAPORATED IS

$$13941.3 - 72.17 = 13869.13 \text{ BTU}$$

STEP 6 DETERMINE THE COMBUSTION PRODUCTS. FROM TABLE 9.11, THE CARBON NEEDS

$.8(2.67) = 2.136$ LBM OXYGEN

AND PRODUCES

$.8(3.67) = 2.936$ LBM CO_2

THE HYDROGEN NEEDS

$(.04375)(8) = .35$ LBM OXYGEN

AND PRODUCES

$(.04375)(9) = .3938$ LBM WATER

THE REMAINING OXYGEN IS

$6.019 - 2.136 - .35 = 3.533$ LBM

STEP 7 THE GASEOUS PRODUCTS MUST BE HEATED FROM 70° TO 550°. THE AVERAGE TEMPERATURE IS $\frac{1}{2}(70 + 550) = 310°F$ OR 770°R. C_p FOR THESE GASES CAN BE FOUND FROM TABLE 9.14.

	C_p
OXYGEN	.228
NITROGEN	.252
WATER	.460
CO_2	.225

$$Q_{HEATING} = \left[(3.533)(.228) + (19.981)(.252) + (.3938)(.460) + (2.936)(.225)\right](550 - 70)$$
$$= 3207.6 \text{ BTU}$$

STEP 8 THE % LOST IS

$$\frac{3207.6 + 72.17}{13941.3} = .235$$

TIMED

1 FROM PAGE 7-33, THE DRILL POWER IS

$$(.25)\text{ HP}(2544.9)\frac{\text{BTU}}{\text{HP-HR}}$$
$$= 636.23 \text{ BTU/HR}$$

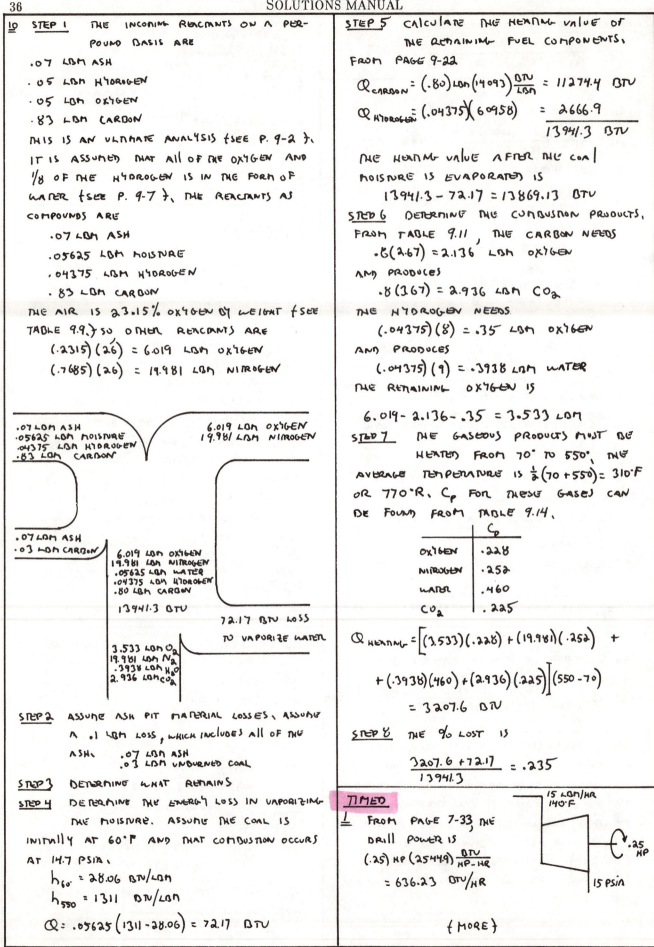

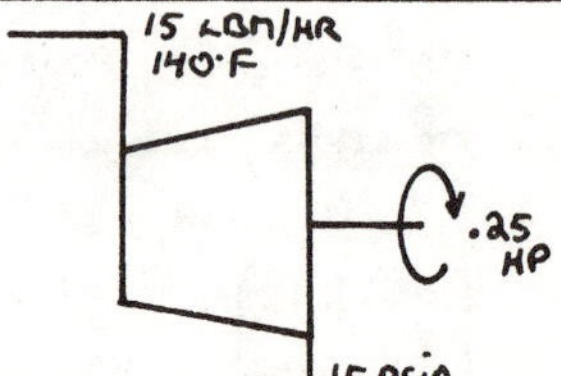

(MORE)

TIMED #1 CONTINUED

FROM PAGE 6-35 FOR $T = (460+140) = 600°R$,

$h_1 = 143.47$ BTU/LBM

$p_{r,1} = 2.005$

$\phi_1 = .62607$

THEN, SINCE $W = \dot{m}(h_1 - h_2)$

$$h_{2,IDEAL} = h_1 - \frac{W}{\dot{m}}$$

Put here since $\eta_{eff} < \Delta h$

$$= (143.47)\frac{BTU}{LBM} - \frac{(636.23)\frac{BTU}{HR}}{(15)\frac{LBM}{HR}(.60)}$$

$= 72.778$ BTU/LBM

SEARCHING KEENAN AND KAYES GAS TABLES FOR $h = 72.778$ GIVES

$T = 305°R$

$p_{r,2} = .18851$

a) HOWEVER, DUE TO THE IRREVERSIBILITY OF THE EXPANSION,

$h_2' = 143.47 - .60(143.47 - 72.778) = 101.05$ BTU/LBM

SEARCHING THE GAS TABLES AGAIN GIVES

$T_2' = 423°R$

$\phi_2 = .54228$

b) SINCE $\frac{P_1}{P_2} = \frac{P_{r,1}}{P_{r,2}}$

$$P_1 = (15)\text{ PSIA}\left(\frac{2.005}{.18851}\right) = 159.5 \text{ PSIA}$$

c) FROM EQN 6.29,

$$S_2 - S_1 = .54228 - .62607 - \left(\frac{53.3}{778}\right)\ln\left(\frac{15}{159.5}\right)$$
$= .07816$ BTU/LBM-°R

2 FOR THE STEAM,

$T_A = 600°F$

$P_A = 200$ PSIG ≈ 215 PSIA

$h_A = 1321$ BTU/LBM

$S_A = 1.6680$ BTU/LBM-°R

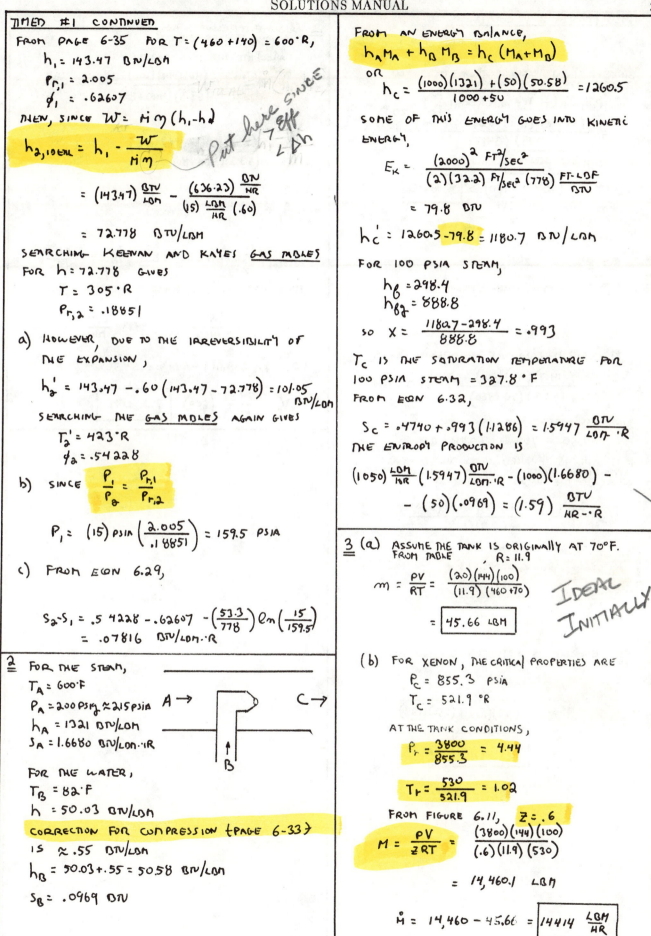

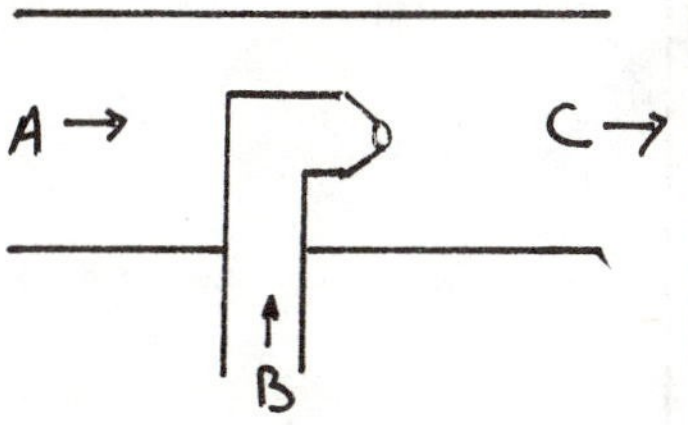

FOR THE WATER,

$T_B = 82°F$

$h = 50.03$ BTU/LBM

CORRECTION FOR COMPRESSION (PAGE 6-33) IS $\approx .55$ BTU/LBM

$h_B = 50.03 + .55 = 50.58$ BTU/LBM

$S_B = .0969$ BTU

FROM AN ENERGY BALANCE,

$$h_A M_A + h_B M_B = h_C(M_A + M_B)$$

OR

$$h_C = \frac{(1000)(1321) + (50)(50.58)}{1000 + 50} = 1260.5$$

SOME OF THIS ENERGY GOES INTO KINETIC ENERGY,

$$E_K = \frac{(2000)^2 \text{ FT}^2/\text{SEC}^2}{(2)(32.2)\frac{\text{FT}}{\text{SEC}^2}(778)\frac{\text{FT-LBF}}{\text{BTU}}}$$

$= 79.8$ BTU

$h_C' = 1260.5 - 79.8 = 1180.7$ BTU/LBM

FOR 100 PSIA STEAM,

$h_f = 298.4$

$h_{fg} = 888.8$

SO $X = \frac{1180.7 - 298.4}{888.8} = .993$

T_C IS THE SATURATION TEMPERATURE FOR 100 PSIA STEAM $= 327.8°F$

FROM EQN 6.32,

$$S_C = .4740 + .993(1.1286) = 1.5947 \frac{\text{BTU}}{\text{LBM-°R}}$$

THE ENTROPY PRODUCTION IS

$$(1050)\frac{\text{LBM}}{\text{HR}}(1.5947)\frac{\text{BTU}}{\text{LBM-°R}} - (1000)(1.6680) - (50)(.0969) = (1.59)\frac{\text{BTU}}{\text{HR-°R}}$$

$\Delta S = \dot{m}S_{FINAL} - \dot{m}S_{INITIAL}$

3 (a) ASSUME THE TANK IS ORIGINALLY AT 70°F. FROM TABLE, $R = 11.9$

$$m = \frac{PV}{RT} = \frac{(20)(144)(100)}{(11.9)(460+70)}$$

$= \boxed{45.66 \text{ LBM}}$

IDEAL INITIALLY

(b) FOR XENON, THE CRITICAL PROPERTIES ARE

$P_c = 855.3$ PSIA

$T_c = 521.9$ °R

AT THE TANK CONDITIONS,

$$P_r = \frac{3800}{855.3} = 4.44$$

$$T_r = \frac{530}{521.9} = 1.02$$

FROM FIGURE 6.11, $Z = .6$

$$M = \frac{PV}{ZRT} = \frac{(3800)(144)(100)}{(.6)(11.9)(530)}$$

$= 14,460.1$ LBM

$$\dot{M} = 14,460 - 45.66 = \boxed{14414 \frac{\text{LBM}}{\text{HR}}}$$

TIMED PROBLEM #3 CONTINUED

SINCE A HEAT EXCHANGER WAS MENTIONED, ASSUME ISOTHERMAL COMPRESSION.

$$W = p_1V_1 \ln\left(\frac{V_2}{V_1}\right) = mRT_1 \ln\left(\frac{p_1}{p_2}\right)$$

$$= \frac{(14414)(11.9)(460+70) \ln\left(\frac{20}{3800}\right)}{(778)(3413)}$$

$$= \boxed{180 \text{ KW-HR}}$$

$$\text{COST} = (.045)(180) = \boxed{\$8.10}$$

4

M_1 = AIR IN TANK WHEN EVACUATED

$$= \frac{PV}{RT} = \frac{(1)(144)(20)}{(53.3)(70+460)} = .102 \text{ LBM}$$

M_2 = AIR IN TANK AFTER OPENING VALVE

$$= \frac{(14.7)(144)(20)}{(53.3)(70+460)} = 1.50 \text{ LBM}$$

THE MASS DIFFERENCE IS

$$1.50 - .102 = 1.398 \text{ LBM}$$

THIS AIR IS PUSHED INTO THE TANK BY ATMOSPHERIC PRESSURE IN A CONSTANT PRESSURE PROCESS. THE VOLUME OF AIR IS

$$V = \frac{mRT}{P} = \frac{(1.398)(53.3)(530)}{(14.7)(144)}$$

$$= 18.66 \text{ FT}^3$$

THE FLOW WORK DONE IS

$$W = P\Delta V = \frac{(14.7)(144)(18.66)}{778}$$

$$= 50.77 \text{ BTU}$$

THE ENERGY IS ABSORBED BY THE AIR AND TANK.

$$\Delta T = \frac{Q}{MC_p} = \frac{50.77}{(40)(.11)+(1.50)(.24)}$$

$$= \boxed{10.67°\text{F}}$$

5

ASSUME PRESSURES ARE LOW ENOUGH TO IGNORE COMPRESSIBILITY.

CALCULATE THE FLOW RATES:

$$\dot{M}_C = \frac{PV}{RT} = \frac{(80)(144)(100)}{(53.3)(460+85)} = 39.66 \frac{\text{LBM}}{\text{HR}}$$

$$\dot{M}_D = \frac{(85)(144)(120)}{(53.3)(460+80)} = 51.03 \frac{\text{LBM}}{\text{HR}}$$

$$\dot{M}_E = 8 \frac{\text{LBM}}{\text{HR}} \quad \text{(GIVEN)}$$

$$\dot{M}_{total} = 39.66 + 51.03 + 8 = 98.69 \frac{\text{LBM}}{\text{HR}}$$

THE INPUT FROM COMPRESSOR A IS

$$\dot{M}_A = \frac{(14.7)(144)(600)}{(53.3)(460+80)} = 44.13 \frac{\text{LBM}}{\text{MIN}}$$

$$\text{SO, } \dot{M}_B = 98.69 - 44.13 = 54.56 \text{ LBM/MIN}$$

$$\dot{V}_B = \frac{54.56}{44.13}(600) = \boxed{742 \text{ CFM}}$$

RESERVED FOR FUTURE USE

Power Cycles

WARM-UPS

1 FROM EQN 7.30,

$$\eta_{th} = \frac{(650+460)-(100+460)}{650+460} = .495$$

2 FROM EQN 7.151

$$COP = \frac{700+460}{(700+460)-(40+460)} = 1.76$$

3 USE THE PROCEDURE ON PAGE 7-9, REFER TO FIGURE 7.12.

AT a: $T_a = 650°F$
$h_a = 696.4$
$s_a = .8833$

AT b: $T_b = 650$
$h_b = 1117.5$
$s_b = 1.2631$

AT c: $T_c = 100°F$
$s_c = s_b = 1.2631$

$$x_c = \frac{1.2631-.1295}{1.8531} = .612$$

$$h_c = 67.97 + .612(1037.2) = 702.7$$

AT d: $T_d = 100°F$
$s_d = s_a = .8833$

$$x_d = \frac{.8833-.1295}{1.8531} = .407$$

$$h_d = 67.97 + .407(1037.2) = 490.1$$

DUE TO THE INEFFICIENCIES (FROM EQUATIONS 7.31 AND 7.32),

$$h_c' = 1117.5 - .9(1117.5-702.7) = 744.2$$

$$h_a' = 490.1 + \frac{696.4-490.1}{.8} = 748.0$$

FROM EQN 7.30

$$\eta_{th} = \frac{(1117.5-744.2)-(748.0-490.1)}{1117.5-748.0} = .312$$

4 FROM EQN 7.150

$$COP = \frac{50+460}{(90+460)-(5+460)} = 5.47$$

$$HP = \frac{(1)\ \text{TON}\ (200)\frac{\text{BTU}}{\text{MIN-TON}}\ (778)\frac{\text{FT-LBF}}{\text{BTU}}}{(5.47)\frac{\text{BTU OUT}}{\text{BTU IN}}\ (33000)\frac{\text{FT-LBF}}{\text{MIN-HP}}} = .862\ \text{HP}$$

FROM EQN 7.148

$$EER = \frac{(200)\ \text{BTU/MIN}\ (60)\ \text{MIN/HR}}{(.862)\ \text{HP}\ (745.7)\ \text{WATTS/hp}} = 18.7$$

5 FROM EQN 3.7, THE SPECIFIC GRAVITY IS

$$SG = \frac{141.5}{131.5+40} = .825$$

FROM EQN. 7.76,

$$HHV = 22320 - 3780(SG)^2$$
$$= 22320 - 3780(.825)^2$$
$$= 19749\ \text{BTU/LBM}$$

6 THE ACTUAL HORSEPOWER IS

$$hp = \frac{(\text{RPM})(\text{TORQUE IN FT-LBF})}{5252}$$

$$= \frac{(200)(600)}{5252} = 22.85\ \text{HP}$$

THE NUMBER OF POWER STROKES PER MINUTE IS

$$N = \frac{(2)(200)(2)}{(4)} = 200$$

THE STROKE IS $\frac{18}{12} = 1.5$ FT

THE BORE AREA IS $\frac{\pi}{4}(10)^2 = 78.54\ \text{IN}^2$

THE IDEAL HORSEPOWER (FROM EQN 7.89) IS

$$hp = \frac{(95)(1.5)(78.54)(200)}{33000} = 67.83$$

THE FRICTION HORSEPOWER IS

$$67.83 - 22.85 = 44.98$$

7 FROM EQUATIONS 7.156 AND 7.157

$$R = \frac{65}{14.7} = 4.42$$

$$\eta_v = 1 - \left((4.42)^{\frac{1}{1.33}} - 1\right)(.07) = .856$$

THE WEIGHT OF AIR DISPLACED PER MINUTE IS $\frac{48}{.856} = 56.07$

THE WEIGHT OF THE AIR AND THE CLEARANCE AIR COMPRESSED IS

$$(1.07)(56.07) = 60\ \text{LBM/MIN}$$

8 THE VALVE IS OPEN

$$180 + 40 = 220°$$

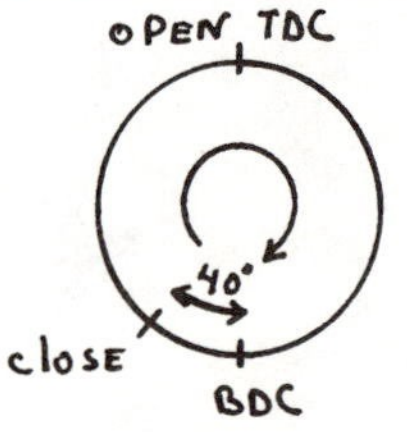

THE TIME THE VALVE IS OPEN IS

$$\left(\frac{220}{360}\right)\left(\text{TIME PER REVOLUTION}\right) =$$

$$\left(\frac{220}{360}\right)\left(\frac{60\ \text{SEC/MIN}}{4000\ \text{RPM}}\right) = (9.167\ \text{EE-3})\ \text{SEC}$$

THE DISPLACEMENT IS

$$\left(\frac{\pi}{4}\right)\left(\frac{3.1}{12}\right)^2\left(\frac{3.8}{12}\right) = .0166\ \text{FT}^3$$

THE ACTUAL VOLUME INCOMING IS

$$V = (.65)(.0166) = .01079\ \text{FT}^3\ \text{PER OPENING}$$

THE AREA IS

$$A = \frac{V}{vt} = \frac{.01079\ \text{FT}^3}{(100)\ \text{FT/SEC}\,(9.167\ \text{EE-3})\ \text{SEC}} = .0118\ \text{FT}^2$$
$$= 1.69\ \text{IN}^2$$

9 METHOD 1: IDEAL GAS RELATIONSHIPS

FROM PAGES 6-24 AND 6-25,

$$V_2 = (10)\frac{FT^3}{sec}\left(\frac{200\ PSIA}{50\ PSIA}\right)^{\frac{1}{1.4}} = 26.918\ CFS$$

$$T_2 = (1500+460)°R\left(\frac{50}{200}\right)^{\frac{1.4-1}{1.4}} = 1319.0\ °R$$

FROM PAGES 6-24 AND 6-25,

$$\Delta h = (.241)\frac{BTU}{LBM\text{-}°R}(1319.0-1960) = -154.5\ \frac{BTU}{LBM}$$

METHOD 2: USING PAGE 6-35,

AT 1960°R,

$h_1 = 493.64$

$P_{r,1} = 160.37$

$v_{r,1} = 4.527$

AFTER EXPANSION,

$$P_{r,2} = 160.37\left(\frac{50}{200}\right) = 40.09$$

SEARCHING THE TABLE FOR THIS VALUE OF P_r,

$T_2 = 1375$

$h_2 = 336.39$

$v_{r,1} = 12.721$

SO $V_2 = (10)\left(\frac{12.721}{4.527}\right) = 28.1$

$\Delta h = 493.64 - 336.39 = 157.25$

10 ALTHOUGH THE IDEAL GAS LAWS COULD BE USED, IT IS EXPEDIENT TO USE AIR TABLES. FROM PAGE 6-35 AT (460 + 500) = 960

$h_1 = 231.06$

$P_{r,1} = 10.610$

$\phi_1 = .7403$

FOR ISENTROPIC COMPRESSION,

$$P_{r,2} = 6P_{r,1} = 6(10.610) = 63.66$$

SEARCHING THE AIR TABLE YIELDS

$T_2 = 1552$

$h_2 = 382.95$

THE ACTUAL ENTHALPY IS

$$h_2' = 231.06 + \frac{382.95-231.06}{.65} = 464.74$$

WHICH CORRESPONDS TO 1855°R, AND

$\phi_2' = .91129$

$$W = \Delta h = 464.74 - 231.06 = 233.68$$

FROM EQUATION 6.29

$$\Delta S = .91129 - .7403 - \left(\frac{53.3}{778}\right)\ln(6) = .04824$$

CONCENTRATES

1 FIRST, ASSUME ISENTROPIC COMPRESSION AND EXPANSION. REFER TO FIGURE 7.14

AT a: $P_a = 100$ PSIA

$h_a = 298.40$

AT b: $P_b = 100$ PSIA

$h_b = 1187.2$

$s_b = 1.6026$

AT c: $P_c = 1$ ATM

$s_c = s_b = 1.6026$

$$x_c = \frac{1.6026-.3120}{1.4446} = .893$$

$$h_c = 180.07 + .893(970.3) = 1046.5$$

AT d: $T_d = 80°F$

$h_d = 48.02$

$P_d = 14.7$ PSIA

$v_d = .01608$

h AND v ARE ESSENTIALLY INDEPENDENT OF PRESSURE

AT e: $P_e = P_a = 100$ PSIA

$$h_e = 48.02 + \frac{.01608(100-14.7)(144)}{778} = 48.2$$

NOW, DUE TO THE INEFFICIENCIES,

$$h_c' = 1187.2 - .80(1187.2-1046.5) = 1074.6$$

{FROM EQN 7.40}

$$h_e' = 48.02 + \frac{48.27-48.02}{.6} = 48.44$$

{FROM EQN 7.41}

FROM EQN 7.39

$$\eta_{th} = \frac{(1187.2-1074.6)-(48.44-48.02)}{(1187.2-48.44)} = .0985$$

2 REFER TO FIGURE 7.16

AT d: $P_d = 500$

$T_d = 1000°F$

$h_d = 1519.6$

$S_d = 1.7363$

AT e: $P_e = 5$ PSIA

$S_e = S_d = 1.7363$

$$x_e = \frac{1.7363-.2347}{1.6094} = .933$$

$$h_e = 130.13 + .933(1001) = 1064.1$$

AT f: $h_f = 130.17$

BUT, BECAUSE THE TURBINE IS 75% EFFICIENT,

$$h_e' = 1519.6 - .75(1519.6-1064.1) = 1178$$

THE MASS FLOW RATE IS

$$M = \frac{(200{,}000)\ KW\ (1000)\ \frac{W}{KW}\ (.05692)\ \frac{BTU}{MIN\text{-}W}}{(1519.6-1178)\ \frac{BTU}{LBM}\ (60)\ \frac{SEC}{MIN}} = 555.7\ \frac{LBM}{SEC}$$

$$Q_{OUT} = (555.7)(1178-130.17) = 5.82\ EE5\ BTU/sec$$

3 USE THE PROCEDURE ON PAGE 7-12 AND REFER TO FIGURE 7.18

AT b: $P_b = 600$ PSIA

$T_b = 486.21\,°F$

$h_b = 471.6$

AT c: $h_c = 1203.2$

AT d: $P_d = 600$ PSIA

$T_d = 600\,°F$

$h_d = 1289.9$

AT e: $P_e = 200$ PSIA

$h_e = 1187$ {FROM MOLLIER, ASSUMING ISENTROPIC EXPANSION}

$h_e' = 1289.9 - .88(1289.9 - 1187)$

$= 1199.3$

AT f: $P_f = 200$ PSIA

$T_f = 600\,°F$

$h_f = 1322.1$

$S_f = 1.6767$

AT g: $T_g = 60\,°F$

$S_g = S_f = 1.6767$

$$x_g = \frac{1.6767 - .0555}{2.0393} = .795$$

$h_g = 28.06 + .795(1059.9) = 870.7$

$h_g' = 1322.1 - .88(1322.1 - 870.7) = 924.9$

AT h: $h_h = 28.06$

$P_h = .2563$

$v_h = .01604$

AT a: $P_a = 600$

$$h_a' = 28.06 + \frac{(.01604)(600 - .2563)(144)}{.96\,(778)}$$

$= 29.9$

FROM EQN 7.59,

$$\eta_{th} = \frac{(1289.9 - 29.9) + (1322.1 - 1199.3) - (924.9 - 28.06)}{(1289.9 - 29.9) + (1322.1 - 1199.3)}$$

$= .351$

4 REFER TO PAGE 8-12 AND THE FOLLOWING DIAGRAM

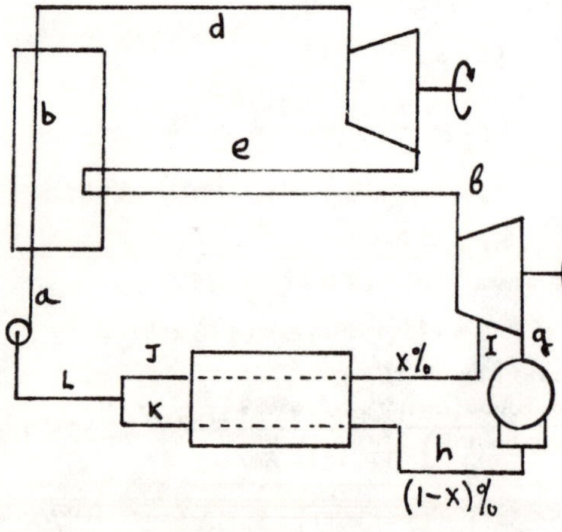

FROM PROBLEM 3

$h_b = 471.6$

$h_d = 1289.9$

$h_e' = 1199.3$

$h_f = 1322.1$

$h_g' = 924.9$

$h_h = 28.04$

AT I: THE TEMPERATURE IS 270°F. USING THE MOLLIER DIAGRAM AND ASSUMING ISENTROPIC EXPANSION TO 270°F,

$h_I \approx 1170$ {SATURATED}

$h_I' = 1322.1 - .88(1322.1 - 1170) = 1188.3$

AT J: THE WATER IS ASSUMED TO BE SATURATED FLUID AT 270°F

$h_J = 238.84$

AT K: THE TEMPERATURE IS $(270 - 6) = 264\,°F$ AND SATURATED FLUID

$h_K = 232.83$

FROM AN ENERGY BALANCE IN THE HEATER,

$(1-x)(h_K - h_h) = x(h_I' - h_J)$

$(1-x)(232.83 - 28.04) = x(1188.3 - 238.84)$

$204.79 = x(1154.25)$

$x = .177$

AT L: $h_L = x(h_J) + (1-x)h_K$

$= .177(238.84) + (1 - .177)232.83$

$= 233.89$

SINCE THIS IS SATURATED LIQUID,

$P_L = 38.5$ PSIA

$v_L = .017132$

$T_L = 265\,°F$

AT a: $P_a = 600$ PSIA

$$h_a = 233.89 + \frac{.017132(600 - 38.5)\,144}{(778)(.96)}$$

$= 235.7$

$$\eta_{th} = \frac{W_{OUT} - W_{IN}}{Q_{IN}}$$

$$= \frac{(h_d - h_e') + (h_f - h_I') + (1-x)(h_I' - h_g') - (h_a - h_L)}{(h_d - h_a) + (h_f - h_e)}$$

$$= \frac{(1289.9 - 1199.3) + (1322.1 - 1188.3) + (1 - .177)(1188.3 - 924.9)}{(1289.9 - 235.7) + (1322.1 - 1199.3)}$$

$$- \frac{(235.7 - 233.89)}{(1289.9 - 235.7) + (1322.1 - 1199.3)} = .376$$

5 REFER TO FIGURE 7.30

AT a: $V = 11\ FT^3$

$T = 460 + 80 = 540\,°R$

$P = 14.2$ PSIA $= 2044.8$ PSFA

$$M = \frac{PV}{RT} = \frac{(2044.8)(11)}{(53.3)(540)} = .781\ LBM$$

{MORE}

CONCENTRATES # 5 CONTINUED

AT b: $V_b = \frac{1}{10} V_a = 1.1 \text{ FT}^3$

AT C: $T_C = 540\left(\frac{11}{1.1}\right)^{1.4-1} + \frac{q_{IN}}{C_V W}$

$= 1356.4 + \frac{160}{(.1724)(.781)}$

$= 2544.7 \text{ °R} = 2084.7 \text{ °F}$

$\eta_{th} = 1 - \frac{1}{(10)^{1.4-1}} = .602$ {EQN 8.71}

6 STEP 1: FIND THE IDEAL WEIGHT OF AIR INGESTED.

THE IDEAL WEIGHT OF AIR TAKEN IN PER SECOND IS

$\dot{V}_i = \left(\text{SWEPT VOLUME}\right)\left(\text{\# INTAKE STROKES PER SECOND}\right)$

FROM EQN 7.80, THE NUMBER OF POWER STROKES PER SECOND IS

$\frac{(2)(1200) \text{ RPM } (6) \text{ CYLINDERS}}{(60) \text{ SEC/HR } (4) \text{ STROKES}} = 60 \text{ 1/SEC}$

THE SWEPT VOLUME IS

$V_S = \left(\frac{\pi}{4}\right)\left(\frac{4.25}{12}\right)^2\left(\frac{6}{12}\right) = .04926 \text{ FT}^3$

SO

$\dot{V}_i = (60)(.04926) = 2.956 \text{ FT}^3\text{/SEC}$

FROM PV=WRT, THE WEIGHT OF THE AIR IS

$W = \frac{(14.7)(144)(2.956)}{(53.3)(530)} = .2215 \text{ LBM AIR/SEC}$

STEP 2: FIND THE CO_2 VOLUME IN THE EXHAUST ASSUMING COMPLETE COMBUSTION WHEN THE AIR/FUEL RATIO IS 15.

FROM TABLE 9.9, AIR IS 76.85% NITROGEN, SO THE NITROGEN/FUEL RATIO IS

$N/F = (.7685)(15) = 11.528$

FROM PV=WRT, THE NITROGEN VOLUME PER POUND OF FUEL BURNED IS

$V_N = \frac{WRT}{P} = \frac{(11.528)(55.2)(530)}{(14.7)(144)} = 159.3 \text{ FT}^3$

THE OXYGEN FUEL RATIO IS

$O/F = (.2315)(15) = 3.472$

THE OXYGEN VOLUME PER POUND OF FUEL BURNED IS

$V_O = \frac{(3.472)(48.3)(530)}{(14.7)(144)} = 41.99 \text{ FT}^3$

WHEN OXYGEN FORMS CARBON DIOXIDE, THE CHEMICAL EQUATION IS

$C + O_2 \longrightarrow CO_2$

SO, IT TAKES 1 VOLUME OF OXYGEN TO FORM 1 VOLUME OF CO_2. THE % CO_2 IN THE EXHAUST IS FOUND FROM

$\% CO_2 = \frac{(\text{vol } CO_2)}{(\text{vol } CO_2) + (\text{vol } O_2) + (\text{vol } N_2)}$

NOW, $(\text{vol } N_2) = 159.3$

$(\text{vol } CO_2) = X$ {UNKNOWN}

$(\text{vol } O_2) = 41.99 -$ OXYGEN USED TO MAKE CO_2

$= 41.99 - X$

OR

$.137 = \frac{X}{X + 41.99 - X + 159.3}$

$X = 27.58 \text{ FT}^3$

ASSUMING COMPLETE COMBUSTION, THIS VOLUME OF CO_2 WILL BE CONSTANT REGARDLESS OF THE AMOUNT OF AIR USED.

STEP 3 CALCULATE THE EXCESS AIR IF $\%CO_2 = 9$

$\% CO_2 = \frac{(\text{vol } CO_2)}{(\text{vol } CO_2) + (\text{vol } O_2 - \text{vol } CO_2) + (\text{vol } N_2) + (\text{vol EXCESS AIR})}$

$.09 = \frac{27.58}{27.58 + 41.99 - 27.58 + 159.3 + (\text{vol EXCESS})}$

$(\text{vol EXCESS AIR}) = 105.2 \text{ FT}^3$

FROM PV=WRT,

$W_{EXCESS} = \frac{(14.7)(144)(105.2)}{(53.3)(530)} = 7.883 \text{ LBM}$

STEP 4: THE ACTUAL AIR FUEL RATIO IS

$15 + 7.883 = 22.883$ LBM AIR/LBM FUEL

THE ACTUAL AIR WEIGHT PER SECOND IS

$\frac{\left(22.883\right)\frac{\text{LBM AIR}}{\text{LBM FUEL}}\left(28\right)\text{ LBM FUEL/HR}}{(3600) \text{ SEC/HR}} = .178 \text{ LBM/SEC}$

STEP 5: $\eta_V = \frac{.178}{.2215} = .804$

7 USE THE PROCEDURE ON PAGE 7-20

STEP 1: 1 - 60°F, 14.7 PSIA

2 - 5000 FT ALTITUDE

STEP 2: $IHP_1 = \frac{1000}{.80} = 1250$

STEP 3: FHP = 1250 - 1000 = 250

STEP 4: $\rho_1 = P/RT = \frac{(14.7)(144)}{(53.3)(520)} = .0764 \text{ LBM/FT}^3$

$\rho_2 = .06592$ AT 5000' {P. 8-20}

STEP 5: $IHP_2 = 1250\left(\frac{.06592}{.0764}\right) = 1078.5$

STEP 6: $BHP_2 = 1078.5 - 250 = 828.5$

STEP 7: THE ORIGINAL FLOW RATE OF FUEL IS

$\dot{W}_{F,1} = (BHP_1)(BSFC_1) = (1000)(.45)$

$= 450$ LBM/HR

THE ORIGINAL AIR WEIGHT IS

$\dot{W}_{A,1} = R_{A/F}(\dot{W}_{F,1}) = (23)(450)$

$= 10350$ LBM/HR

THIS IS A VOLUME OF

$V = \frac{WRT}{P} = \frac{(10350)(53.3)(520)}{(14.7)(144)}$ {MORE}

CONCENTRATE #7 CONTINUED

THIS VOLUME IS THE SAME AT 5000'

STEP 8: $\dot{W}_{A,2} = (1.355\text{ EE }5)(.06592) = 8932$

STEP 9: $\dot{W}_{F,2} = \frac{\dot{W}_{A,2}}{R_{A/F}} = \frac{8932}{23} = 388$

STEP 10: $BSFC_2 = \frac{388}{828.5} = .469$

8 REFER TO FIGURE 7.39, AS WITH OTHER IC ENGINES THE 'COMPRESSION RATIO' IS A RATIO OF VOLUMES. FIRST, ASSUME ISENTROPIC OPERATION

AT a: $P_a = 14.7\text{ PSIA} = 2116.8\text{ PSFA}$

$T_a = 60°F = 520°R$

FOR 1 POUND,

$V_a = \frac{(1)(53.3)(520)}{2116.8} = 13.09\text{ FT}^3$

AT b: $V_b = \frac{13.09}{5} = 2.618$

$T_b = 520\left(\frac{13.09}{2.618}\right)^{1.4-1} = 989.9°R$

$P_b = \frac{(1)(53.3)(989.9)}{2.618} = 20153\text{ PSF}$

AT c: $T_c = 1500°F = 1960°R$

$P_c = 20153\text{ PSF}$

AT d: $P_d = 14.7\text{ PSIA} = 2116.8\text{ PSF}$

$T_d = 1960\left(\frac{2116.8}{20153}\right)^{\frac{1.4-1}{1.4}} = 1029.5$

NOW, INCLUDE THE INEFFICIENCIES

$T_a = 520°R$

$T_b' = 520 + \frac{989.9-520}{.83} = 1086°R$

$T_c = 1960°R$

$T_d' = 1960 - .92(1960-1029.5) = 1103.9$

FROM EQN 7.124

$\eta_{th} = \frac{(1960-1086)-(1102.9-520)}{1960-1086} = .332$

9 REFER TO FIGURE 7.38

FROM PROBLEM 8

$T_a = 520°R$

$T_b' = 1086$

$T_d = 1960$

$T_e' = 1103.9$

FROM EQN 7.125, ASSUMING AN IDEAL GAS

$.65 = \frac{T_c - T_b'}{T_e - T_b'} = \frac{T_c - 1086}{1103.9 - 1086}$

OR $T_c = 1097.6$

THEN FROM EQN 7.126 ASSUMING AN IDEAL GAS,

$\eta_{th} = \frac{(1960-1103.9)-(1086-520)}{(1960-1097.6)}$

$= .336$

10 THIS IS A HARD ONE TO VISUALIZE. SHOWN BELOW IS ONE OF N LAYERS. EACH LAYER CONSISTS OF 24 TUBES, ONLY 3 OF WHICH ARE SHOWN.

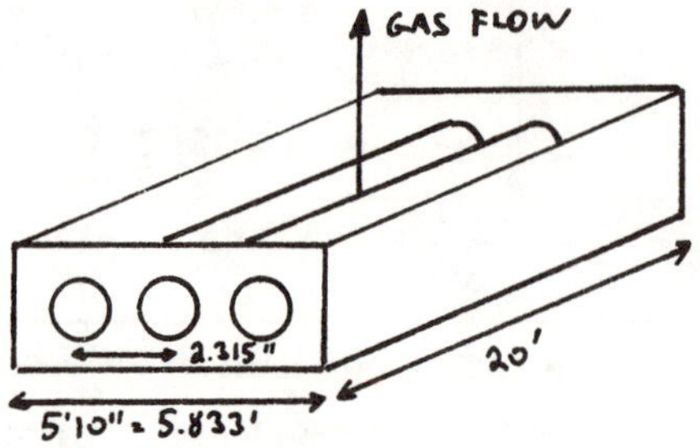

APPROACH: $q = UA\Delta T_M$

THE LOG MEAN TEMPERATURE DIFFERENCE (ASSUMING COUNTER FLOW OPERATION) IS

635 — GASES → 470

$\Delta T = 350$ $\Delta T = 258$

285 ← WATER — 212

$\Delta T_M = \frac{350-258}{\ln\left(\frac{350}{258}\right)} = 301.7°F$

SINCE NO INFORMATION IS GIVEN ABOUT THE STACK GASES, ASSUME THAT THEY CONSIST OF PRIMARILY NITROGEN. THE AVERAGE GAS TEMPERATURE IS

$\frac{1}{2}(635+470) = 552.5°F = 1012.5°R$

FROM TABLE 9.14

$\bar{C}_{p,\text{NITROGEN}} \approx .255\text{ BTU/LBM-°F}$

SO

$q = \dot{m}C_p\Delta T = (191{,}000)\frac{\text{LBM}}{\text{HR}}(.255)\frac{\text{BTU}}{\text{LBM-°F}} \times (635-470) = 8.036\text{EE}6\text{ BTU/HR}$

SINCE NOT ENOUGH INFORMATION IS GIVEN, h_i AND h_o CANNOT BE EVALUATED, SO U MUST BE ASSUMED. ASSUME $U_o = 10\text{ BTU/HR-FT}^2\text{-°F}$

THEN,

$A_o = \frac{q}{U_o\Delta T_M} = \frac{8.036\text{EE}6}{(10)(301.7)} = 2663.6\text{ FT}^2$

THE TUBE AREA PER BANK IS

$(24)\text{TUBES}\,(\pi)\left(\frac{1.315}{12}\right)\text{FT}\,(20)\text{FT} = 165.2\text{ FT}^2$, SO

$\#\text{ LAYERS} = \frac{2663.6}{165.2} = 16.1$ (SAY 17)

TIMED

1 a) FROM EQN 7.80, THE NUMBER OF POWER STROKES PER MINUTE IS

$\frac{(2)(4600)\text{ RPM}\,(8)\text{ CYLINDERS}}{(4)\text{ STROKES/CYCLE}} = 18400\ 1/\text{MIN}$

THE NET WORK PER CYCLE IS

$(1500-1200) = 300\text{ FT-LBF/MIN}$

(MORE)

TIMED #1, CONTINUED

THE HORSEPOWER IS

$$IHP = \frac{(18400)\ 1/\text{MIN}\ (300)\ \text{FT-LBF}}{33000\ \frac{\text{FT-LBF}}{\text{HP-MIN}}} = 167.27\ \text{HP}$$

b) THE THERMAL EFFICIENCY IS

$$\frac{(300)\ \text{FT-LBF}}{(1.27)\ \text{BTU}\ (778)\ \text{FT-LBF}} = .304$$

c) METHOD 1 ASSUME AN AIR FUEL RATIO OF 15 AND A VOLUMETRIC EFFICIENCY OF 90%. ASSUME AIR AT 70°F AND 14.7 PSIA. THE SWEPT VOLUME PER CYLINDER IS

$$\frac{(265)\ \text{IN}^3}{(8)\ (12)^3} = .01917\ \text{FT}^3$$

THE AIR DENSITY IS

$$\rho = P/RT = \frac{(14.7)(144)}{(53.3)(460+70)} = .07493\ \text{LBM/FT}^3$$

THE AIR WEIGHT PER HOUR IS

$$\dot{w}_A = (.01917)\ \text{FT}^3\ (18400)\ 1/\text{MIN}\ (60)\frac{\text{MIN}}{\text{HR}}\ (.07493)\frac{\text{LBM}}{\text{FT}^3}$$

$$= 1585.8\ \text{LBM/HR}$$

THE FUEL WEIGHT PER HOUR IS

$$\dot{w}_F = \frac{\dot{w}_A}{R_{A/F}} = \frac{1585.8}{15} = 105.7\ \text{LBM/HR}$$

d) THE SPECIFIC FUEL CONSUMPTION IS

$$ISFC = \frac{105.7\ \text{LBM/HR}}{167.27\ \text{HP}} = .632\ \text{LBM/HP-HR}$$

METHOD 2

c) ASSUME THE HEATING VALUE OF GASOLINE IS

HHV = 18,700 BTU/LBM

THEN, THE FUEL CONSUMPTION IS

$$\dot{w}_F = \frac{(1.27)\ \text{BTU/cycle}\ (18400)\ 1/\text{MIN}\ (60)\ \text{MIN/HR}}{(18,700)\ \text{BTU/LBM}}$$

$$= 74.98\ \text{LBM/HR}$$

d) $$ISFC = \frac{74.98}{167.27} = .448\ \text{LBM/HP-HR}$$

2 collect all enthalpies

AT 1: 1393.9

AT 2: 1270

AT 3: 1425.2

AT 4: 1280

AT 5: 1075

AT 6: 69.73

AT 7: 69.73 + .15 = 69.88

AT 8: 250.2

AT 9: 253.1

a) IF THE EXPANSION HAD BEEN ISENTROPIC TO 200 PSIA,

$h_2 = 1230$ BTU/LBM (FROM MOLLIER)

SO $$\eta_{ISEN} = \frac{1393.9-1270}{1393.9-1230} = .756$$

b) LET X BE THE BLEED FRACTION. FROM AN ENERGY BALANCE IN THE HEATER,

$$h_8 = x h_4 + (1-x) h_7$$

$$250.2 = x(1280) + (1-x)69.88$$

$$x = .149$$

THE THERMAL EFFICIENCY IS

$$\eta_{th} = \frac{Q_{IN}-Q_{OUT}}{Q_{IN}} = \frac{(h_1-h_9)+(h_3-h_2)-(1-x)(h_5-h_6)}{(h_1-h_9)+(h_3-h_2)}$$

$$= \frac{(1393.9-253.1)+(1425.2-1270)-(1-.149)(1075-69.73)}{(1393.9-253.1)+(1425.2-1270)}$$

$$= .34$$

3 a) ASSUME THE ISENTROPIC EFFICIENCY IS WANTED. FROM AIR TABLES,

AT 1 $T_1 = -10°F = 450°R$

$h_1 = 107.5$

$p_{r,1} = .7329$

$p_1 = 8$ PSIA

AT 2 $$p_{r,2} = \frac{p_2}{p_1}(p_{r,1}) = \left(\frac{40}{8}\right)(.7329) = 3.6645$$

(IF THE PROCESS WAS ISENTROPIC,

$T_2 = 712°R$

$h_2 = 170.47$

HOWEVER,

$T_2' = 315°F = 775°R$

SO $h_2' = 185.75$

$$\eta_{ISENTROPIC} = \frac{170.47-107.50}{185.73-107.50} = .805$$

b) FOR 35.7 PSIA, $T_{SAT} \approx 20°F = 480°R$

FOR 172.4 PSIA, $T_{SAT} \approx 120°F = 580°R$

FROM EQN 7.151 FOR AN IDEAL HEAT PUMP,

$$COP = \frac{580}{580-480} = 5.8$$

c) $$COP = \frac{(450)\ \text{BTUH}\ (1000)\ \text{W/KW}}{(585)\ \text{W}\ (3413)\ \text{BTUH/KW}} = .225$$

d) $$W_{OUT} = \frac{(600\ \text{EE}6)\ \text{W}\ (3413)\ \text{BTU/KW}}{(1000)\ \text{W/KW}} = 2.048\ \text{EE}9\ \text{BTUH}$$

NOW, $Q_{IN} - Q_{OUT} = W_{OUT} - W_{IN}$

BUT $W_{IN} \approx 0$, SO $Q_{IN} - Q_{OUT} = W_{OUT}$

OR $$Q_{IN} = W_{OUT} + Q_{OUT}$$

$$= 2.048\ \text{EE}9 + 3.07\ \text{EE}9$$

$$= 5.118\ \text{EE}9$$

{MORE}

TIMED #3 CONTINUED

THEN

$$\eta_{th} = \frac{Q_{IN} - Q_{OUT}}{Q_{IN}} = \frac{(5.118\,EE9) - (3.07\,EE9)}{5.118\,EE9}$$

$$= .400$$

e) ASSUME CARNOT CYCLE

$$\eta_{th} = \frac{(82° + 460) - (40 + 460)}{82 + 460} = .0775$$

4) a) IF THE HORSEPOWERS ARE THE SAME,

$$HP_1 = HP_2$$

$$(\dot{W}_{F,1})(HV_1) = (\dot{W}_{F,2})(HV_2)$$

BUT $\dot{W}_F = (SFC)(HP)$. SO

$$(SFC_1)(HV_1) = (SFC_2)(HV_2)$$

$$\frac{(SFC_2)}{(SFC_1)} = \frac{(HV_1)}{(HV_2)} = \frac{23{,}200}{11{,}930} = 1.945$$

THEN

$$\frac{(SFC_2) - (SFC_1)}{(SFC_1)} = \frac{(1.945)(SFC_1) - (SFC_1)}{(SFC_1)}$$

$$= .945 \quad (94.5\% \text{ INCREASE})$$

b) $\dot{W} = VA\rho$, SO $A = \frac{\dot{W}}{V\rho}$

AND $\dot{W}_2 = 1.945\,\dot{W}_1$

$$\frac{A_2 - A_1}{A_1} = \frac{\frac{\dot{W}_2}{V\rho_2} - \frac{\dot{W}_1}{V\rho_1}}{\frac{\dot{W}_1}{V\rho_1}} = \frac{\frac{\dot{W}_2}{\rho_2} - \frac{\dot{W}_1}{\rho_1}}{\frac{\dot{W}_1}{\rho_1}}$$

$$= \frac{\left(\frac{1.945}{\rho_2} - \frac{1}{\rho_1}\right)}{\frac{1}{\rho_1}} = 1.945\left(\rho_1/\rho_2\right) - 1$$

USING INTERPOLATED SPECIFIC GRAVITIES AT 68°F FROM PAGE 5-45

$$\frac{A_2 - A_1}{A_1} = 1.945\left(\frac{.724}{.789}\right) - 1 = .785$$

c) POWER IS PROPORTIONAL TO THE WEIGHT FLOW AND HEATING VALUE,

$$\frac{P_2 - P_1}{P_1} = \frac{V_2A_2\rho_2(HV)_2 - V_1A_1\rho_1(HV)_1}{V_1A_1\rho_1(HV)_1}$$

$$= \frac{\rho_2(HV)_2 - \rho_1(HV)_1}{\rho_1(HV)_1}$$

$$= \frac{(.789)(11930) - (.724)(23200)}{(.724)(23200)} = -.44$$

5 WORK WITH 1 LBM. ASSUME IDEAL GASES.

FOR THE 3→1 PROCESS

FIND SOME COMPOSITE PROPERTY OF THE GAS MIXTURE

$$P_1 = P_3\left(\frac{T_1}{T_3}\right)^{\frac{k}{k-1}}$$

$$14.7 = 568.6\left(\frac{520}{1600}\right)^{\frac{k}{k-1}}$$

$$.02585 = (.325)^{\frac{k}{k-1}}$$

$$\ln(.02585) = \frac{k}{k-1}\ln(.325)$$

$$\frac{k}{k-1} = 3.252 \longrightarrow k = 1.444$$

NOW, GET THE PROPERTIES OF EACH GAS.

$$C_{v,He} = .754$$

$$C_{v,CO_2} = .1599$$

$$k_{He} = 1.66$$

$$k_{CO_2} = 1.28$$

SINCE COMPOSIT $k = 1.444$, LET X = FRACTION He

$$1.444 = x(1.66) + (1-x)(1.28)$$

$$x = .432 \quad \boxed{(43.2\%\ \text{Helium})}$$

SIMILARLY,

$$C_v = (.432)(.754) + (1-.432)(.1599)$$

$$= .4166\ \frac{BTU}{LBM\text{-}°R}$$

$$W_{3-1} = C_v(T_3 - T_1)$$

$$= .4166(1600 - 520) = \boxed{449.9\ \frac{BTU}{LBM}}$$

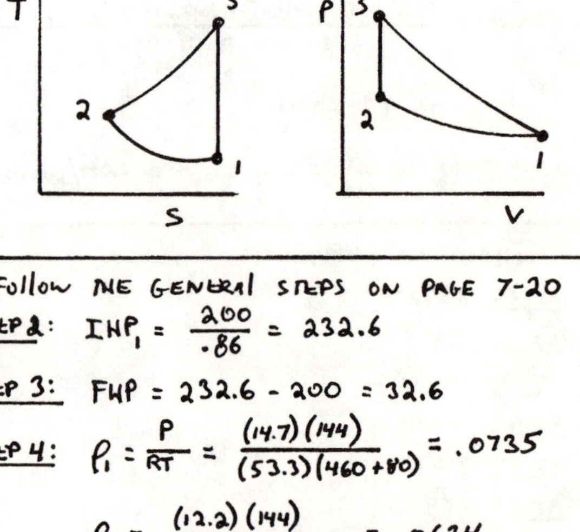

6 FOLLOW THE GENERAL STEPS ON PAGE 7-20

STEP 2: $IHP_1 = \frac{200}{.86} = 232.6$

STEP 3: $FHP = 232.6 - 200 = 32.6$

STEP 4: $\rho_1 = \frac{P}{RT} = \frac{(14.7)(144)}{(53.3)(460+80)} = .0735$

$$\rho_2 = \frac{(12.2)(144)}{(53.3)(460+60)} = .0634$$

{MORE}

TIMED #6 CONTINUED

STEP 5: $IHP_2 = 232.6\left(\frac{.0634}{.0735}\right) = 200.6$

STEP 6: $BHP_2 = 200.6 - 32.6 = \boxed{168}$

$$\eta = \frac{168}{200.6} = \boxed{.837}$$

STEP 7: $\dot{W}_{f1} = (.48)(200) = 96 \frac{LBM}{HR}$

$$\dot{W}_{a1} = (22)(96) = 2112 \frac{LBM}{HR}$$

$$\dot{V}_{a1} = \frac{2112}{.0735} = 28735 \frac{FT^3}{HR}$$

SINCE THE ENGINE SPEED IS CONSTANT,

$$\dot{V}_{a2} = \dot{V}_{a1} = 28735 \frac{FT^3}{HR}$$

STEP 8: $\dot{W}_{a2} = (28735)(.0634) = 1821.8 \frac{LBM}{HR}$

ASSUME THE METERED FUEL INJECTION VOLUME IS THE SAME.

$$\dot{W}_{f2} = \dot{W}_{f1} = 96 \frac{LBM}{HR}$$

$$R_{a/f,2} = \frac{1821.8}{96} = \boxed{18.98}$$

$$BSFC_2 = \frac{96}{168} = \boxed{.57 \frac{LBM}{HP\text{-}HR}}$$

7 FIRST WORK WITH THE ORIGINAL SYSTEM TO FIND THE STEAM FLOW.

$h_1 = 1227.6$ BTU/LBM (SUPERHEATED)

$h_2 = 38.04$ (SUBCOOLED)

$h_3 = 147.92$ (SUBCOOLED)

LET X = FRACTION OF STEAM IN MIXTURE

$$147.92 = x(1227.6) + (1-x)(38.04)$$

$$x = .0924$$

SO, THE STEAM FLOW IS

$$(.0924)(2000) = 184.8 \frac{LBM}{HR}$$

SINCE THE PRESSURE DROP ACROSS THE HEATER IS 5 PSI,

$$P_4 = P_6 + 5 = 20 + 5 = 25 \text{ PSIA}$$

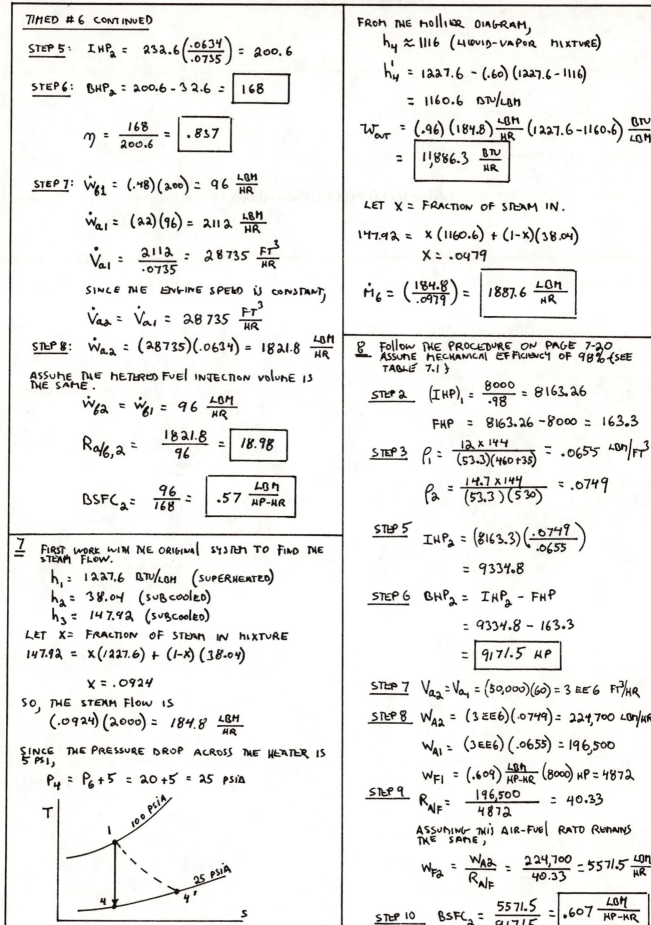

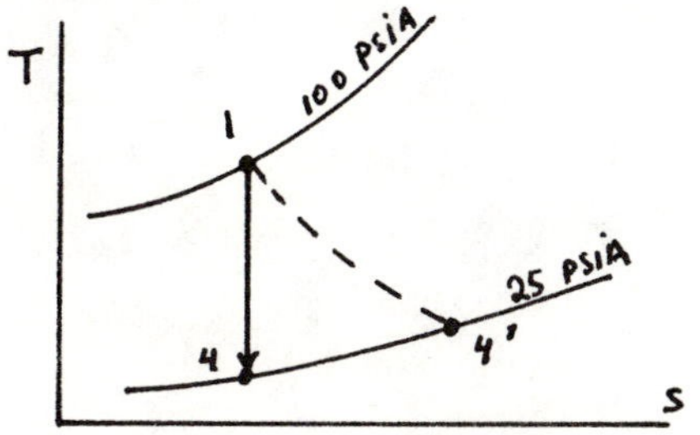

FROM THE MOLLIER DIAGRAM,

$h_4 \approx 1116$ (LIQUID-VAPOR MIXTURE)

$$h_4' = 1227.6 - (.60)(1227.6 - 1116)$$

$$= 1160.6 \text{ BTU/LBM}$$

$$W_{out} = (.96)(184.8)\frac{LBM}{HR}(1227.6 - 1160.6)\frac{BTU}{LBM}$$

$$= \boxed{11{,}886.3 \frac{BTU}{HR}}$$

LET X = FRACTION OF STEAM IN.

$$147.92 = x(1160.6) + (1-x)(38.04)$$

$$x = .0479$$

$$\dot{M}_6 = \left(\frac{184.8}{.0979}\right) = \boxed{1887.6 \frac{LBM}{HR}}$$

8 FOLLOW THE PROCEDURE ON PAGE 7-20. ASSUME MECHANICAL EFFICIENCY OF 98% (SEE TABLE 7.1)

STEP 2 $(IHP)_1 = \frac{8000}{.98} = 8163.26$

$FHP = 8163.26 - 8000 = 163.3$

STEP 3 $\rho_1 = \frac{12 \times 144}{(53.3)(460+35)} = .0655 \text{ LBM/FT}^3$

$$\rho_2 = \frac{14.7 \times 144}{(53.3)(530)} = .0749$$

STEP 5 $IHP_2 = (8163.3)\left(\frac{.0749}{.0655}\right)$

$$= 9334.8$$

STEP 6 $BHP_2 = IHP_2 - FHP$

$$= 9334.8 - 163.3$$

$$= \boxed{9171.5 \text{ HP}}$$

STEP 7 $V_{a2} = V_{a1} = (50{,}000)(60) = 3\text{ EE}6 \text{ FT}^3\text{/HR}$

STEP 8 $W_{A2} = (3\text{EE}6)(.0749) = 224{,}700$ LBM/HR

$W_{A1} = (3\text{EE}6)(.0655) = 196{,}500$

$W_{F1} = (.609)\frac{LBM}{HP\text{-}HR}(8000)\text{ HP} = 4872$

STEP 9 $R_{A/F} = \frac{196{,}500}{4872} = 40.33$

ASSUMING THIS AIR-FUEL RATIO REMAINS THE SAME,

$$W_{F2} = \frac{W_{A2}}{R_{A/F}} = \frac{224{,}700}{40.33} = 5571.5 \frac{LBM}{HR}$$

STEP 10 $BSFC_2 = \frac{5571.5}{9171.5} = \boxed{.607 \frac{LBM}{HP\text{-}HR}}$

RESERVED FOR FUTURE USE

Compressible Fluid Dynamics

WARM-UPS

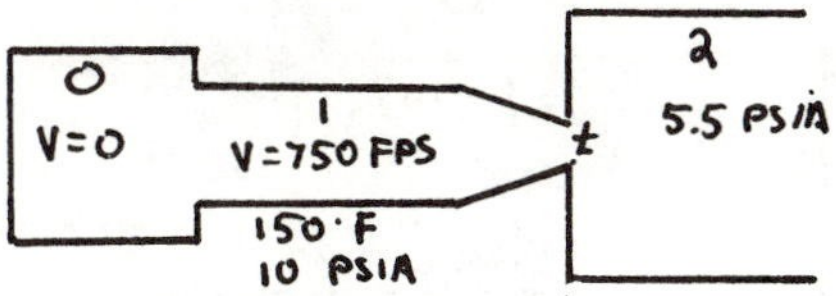

1 $T_1 = 460 + 150 = 610°R$

THE VELOCITY OF SOUND AT POINT 1 IS

$$C_1 = \sqrt{kg_cRT} = \sqrt{(1.4)(32.2)(53.3)610} = 1210.7$$

THE ENTRANCE MACH NUMBER IS

$$M_1 = \frac{V_1}{C_1} = \frac{750}{1210.7} = .619 \ \{\text{SAY } .62\}$$

SO, FROM PAGE 8-17,

$$\frac{T_1}{T_0} = \frac{T}{TT} = .9286$$

$$\frac{P_1}{P_0} = \frac{P}{TP} = .7716$$

$$\text{SO } T_0 = \frac{610}{.9286} = 656.9\ °R$$

$$P_0 = \frac{10}{.7716} = 12.96 \text{ PSIA}$$

$$\frac{P_2}{P_0} = \frac{5.5}{12.96} = .424$$

FROM PAGE 8-5, THE CRITICAL PRESSURE RATIO IS .5283. SINCE $\frac{P_2}{P_0} < .5283$, THE FLOW IS CHOKED AND $M_t = 1$. FROM PAGE 8-18 FOR M=1

$$\frac{P}{P_0} = \frac{P}{TP} = .5283$$

$$\frac{T}{T_0} = \frac{T}{TT} = .8333$$

$$\text{SO } P = (.5283)(12.96) = 6.85 \text{ PSIA}$$

$$T = (.8333)(656.9) = 547.4\ °R$$

2 ASSUME STP IS 14.7 PSIA AND 32°F

THUS, $T_x = 460 + 32 = 492°R$

$P_x = 14.7$ PSIA

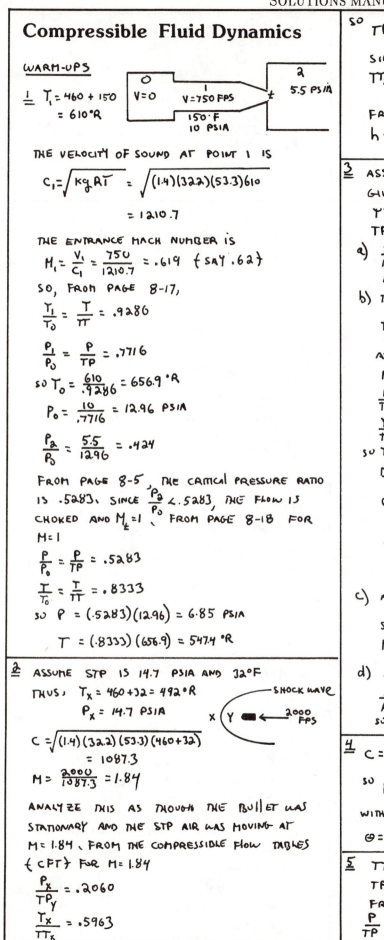

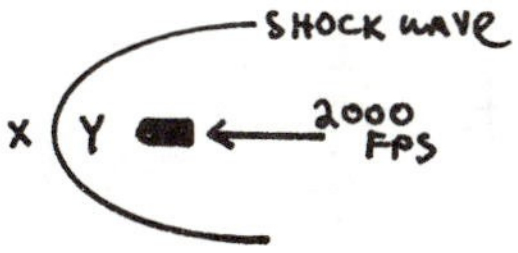

$$C = \sqrt{(1.4)(32.2)(53.3)(460+32)} = 1087.3$$

$$M = \frac{2000}{1087.3} = 1.84$$

ANALYZE THIS AS THOUGH THE BULLET WAS STATIONARY AND THE STP AIR WAS MOVING AT M= 1.84. FROM THE COMPRESSIBLE FLOW TABLES {CFT} FOR M= 1.84

$$\frac{P_x}{TP_y} = .2060$$

$$\frac{T_x}{TT_x} = .5963$$

SO

$$TP_y = \frac{14.7}{.2060} = 71.36 \text{ PSIA}$$

SINCE A SHOCK WAVE IS ADIABATIC,

$$TT_x = TT_y = \frac{492}{.5963} = 825.1$$

FROM PAGE 6-35

$$h \approx 197.9 \ \frac{\text{BTU}}{\text{LBM}}$$

3 ASSUME THE PRESSURE AND TEMPERATURE GIVEN ARE THE CHAMBER PROPERTIES, SO

$TT = 240°F = 700°R$

$TP = 160$ PSIA

a) $\frac{20}{160} < .5283$, SO THE NOZZLE IS SUPERSONIC AND THE THROAT FLOW IS SONIC

b) THE TOTAL DENSITY IS

$$TD = \frac{(TP)}{R(TT)} = \frac{(160)(144)}{(53.3)(700)} = .6175 \ \text{LBM/FT}^3$$

AT THE THROAT,

$M=1$

$$\frac{D}{TD} = .6339$$

$$\frac{T}{TT} = .8333$$

$$\text{SO } T^* = (700)(.8333) = 583.3\ °R$$

$$D^* = (.6175)(.6339) = .3914 \ \text{LBM/FT}^3$$

$$C^* = \sqrt{(1.4)(53.3)(32.2)(583.3)} = 1183.9 \ \text{FT/SEC}$$

$$A^* = \frac{\dot{M}}{D^*C^*} = \frac{(4.5)\ \text{LBM/SEC}}{(.3914)\frac{\text{LBM}}{\text{FT}^3}(1183.9)\frac{\text{FT}}{\text{SEC}}} = .00971 \ \text{FT}^2$$

c) AT EXIT, $\frac{P}{TP} = \frac{20}{160} = .125$

SEARCHING THE CFT GIVES

$M \approx 2.01$

d) AT M= 2.01

$$\frac{A}{A^*} = 1.7016$$

$$\text{SO } A_e = (1.7016)(.00971) = .01652 \ \text{FT}^2$$

4 $C = \sqrt{(1.4)(32.2)(53.3)(460+60)} = 1117.8$ FT/SEC

$$\text{SO } M = \frac{2700}{1117.8} = 2.415$$

WITHOUT KNOWING δ, WE CAN ONLY APPROXIMATE θ

$$\theta = \text{ARCSIN}\left(\frac{1}{2.415}\right) = 24.46°$$

5 $TT = 460 + 80 = 540°R$

$TP = 100$ PSIA

FROM THE CFT AT M=2,

$$\frac{P}{TP} = .1278; \ \frac{T}{TT} = .5556; \ \frac{A}{A^*} = 1.6875$$

{MORE}

WARM-UP #5 CONTINUED

SO, $T = (.5556)(540) = 300°R$
$A = (1.6875)(1) = 1.6875$
$p = (.1278)(100) = 12.78$ PSIA

$C = \sqrt{(1.4)(32.2)(53.3)(300)} = 849$ FT/SEC

$V = Mc = (2)(849) = 1698$ FT/SEC
$D = \rho = P/RT = \frac{(12.78)(144)}{(53.3)(300)} = .1151$ LBM/FT³

$\dot{m} = VA\rho = (1698)\frac{1.6875}{144}(.1151) = 2.29$ LBM/SEC

6

FROM PAGE 8-18 OF THE CFT AT $M_x = 2$

$M_y = .5744$
$\frac{T_y}{T_x} = 1.687$

SO $T_y = (1.687)(500) = 843.5°R$

$C_y = \sqrt{(1.4)(32.2)(53.3)(843.5)} = 1423.6$ FT/SEC

$V_y = MC = (.5744)(1423.6) = 822$ FT/SEC

7

$TP = 100$ PSIA
$TT = (460+70) = 530°R$
SEARCHING THE CFT FOR $\left(\frac{A}{A^*}\right) = 1.555$,
$M = 1.9$
$\frac{T}{TT} = .5807$
$\frac{P}{TP} = .1492$
SO $T = (.5807)(530) = 307.8$
$P = (.1492)(100) = 14.92$

8

SINCE V IS UNKNOWN, ASSUME THAT THE STATIC TEMPERATURE IS $(40+460) = 500°F$. THEN

$\rho = \frac{P}{RT} = \frac{(10)(144)}{(53.3)(500)} = .054$ LBM/FT³

SO $V = \frac{\dot{m}}{A\rho} = \frac{(20) \text{ LBM/SEC}}{(1) \text{ FT}^2 (.054) \frac{\text{LBM}}{\text{FT}^3}} = 370.4$ FT/SEC

AT 500°F,
$C = \sqrt{(1.4)(32.2)(53.3)(500)} = 1096.1$

SO $M = \frac{370.4}{1096.1} = .338$ (SAY .34)

AT $M = .34$, $\frac{T}{TT} = .9774$

SO, A CLOSER APPROXIMATION TO T WOULD BE
$T = (.9774)(500) = 488.7$
$\rho = \frac{P}{RT} = \frac{(10)(144)}{(53.3)(488.7)} = .0553$

$V = \frac{\dot{m}}{A\rho} = \frac{20}{(1)(.0553)} = 361.7$

$C = \sqrt{(1.4)(32.2)(53.3)(488.7)} = 1083.6$

$M = \frac{361.7}{1083.6} = .334$ (SAY .33)

AT $M = .33$,
$\frac{A}{A^*} = 1.8707$, SO
$A_{smallest} = \frac{1}{1.8707} = .535$

9

$\frac{P_x}{TP_y} = \frac{1.38}{20} = .069$

SEARCHING THE CFT, $M = 3.3$

10

$h_1 = 1428.9$
USING THE MOLLIER DIAGRAM ASSUMING ISENTROPIC EXPANSION,
$h_2 = 1362$
FROM EQN 26.67

$V = \sqrt{(2)(32.2)(778)(1428.9-1362)}$
$= 1830.8$ FT/SEC

CONCENTRATES

1

AT THAT POINT,
$C = \sqrt{(1.4)(32.2)(53.3)(1000)} = 1550.1$

SO $M = \frac{600}{1550.1} = .387$

AT $M = .39$
$\frac{T}{TT} = .9705$
$\frac{P}{TP} = .9004$
$\frac{A}{A^*} = 1.6234$

SO $TT = \frac{1000}{.9705} = 1030.4$

$TP = \frac{50}{.9004} = 55.5$

$A^* = \frac{.1}{1.6234} = .0616$

AT $M = 1$,
$\frac{P}{TP} = .5283$
$\frac{T}{TT} = .8333$

SO $P^* = (55.5)(.5283) = 29.32$ PSIA
$T^* = (1030.4)(.8333) = 858.6$

2

THE EXPANSION IS NOT ISENTROPIC AND $V_i \neq 0$, SO EQN 8.69 SHOULD NOT BE USED.
$h_1 = 1322.1$
FROM THE MOLLIER DIAGRAM, ASSUMING ISENTROPIC EXPANSION,
$h_2 = 1228$

{MORE}

CONCENTRATES # 2 CONTINUED

$h_2' = 1322.1 - .85(1322.1 - 1228)$
$= 1242.1$

KNOWING $h = 1242.1$ AND $p = 80$ PSI ESTABLISHES

$T_2' = 420°F$
$v_2' = 6.383$ } FROM DETAILED SUPERHEAT TABLES

SO $\rho_2 = 1/v_2 = 1/6.383 = .1567$ LBM/FT³

FROM EQUATION 26.62

$$v_2' = \sqrt{(2)(32.2)(778)(1322.1-1242.1)+300^2}$$

$= 2024.4$ FT/SEC AT EXIT

$$A_e = \frac{\dot{m}}{v\rho} = \frac{(3)\ \text{LBM/SEC}}{(2024.4)\frac{\text{FT}}{\text{SEC}}(.1567)\frac{\text{LBM}}{\text{FT}^3}} = .009457$$

FROM FIGURE 8.13 FOR $(420+460 = 880°R)$

STEAM, $k_{STEAM} = 1.31$

FROM TABLE 6.4

$R_{STEAM} = 85.8$

$$c = \sqrt{(1.31)(32.2)(85.8)(880)} = 1784.6$$

$$\text{SO}\quad M = \frac{2024.4}{1784.6} = 1.13$$

FROM THE CFT AT $M = 1.13$ AND $k = 1.30$

$$\frac{A}{A^*} = 1.0139$$

$$\text{SO}\quad A^* = \frac{A_e}{1.0139} = \frac{.009457}{1.0139} = .00933$$

TIMED

1 THIS IS A FANNO FLOW PROBLEM. IT'S NOT CLEAR WHAT THE PROBLEM WANTS, BUT ASSUME WE ARE TO CHECK FOR CHOKED FLOW.

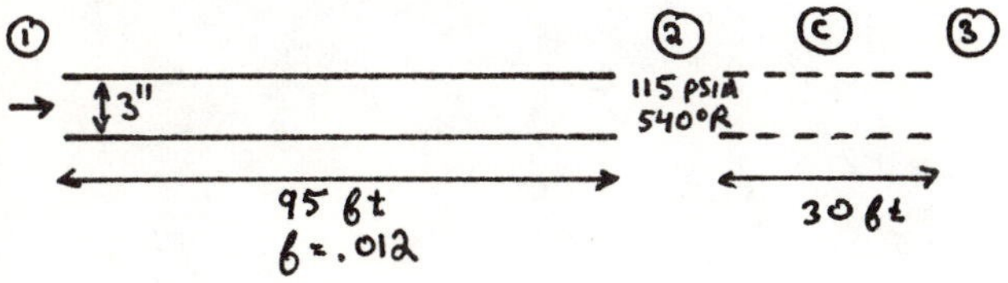

SINCE P IS DECREASING, THE FLOW IS INITIALLY SUB-SONIC.

METHOD 1

AT POINT 2, FROM THE SUPERHEAT TABLES,

$v_2 = 5.066$ FT³/LB

$$V = \frac{\dot{m}}{A\rho} = \frac{\dot{m}v}{A}$$

$$= \frac{(35200)(5.066)}{(3600)\left(\frac{\pi}{4}\right)\left(\frac{3}{12}\right)^2} = 1009 \text{ FT/SEC}$$

ASSUME $k = 1.33$ (SEE FIGURE 8.13)

$$c = \sqrt{kgRT} = \sqrt{(1.33)(32.2)(85.8)(540+460)}$$

$= 1917$ FT/SEC

SO, THE MACH NUMBER AT POINT 2 IS

$$M_2 = \frac{1009}{1917} = .526$$

FROM EQUATION 8.38, THE DISTANCE FROM POINT 2 TO WHERE THE FLOW BECOMES CHOKED IS

$$X_{MAX} = \frac{3/12}{(4)(.012)}\left[\frac{1-(.526)^2}{(1.33)(.526)^2} + \frac{1.33+1}{2(1.33)}\ln\left(\frac{(1.33+1)(.526)^2}{2\left[1+\left(\frac{1.33-1}{2}\right)(.526)^2\right]}\right)\right]$$

$$= 5.208\left[1.966 + .876\ln\left(\frac{.6447}{2.091}\right)\right]$$

$= 4.87$ FT

SO THE FLOW WILL BE CHOKED IN LESS THAN 30 FT

METHOD 2

USE FANNO FLOW TABLE FOR $k = 1.3$

AT $M_2 = .53$, $\frac{4fL_{max}}{D} = .949$

AT $M_c = 1$, $\frac{4fL_{max}}{D} = 0$

$$L_{max} = \frac{(.949-0)(3/12)}{(4)(.012)} = 4.94 \text{ FT}$$

2 SINCE THE PRESSURE DROPS TO 14.7 PSIA, THIS MUST BE A SUPERSONIC NOZZLE.

ASSUME THE GIVEN PROPERTIES ARE TOTAL PROPERTIES.

AT THE THROAT, $M = 1$, AND

$$\frac{T}{TT} = .8333$$

$$\frac{P}{TP} = .5283$$

SO

$T_{throat} = (.8333)(660) = 550°R$

$P_{throat} = (.5283)(160) = 84.53$ PSIA

$$c^* = \sqrt{kgRT} = \sqrt{(1.4)(32.2)(53.3)(550)}$$

$= 1150$ ft/sec

(MORE)

SINCE AIR IS AN IDEAL GAS,

$$\rho_{throat} = \frac{P}{RT} = \frac{(84.53)(144)}{(53.3)(550)}$$

$$= .415 \text{ LBM/FT}^3$$

FROM $\dot{m} = Av\rho$,

$$A^* = \frac{\dot{m}}{\eta v \rho} = \frac{3600}{(3600)(1150)(.415)(.90)}$$

$$= \boxed{.002328 \text{ FT}^2}$$

$$D_{throat} = \sqrt{\frac{4A}{\pi}} = \sqrt{\frac{(4)(.002328)}{\pi}}$$

$$= .0544''$$

AT THE EXIT, $P = 14.7$

$$\frac{P}{TP} = \frac{14.7}{160} = .0919$$

FINDING THIS IN THE ISENTROPIC FLOW TABLES YIELDS

$$M = 2.22$$

$$\frac{A}{A^*} = 2.041$$

SO,

$$A_{exit} = (2.041)(.002328) = .004751$$

$$D_{exit} = \sqrt{\frac{4(.004751)}{\pi}} = .0778$$

THE LONGITUDINAL DISTANCE FROM THROAT TO EXIT IS

$$x = \frac{.0778 - .0544}{(2)(\text{TAN } 3°)} = .223''$$

THE ENTRANCE VELOCITY IS NOT KNOWN, SO THE ENTRANCE AREA CANNOT BE FOUND. HOWEVER, THE LONGITUDINAL DISTANCE FROM ENTRANCE TO THROAT IS

$$.05(.223) = .0112'$$

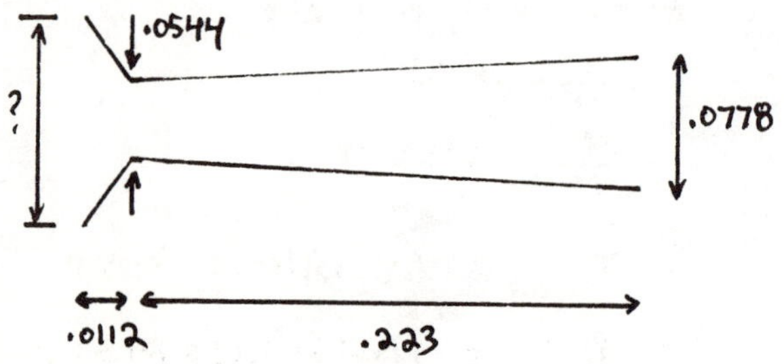

RESERVED FOR FUTURE USE

Combustion

1

	G	AW	G/AW
C	40.0	12	3.33
H	6.7	1	6.7
O	53.3	16	3.33

CH_2O

2 # g MOLES of $CH_4 = \frac{200}{22.4} = 8.929$

1 MOLE OF METHANE HAS A MOLECULAR WEIGHT OF 16 g, SO, THE METHANE HAS A MASS OF

$(16)(8.929) = 142.86$ g

IN POUNDS,

$(142.86)\text{g}(.0022046)\text{LB/g} = .315$ LB

THE AVAILABLE HEAT IS

$(.315)(24{,}000)$ BTU/LB = 7560 BTU

$\Delta T = (95-15)\left(\frac{9}{5}\right) = 144°F$

$q = MC_p\Delta T/\eta$

$$M = \frac{(.5)(7560)(.4536)\text{ kg/lbm}}{(1)(144)} = 11.91 \text{ kg}$$

3 $C_3H_8 + 5O_2 \rightarrow 3CO_2 + 4H_2O$

(MW) 44 + 160 → 132 + 72

$$\frac{CO_2}{C_3H_8} = \frac{132}{44} = \frac{X}{15}$$

X = 45 LBM/HR

$$V = \frac{WRT}{P} = \frac{(45)(35.1)(530)°R}{(14.7)(144)} = 395.5 \text{ FT}^3\text{/HR}$$

4 $CH_4 + 2O_2 \rightarrow CO_2 + 2H_2O$

(volumes) 1 2 1 2

SO, IDEAL OXYGEN VOLUME IS

$2(4000) = 8000$ CFH

ACTUAL VOLUME = $(1.3)(8000) = 10400$ CFH

ACTUAL OXYGEN WEIGHT

$$= \frac{PV}{RT} = \frac{(15)(144)(10400)}{(48.3)(460+100)} = 830.5$$

BUT AIR IS .2315 OXYGEN BY WEIGHT, SO NITROGEN WEIGHT IS

$$\frac{1-.2315}{.2315}(830.5) = 2757 \text{ LBM/HR}$$

5 $C + O_2 \rightarrow CO_2$

12 + 32 → 48

SO $\frac{32}{12} = 2.67$ LBM OXYGEN REQ'D PER POUND CARBON

$2H_2 + O_2 \rightarrow 2H_2O$

4 + 32 → 36

SO $\frac{32}{4} = 8 \frac{\text{LBM } O_2}{\text{LBM } H_2}$

$S + O_2 \rightarrow SO_2$

32.1 + 32 → 64.1

SO $\frac{32}{32.1} = 1 \frac{\text{LBM } O_2}{\text{LBM } S}$

NITROGEN DOES NOT BURN

$$(.84)(2.67) + (.153)(8) + (.003)(1) = 3.47 \frac{\text{LBM } O_2}{\text{LBM FUEL}}$$

BUT AIR IS .2315 OXYGEN

$$\text{AIR} = \frac{3.47}{.2315} = 15 \frac{\text{LBM AIR}}{\text{LBM FUEL}}$$

6 IDEALLY,

$C_3H_8 + 5O_2 \rightarrow 3CO_2 + 4H_2O$

44 + 160 → 132 + 72

THE EXCESS OXYGEN IS $(160)(.2) = 32$

SINCE AIR IS .2315 OXYGEN BY WEIGHT, THE NITROGEN IS

$$\frac{(1-.2315)}{.2315}(160+32) = 637.4$$

$\%CO_2$ BY WEIGHT =

$$\frac{132}{132+72+32+637.4} = .151 \text{ (WET)}$$

CONCENTRATES

1 STEP 1 FIND THE WEIGHT OF OXYGEN IN THE STACK GASES. ASSUME THE STACK GASES ARE

AT 60°F AND 14.7 PSIA WHEN SAMPLED, FROM TABLE 6.4

$R_{CO_2} = 35.1 \quad R_{CO} = 55.2 \quad R_{O_2} = 48.3$

SO

$$\rho_{CO_2} = \frac{P}{RT} = \frac{(14.7)(144)}{(35.1)(460+60)} = .1160$$

$$\rho_{CO} = \frac{(14.7)(144)}{(55.2)(520)} = .07374$$

$$\rho_{O_2} = \frac{(14.7)(144)}{(48.3)(520)} = .08428$$

CO_2 IS $\frac{32}{44} = .7273$ OXYGEN

CO IS $\frac{16}{28} = .5714$ OXYGEN

O_2 IS ALL OXYGEN

IN 100 FT^3 OF STACK GASES, THE TOTAL OXYGEN WEIGHT IS

$$(.12)(100)(.7273)(.1160) + (.01)(100)(.07374)(.5714) + (.07)(100)(.08428)(1.00) = 1.644 \frac{LBM\ O_2}{100\ FT^3}$$

STEP 2 SINCE AIR IS 23.15% OXYGEN BY WEIGHT, THE AIR PER 100 FT^3 IS

$$\frac{1.644}{.2315} = 7.102 \frac{LBM\ AIR}{100\ FT^3}$$

STEP 3 FIND THE WEIGHT OF CARBON IN THE STACK GASES.

CO_2 IS $\frac{12}{44} = .2727$ CARBON

CO IS $\frac{12}{28} = .4286$ CARBON

THEN

$$(.12)(100)(.1160)(.2727) + (.01)(100)(.07374)(.4286) = .4112 \frac{LBM\ CARBON}{100\ FT^3}$$

STEP 4 THE COAL IS 80% CARBON, SO THE AIR PER POUND OF COAL IS

$$\frac{(.80)\frac{LBM\ CARBON}{LBM\ COAL}}{(.4112)\frac{LBM\ CARBON}{100\ FT^3}}\left(7.102\right)\frac{LBM\ AIR}{100\ FT^3} = 13.82 \frac{LBM\ AIR}{LBM\ COAL}$$

THIS DOES NOT INCLUDE AIR TO BURN THE HYDROGEN, SINCE ORSAT IS A DRY ANALYSIS,

STEP 5. FROM EQN 9.3

THE THEORETICAL AIR FOR THE HYDROGEN IS

$$34.34\left(.04 - \frac{.02}{8}\right) = 1.288 \frac{LBM\ AIR}{LBM\ COAL}$$

STEP 6 IGNORING ANY EXCESS AIR FOR THE HYDROGEN, THE AIR PER POUND COAL IS

$$13.82 + 1.29 = 15.11 \frac{LBM\ AIR}{LBM\ COAL}$$

---

2 STEP 1 FROM TABLE 9.22

THE HEATING VALUE PER POUND OF COAL IS

$$(.75)(14093) + \left(.05 - \frac{.03}{8}\right)(60958) = 13389 \frac{BTU}{LBM}$$

STEP 2 THE GRAVIMETRIC ANALYSIS OF 1 LBM OF FUEL IS

CARBON: .... .75 LBM

FREE HYDROGEN $\left(.05 - \frac{.03}{8}\right)$ = ... .0463

WATER $9(.05 - .0463)$ .... .0333

NITROGEN .... .02

STEP 3 THE THEORETICAL STACK GASES PER POUND OF COAL (TABLE 9.11)

FOR .75 LBM COAL ARE

CO_2: $(.75)(3.67) = 2.753$

N_2: $(.75)(8.78) = 6.585$

ALL PRODUCTS ARE SUMMARIZED IN THE FOLLOWING TABLE:

	CO_2	N_2	H_2O
FROM CARBON	2.753	6.583	
FROM H_2		1.218	.417
FROM H_2O			.0333
FROM O_2	SHOWS UP IN CO_2, H_2O		
FROM N_2		.02	
TOTALS:	2.753	7.821	.4503

STEP 4 ASSUME THE STACK GASES LEAVE AT 1000°F. THEN, $T_{AVE} = \frac{1}{2}\left((60+460) + (1000+460)\right) = 990$ (SAY 1000°R)

FROM TABLE 9.14

$\bar{C}_p(CO_2) = .251$

$\bar{C}_p(N_2) = .255$

$\bar{C}_p(H_2O) = .475$

THE HEAT REQUIRED TO RAISE THE COMBUSTION PRODUCTS FROM 1 LBM OF COAL 1°F IS

$$(2.753)(.251) + (7.821)(.255) + (.4503)(.475) = 2.9\ BTU$$

STEP 5 ASSUMING ALL COMBUSTION HEAT GOES INTO THE STACK GASES, THE FINAL TEMPERATURE IS

$$T_2 = T_1 + \frac{HHV}{2.9} = 60 + \frac{13389}{2.9} = 4677°F$$

STEP 6 IN REALITY, THERE WILL BE APPROXIMATELY 40% EXCESS AIR, AND 75% OF THE HEAT WILL BE ABSORBED BY THE BOILER

$$\text{EXCESS AIR} = \frac{(.40)(7.821)}{.7685} = 4.071 \text{ LBM}$$

FROM TABLE 9.14, $\overline{C}_p$ FOR AIR AT 1000°R IS

$$(.7685)(.255) + (.2315)(.236) = .251$$

$$\text{THEN, } T_2 = 60 + \frac{13389(1-.75)}{2.9 + (4.071)(.251)} = 913.5°$$

3 STEP 1 USE TABLE 9.11 TO FIND THE STOICHIOMETRIC OXYGEN REQUIRED PER POUND OF FUEL OIL

$C \rightarrow CO_2$:	$(.8543)(2.67) =$	2.2810
$H \rightarrow H_2O$:	$(.1131)(8) =$	.9048
$S \rightarrow SO_2$:	$(.0034)(1) =$	.0034
LESS OXYGEN IN FUEL		−.0270
		3.1622 LBM OXY / LBM FUEL

STEP 2 THE THEORETICAL NITROGEN IS $\frac{3.1622}{.2315}(.7685) = 10.497$

THE ACTUAL NITROGEN WITH EXCESS AIR AND THE NITROGEN IN THE FUEL IS

$$.0022 + (1.6)(10.497) = 16.8$$

THIS IS A VOLUME OF

$$V = \frac{WRT}{P} = \frac{(16.8)(55.2)(460+60)}{(14.7)(144)} = 227.81 \text{ FT}^3$$

STEP 3 THE EXCESS OXYGEN

$$(.6)(3.1622) = 1.897 \ \frac{\text{LBM AIR}}{\text{LBM FUEL}}$$

THIS IS A VOLUME OF

$$V = \frac{WRT}{P} = \frac{(1.897)(48.3)(460+60)}{(14.7)(144)} = 22.51 \text{ FT}^3$$

STEP 4 FROM TABLE 9.11, THE 60°F COMBUSTION PRODUCT VOLUMES PER POUND OF FUEL WILL BE

CO_2 :	$(.8543)(316)$ =	27.0 FT^3
H_2O :	$(.1131)(189.5)$ =	21.43
SO_2 :	$(.0034)(1184)$ =	.04
N_2 :	FROM STEP 2 =	227.81
O_2 :	FROM STEP 3	22.51
		298.79 FT^3

STEP 5 AT 600°F, THE WET VOLUME WILL BE

$$V_{WET} = \left(\frac{460+600}{460+60}\right)(298.79) = 609 \text{ FT}^3$$

AT 600°F, THE DRY VOLUME WILL BE

$$V_{DRY} = \left(\frac{460+600}{460+60}\right)(298.79 - 21.43) = 565.4 \text{ FT}^3$$

STEP 6 % CO_2 BY VOLUME (DRY)

$$= \frac{27}{27 + .04 + 227.81 + 22.51} = .097$$

4 NOTE THAT THE HYDROGEN AND OXYGEN LISTED MUST BOTH BE FREE SINCE THE WATER VAPOR IS LISTED SEPARATELY.

STEP 1 THE USEABLE % OF CARBON PER POUND OF FUEL IS

$$.5145 - \frac{2816(.209)}{15395} = .4763$$

STEP 2 THE THEORETICAL OXYGEN REQUIRED PER POUND OF FUEL IS

$C \rightarrow CO_2$:	$(.4763)(2.67)$ =	1.2717
$H_2 \rightarrow H_2O$:	$(.0402)(8.0)$ =	.3216
$S \rightarrow SO_2$:	$(.0392)(1.0)$ =	.0392
MINUS O_2 IN FUEL		−.0728
		1.5597

STEP 3 THE THEORETICAL AIR IS

$$\frac{1.5597}{.2315} = 6.737 \ \frac{\text{LBM AIR}}{\text{LBM FUEL}}$$

STEP 4 IGNORING FLY ASH, THE THEORETICAL DRY PRODUCTS ARE

CO_2 :	$(.4763)(3.67)$	= 1.748
SO_2 :	$(.0392)(2.0)$	= .0784
N_2 :	$(.0093 + (.7685)(6.737))$	= 5.187
		7.013

STEP 5 THE EXCESS AIR IS

$$13.3 - 7.013 = 6.287$$

STEP 6 THE TOTAL AIR SUPPLIED WAS

$$6.287 + 6.737 = 13.024 \ \frac{\text{LBM AIR}}{\text{LBM FUEL}}$$

5 STEP 1 THE DENSITY OF CARBON DIOXIDE AT 60°F IS

$$\rho = \frac{P}{RT} = \frac{(14.7)(144)}{(35.1)(460+60)} = .1160$$

THE CARBON DIOXIDE IS $\frac{12}{44} = .2727$ CARBON, SO THE WEIGHT OF CARBON PER 100 FT^3 OF STACK GASES IS $[(.2727)(.1160)(100)(.095)] = .3005$

STEP 2 THE DENSITY OF NITROGEN IS

$$\rho = \frac{(14.7)(144)}{(55.2)(460+60)} = .07375$$

SO, THE WEIGHT OF NITROGEN IN 100 FT^3 IS

$(.815)(100)(.07375) = 6.0106$

STEP 3 THE ACTUAL NITROGEN PER POUND OF COAL IS

$$\frac{(.65)(1-.03)\ \frac{LBM\ CARBON}{LBM\ FUEL}}{(.3005)\ \frac{LBM\ CARBON}{100\ FT^3}}\ (6.0106)\ \frac{LBM\ N_2}{100\ FT^3}$$

$$= 12.611\ \frac{LBM\ N_2}{LBM\ FUEL}$$

STEP 4 ASSUMING THE ASH PIT LOSS COAL HAS THE SAME COMPOSITION AS THE UNBURNED COAL, THE THEORETICAL NITROGEN PER POUND OF FUEL BURNED IS

$(.7685)(9.45)(1-.03) = 7.0444$

STEP 5 THE % EXCESS AIR IS

$$\frac{12.611 - 7.0444}{7.0444} = .790$$

6 FROM EQN 9.20

$[C] = (1-.03)(.6734) = .6532$

$$q_4 = \frac{(10,143)[.6532](1.6)}{15.5+1.6} = 619.9\ \frac{BTU}{LBM}$$

7 FROM TABLE 6.5, DENSITY IS VOLUMETRICALLY WEIGHTED.

METHANE

$B = .93$

$$\rho = \frac{P}{RT} = \frac{(14.7)(144)}{(96.4)(460+60)} = .0422$$

$$\frac{FT^3\ AIR}{FT^3\ FUEL} = 9.52 \quad \{\text{TABLE}$$

PRODUCTS: 1 FT^3 CO_2, 2 FT^3 H_2O

HHV = 1013 {TABLE 23.15

THE RESULTS FOR ALL THE FUEL COMPONENTS IS TABULATED IN THE FOLLOWING TABLE

					VOLUMES OF PRODUCTS		
GAS	B	ρ	FT^3 AIR	HHV	CO_2	H_2O	OTHER
CH_4	.93	.0422	9.52	1013.	1	2	
N_2	.034	.0737	—	—	—	—	1 N_2
CO	.0045	.0737	2.38	322	1	—	—
H_2	.0182	.0053	2.39	325	—	1	—
C_2H_4	.0025	.0739	14.29	1614	2	2	—
H_2S	.0018	.0900	7.15	647	—	1	1 SO_2
O_2	.0035	.0843	—	—	—	—	—
CO_2	.0022	.1160	—	—	1	—	—

a) THE COMPOSITE DENSITY IS

$\sum B\rho = .0431$

b) THE THEORETICAL AIR REQUIREMENTS ARE

$$\sum B_i\,(FT^3\ AIR) - \frac{OXYGEN\ IN\ FUEL}{.209}$$

$$= 8.9564 - \frac{.0035}{.209} = 8.94\ \frac{FT^3\ AIR}{FT^3\ FUEL}$$

c) THE AIR IS 20.9% OXYGEN BY VOLUME, SO THE THEORETICAL OXYGEN WILL BE

$(8.94)(.209) = 1.868\ FT^3/FT^3$

THE EXCESS OXYGEN WILL BE

$(.4)(1.868) = .747\ FT^3/FT^3$

SIMILARLY, THE TOTAL NITROGEN IN THE STACK GASES IS

$(1.4)(.791)(8.94) + .034 = 9.934$

THE STACK GASES PER FT^3 OF FUEL ARE

EXCESS O_2: = .747

NITROGEN: = 9.934

SO_2: = .0018

CO_2: $(.93)(1) + (.0045)(1) + (.0025)(2) + (.0022)(1) = .9417$

H_2O: $(.93)(2) + (.0182)(1) + (.0025)(2) + (.0018)(1) = 1.885$

THE TOTAL WET VOLUME IS 13.51

THE TOTAL DRY VOLUME IS 11.62

THE VOLUMETRIC ANALYSES ARE

	O_2	N_2	SO_2	CO_2	H_2O
WET	.055	.735	—	.070	.140
DRY	.064	.855	—	.081	—

8 REFER TO THE PROCEDURE STARTING ON PAGE 9-15 {WHICH IS LIBERALLY INTERPRETED}

STEP 1 HEAT ABSORBED IN THE BOILER IS

$(11.12)\ LBM\ WATER\ (970.3)\ \frac{BTU}{LBM} = 10789.7\ \frac{BTU}{LBM}$

STEP 2 THE LOSSES ARE

a) HEATING STACK GASES: THE BURNED CARBON PER POUND OF FUEL IS

$.7842 - (.315)(.0703) = .7621$

FROM EQUATION 23.15, THE WEIGHT OF DRY FUEL GAS IS

$$\frac{\{11(14.0) + 8(5.5) + 7(.42 + 80.08)\}\{.7621 + \frac{.01}{1.833}\}}{3(14 + .42)}$$

$$= 13.51\ \frac{LBM\ STACK\ GASES}{LBM\ FUEL}$$

ASSUME $\bar{C}_p = .245\ BTU/LBM\text{-}°F$

ASSUME THE AIR IS 73°F INITIALLY.

THEN

$q_1 = (13.51)(.245)(575-73) = 1661.6$ BTU/LBM

b) THE COAL ANALYSIS GIVEN ADDS TO 100% WITHOUT THE 1.91% MOISTURE, WE MUST ASSUME THAT THE H_2 AND O_2 ARE FREE GASES, WITH LITTLE ERROR, THE AMOUNT OF MOISTURE CAN BE TAKEN AS

$.0191 \frac{\text{LBM}}{\text{LBM COAL}}$

b1) TO HEAT, EVAPORATE, AND SUPERHEAT THE WATER FORMED FROM COMBUSTION:

FROM FIGURE 8.13, ASSUME $\bar{C}_p = .46$ FOR SUPERHEATED STEAM

$(.0556)$ LBM HYDROGEN $(9) \frac{\text{LBM WATER}}{\text{LBM HYDROGEN}} \times$

$\times \left((1)(212-73) + (970.3) + .46(575-212)\right)$

$= 638.7$ BTU/LBM

b2) TO EVAPORATE THE MOISTURE IN THE COAL

$(.0191)\left((1)(212-73) + (970.3) + .46(575-212)\right)$

$= 24.4$ BTU/LBM

c) FROM THE PSYCHROMETRIC CHART,

$\omega = 90 \frac{\text{GRAINS}}{\text{LBM AIR}}$

$\frac{90}{7000} = .0129 \frac{\text{LBM WATER}}{\text{LBM AIR}}$

SINCE THE CARBON BURNED PER POUND OF FUEL IS

$.7842 - (.315)(.0703) = .7621$

EQN 9.13 {MODIFIED FOR SULFUR CONTENT} IS

$$\frac{\text{LBM AIR}}{\text{LBM FUEL}} = \frac{3.04(80.08)\left[.7621 + \frac{.01}{1.833}\right]}{(14+.42)}$$

$= 12.96$

THE ENERGY TO SUPERHEAT THE MOISTURE IN THE AIR IS APPROXIMATELY

$(.0129) \frac{\text{LBM WATER}}{\text{LBM AIR}} (12.96) \frac{\text{LBM}}{\text{AIR}} (.46) \frac{\text{BTU}}{\text{LBM-°F}}$

$\times (575-73)$°F

$= 38.6$ BTU/LBM

d) IN INCOMPLETE COMBUSTION OF CARBON {EQN 9.20}

$$q_4 = \frac{10,143[.7621](.42)}{14+.42} = 225.1$$

e) IN UNBURNED CARBON {SEE EQN 9.21}

$$q_5 = \frac{(14,093)(.0703)(31.5)}{100} = 312.1$$

f) RADIATION AND UNACCOUNTED FOR:

$14000 - 10789.7 - 1661.6 - 638.7 - 24.4 - 38.6 - 225.1 - 312.1$

$= 309.8$

9 (a) ONLY THE DOUBLY-CHARGED METALLIC IONS CONTRIBUTE TO HARDNESS.

Ca^{++} 80.2 mg/l

Mg^{++} 24.3 mg/l

Fe^{++} 1.0 mg/l

PROCEED AS IN EXAMPLE 9.19. THE EQUIVALENT WEIGHTS ARE:

Ca $40.1/2 = 20.05$

Mg $24.3/2 = 12.15$

Fe $55.8/2 = 27.9$

$CaCO_3$ $[40.1 + 12 + (3)(16)]/2 = 50.05$

$$\text{HARDNESS} = (80.2)\left(\frac{50.05}{20.05}\right) + (24.3)\left(\frac{50.05}{12.15}\right) + (1.0)\left(\frac{50.05}{27.9}\right)$$

$= 302.1$ mg/l AS $CaCO_3$

(b) ASSUME ZERO HARDNESS IS POSSIBLE. USE STOICHIOMETRIC RELATIONSHIPS.

$$CO_2 + Ca(OH)_2 \rightarrow CaCO_3\downarrow + H_2O$$

(MW) 44 + 74.1 = 100.1 + 18

(mg/l) 19.0 + X

THE LIME REQUIRED TO REMOVE THE CO_2 IS

$$\frac{X}{19.0} = \frac{74.1}{44}$$

$X = 32.0$ mg/l $Ca(OH)_2$

SINCE THE ACTUAL COMPOUNDS IN THE WATER SUPPLY ARE NOT KNOWN, WE HAVE TO MAKE SOME ASSUMPTIONS ABOUT THE DISTRIBUTION OF RADICALS.

(1) ASSUME ALL OF THE $(HCO_3)^-$ IS IN THE FORM OF $Ca(HCO_3)_2$

Ca^{++} HAS A MW OF 40.1

$(HCO_3)_2^{--}$ HAS A MW OF 122.0

SO, THE 185 mg/l OF HCO_3^- WILL ACCOUNT FOR

$(40.1)\left(\frac{185}{122}\right) = 60.8$ mg/l Ca^+

THE TOTAL $Ca(HCO_3)_2$ IS

$185 + 60.8 = 245.8$ mg/l

(2) ASSUME THE REMAINING $(80.2-60.8)=19.4$ mg/ℓ OF Ca^{++} IS IN THE FORM $CaSO_4$

Ca^{++} HAS A MW OF 40.1

$SO_4^{=}$ HAS A MW OF 96.1

SO, THE 19.4 mg/ℓ OF Ca^{++} will ACCOUNT FOR

$$(96.1)\left(\frac{19.4}{40.1}\right)=46.5 \text{ mg/ℓ } SO_4^{--}$$

THE TOTAL $CaSO_4$ IS

$$19.4 + 46.5 = 65.9 \text{ mg/ℓ}$$

$$Ca(HCO_3)_2 + Ca(OH)_2 \longrightarrow 2CaCO_3\downarrow + 2H_2O$$

(MW) 162.1 + 74.1 = 200.2 + 36

(mg/ℓ) 245.8 X

THE REQUIRED LIME IS

$$(74.1)\left(\frac{245.8}{162.1}\right) = 112.4 \text{ mg/ℓ } Ca(OH)_2$$

$$CaSO_4 + Na_2CO_3 \longrightarrow CaCO_3\downarrow + Na_2SO_4$$

(MW) 136.2 + 106 = 100.1 + 142.1

(mg/ℓ) 65.9 X

THE REQUIRED SODA ASH IS

$$106\left(\frac{65.9}{136.2}\right) = 51.3 \text{ mg/ℓ } Na_2CO_3$$

TOTAL LIME: $32.0 + 112.4 = 144.4$ mg/ℓ $Ca(OH)_2$

Total SODA ASH: 51.3 mg/ℓ Na_2CO_3

10 $Ca(HCO_3)_2$ AND $MgSO_4$ BOTH CONTRIBUTE TO HARDNESS. SINCE 100 mg/ℓ OF HARDNESS IS THE GOAL, LEAVE ALL $MgSO_4$ IN THE WATER. SO, NO SODA ASH IS REQUIRED. PLAN ON TAKING OUT $(137+72-100)=109$ mg/ℓ of $Ca(HCO_3)_2$.

FROM EQUATION 9.31 (INCLUDING THE EXCESS),

$$\text{pure } CaO = 30 + \frac{56}{100}(109) = 91.04 \text{ mg/ℓ}$$

$$(91.04)(8.345) = 759.7 \text{ LB/million gallons}$$

(b) HARDNESS REMOVED = $(137+72-100)=109$ mg/ℓ

THE CONVERSION FROM mg/ℓ TO LB/MG is 8.345.

$$(109) \text{ mg/ℓ } (8.345) = 909.6 \frac{\text{LB HARDNESS}}{\text{MILLION GALLONS}}$$

$$\left(\frac{.5}{1000}\right)\frac{\text{LB}}{\text{GRAIN}}(909.6)\frac{\text{LB}}{\text{MIL. GAL}}(7000)\frac{\text{GRAIN}}{\text{LB}} =$$

$= 3.18$ EE LB/MILLION gallons

TIMED

1a) SILICON IN ASH IS SiO_2, WITH A MOLECULAR WEIGHT OF

$$28.09 + 2(16) = 60.09.$$

THE OXYGEN TIED UP IS

$$\frac{2(16)}{28.09}(.061) = .0695$$

THE SILICON ASH PER HOUR IS

$(15300)(.061 + .0695) = 1996.7$ LBM/HR

THE REFUSE SILICON IS

$410(1-.30) = 287$ LBM/HR

THE EMISSION RATE IS

$1996.7 - 287 = 1709.7$ LBM/HR

b) FROM TABLE 9.11 EACH POUND OF SULFUR PRODUCES 2 LBM SO_2

$(15300)(.0244)(2) = 7466$ LBM/HR

c) THE COOLING WATER WOULD BE USED IN THE CONDENSER. INSUFFICIENT INFORMATION IS GIVEN TO EVALUATE THE CONDENSER LOAD

d) FROM EQUATION 9.12, THE HEATING VALUE OF THE FUEL IS

$$14093(.7656) + 60{,}958\left(.055 - \frac{.077}{8}\right) + 3983(.0244)$$

$$= 13653 \text{ BTU/LBM}$$

THE ALLOWABLE EMISSION RATE IS

$$\frac{(1)\frac{\text{LBM}}{\text{MBTU}}(15300)\frac{\text{LBM}}{\text{HR}}(13653)\frac{\text{BTU}}{\text{LBM}}}{1{,}000{,}000\frac{\text{BTU}}{\text{MBTU}}} = 20.89 \frac{\text{LBM}}{\text{HR}}$$

$$\eta = \frac{1709.7 - 20.89}{1709.7} = .988$$

2 FOR PROPANE

$$C_3H_8 + 5O_2 \longrightarrow 3CO_2 + 4H_2O$$

MW: 44.09 + 160 ⟶ 132.03 + 72.06

OXYGEN IS

$$(160)(1.4) = 224$$

EXCESS: $224 - 160 = 64$

THE WEIGHT RATIO OF NITROGEN TO OXYGEN IS

$$\frac{G_N}{G_O} = \frac{B_N R_O}{B_O R_N} = \frac{(.40)(48.3)}{(.60)(55.2)} = .583$$

ACCOMPANYING NITROGEN

$$(224)(.583) = 130.6$$

SO, THE MASS BALANCE PER MOLE OF PROPANE IS

$$C_3H_8 + O_2 + N_2 \longrightarrow CO_2 + H_2O + O_2 + N_2$$
$$44.09 + 224 + 130.6 \longrightarrow 132.03 + 72.06 + 64 + 130.6$$

AT STANDARD INDUSTRIAL CONDITIONS {60°F, 1 ATMOS}
THE PROPANE DENSITY IS

$$\rho = \frac{P}{RT} = \frac{(14.7)(144)}{(35.0)(460+60)} = .1163 \frac{LBM}{FT^3}$$

SINCE 250 SCFM FLOW, THE MASS FLOW RATE OF THE PROPANE IS

$$(250)(.1163) = 29.075 \frac{LBM}{MIN}$$

SCALING THE OTHER MASS BALANCE FACTORS DOWNWARD,

$$C_3H_8 + O_2 + N_2 \longrightarrow CO_2 + H_2O + O_2 + N_2$$
$$29.08 + 147.72 + 86.1 \longrightarrow 87.07 + 47.52 + 42.20 + 86.1$$

THE OXYGEN FLOW IS

$$\boxed{147.72 \frac{LBM}{MIN}}$$

THE SPECIFIC VOLUMES OF THE REACTANTS ARE

$$v_{C_3H_8} = \frac{RT}{P} = \frac{(35.0)(460+80)}{(14.7)(144)} = 8.929 \frac{FT^3}{LBM}$$

$$v_{O_2} = \frac{(48.3)(540)}{(14.7)(144)} = 12.321$$

$$v_{N_2} = \frac{(55.2)(540)}{(14.7)(144)} = 14.082$$

THE TOTAL INCOMING VOLUME IS

$$(29.08)(8.929) + (147.72)(12.321) + (86.1)(14.082) =$$
$$= 3292 \text{ CFM}$$

SINCE VELOCITY MUST BE KEPT BELOW 400 FPM,

$$A = \frac{Q}{V} = \frac{3292}{400} = \boxed{8.23 \text{ FT}^2}$$

SIMILARLY, THE SPECIFIC VOLUMES OF THE PRODUCTS ARE

$$v_{CO_2} = \frac{(44)(460+460)}{(8)(144)} = 35.14 \frac{FT^3}{LBM}$$

$$v_{H_2O} = \frac{(85.8)(920)}{1152} = 68.52$$

$$v_{O_2} = \frac{(48.3)(920)}{1152} = 38.57$$

$$v_{N_2} = \frac{(55.2)(920)}{1152} = 44.08$$

THE TOTAL EXHAUST VOLUME IS

$$(87.07)(35.14) + (47.52)(68.52) + (42.20)(38.57) +$$
$$(86.1)(44.08)$$
$$= \boxed{11739 \text{ CFM}}$$

$$A_{stack} = \frac{11739}{800} = \boxed{14.67 \text{ FT}^2}$$

FOR IDEAL GASES, THE PARTIAL PRESSURE IS VOLUMETRICALLY WEIGHTED.

THE WATER VAPOR PARTIAL PRESSURE IS

$$8\left[\frac{(47.52)(68.52)}{11739}\right] = 2.22 \text{ PSIA}$$

THE SATURATION TEMPERATURE CORRESPONDING TO 2.22 PSIA IS

$$\boxed{130°F = T_{dewpoint}}$$

3 ASSUME THE NITROGEN AND OXYGEN CAN BE VARIED INDEPENDENTLY.

	C_3H_8	$+ 5O_2 \longrightarrow$	$3CO_2$	$+ 4H_2O$
WTS:	44.09	160	132.03	72.06
moles:	(1)	(5)	(3)	(4)

ASSUME 0% EXCESS OXYGEN. (JUSTIFIED SINCE ENTHALPY INCREASE INFORMATION IS NOT GIVEN FOR OXYGEN.)

USE THE ENTHALPY OF FORMATION DATA TO CALCULATE THE HEAT OF REACTION (MAXIMUM FLAME TEMPERATURE).

OXYGEN AND NITROGEN HAVE $\Delta H_f = 0$. SO,

$$3(-169{,}300) + 4(-104{,}040) - (28{,}800) = -952{,}860 \frac{BTU}{mole}$$

THE NEGATIVE SIGN SIGN INDICATES EXOTHERMIC REACTION.

LET X BE THE # OF MOLES OF NITROGEN PER MOLE OF PROPANE. USE THE NITROGEN TO COOL THE COMBUSTION.

$$952{,}860 = 3(39{,}791) + 4(31{,}658) + x(24{,}471)$$

$$x = 28.89 \text{ moles}$$

$$M_{N_2} = (28.89)(28.016) = \boxed{809.4 \frac{LBM}{MIN}}$$

$$M_{O_2} = \boxed{160 \frac{LBM}{MIN}}$$

EACH MOLE OF PROPANE PRODUCES 4 MOLES OF H_2O

$$M_{H_2O} = (4)(18) = \boxed{72 \frac{LBM}{MIN}}$$

ALL FLOWS ARE PER MOLE PROPANE

RESERVED FOR FUTURE USE

Heat Transfer

WARM-UPS

1 USE THE VALUE OF K AT THE AVERAGE TEMPERATURE $\frac{1}{2}(150+350) = 250$

$$K = .030(1+.0015(250)) = .04125$$

2 USE EQUATION 10.3

$$q = \frac{KA\Delta T}{L}$$

$$= \frac{(.038)\frac{\text{BTU-FT}}{\text{FT}^2\text{-HR-}^\circ\text{R}}(350^\circ)\text{F}}{(1)\ \text{FT}} = 13.3\ \frac{\text{BTU}}{\text{HR-FT}^2}$$

3 $\Delta T_A = 190^\circ F - 85^\circ F = 105^\circ F$

$\Delta T_B = 160 - 85 = 75^\circ F$

FROM EQUATION 10.66

$$\Delta T_M = \frac{105-75}{\ln\left(\frac{105}{75}\right)} = 89.16^\circ F$$

4 a) h SHOULD BE EVALUATED AT $\frac{1}{2}(T_{PIPE} + T_{FLUID})$. THE MIDPOINT PIPE TEMPERATURE IS $\frac{1}{2}(160+190) = 175^\circ F$. AT THE START OF THE HEATING PROCESS,

$$T_{FILM} = \frac{1}{2}(85+175) = 130^\circ F$$

SINCE THE STEAM OUTLET TEMPERATURE WILL INCREASE AS THE FLUID TEMPERATURE INCREASES, WE DON'T HAVE ENOUGH INFORMATION TO EVALUATE THE FINAL FILM COEFFICIENT.

b) SINCE THE STEAM IS COOLING, IT MUST BE SUPERHEATED. THE INITIAL FILM SHOULD BE EVALUATED AT $\frac{1}{2}(190+160) = 175^\circ F$

5 FROM PAGE 10-14, ASSUME COUNTER-FLOW OPERATION TO FIND ΔT_M

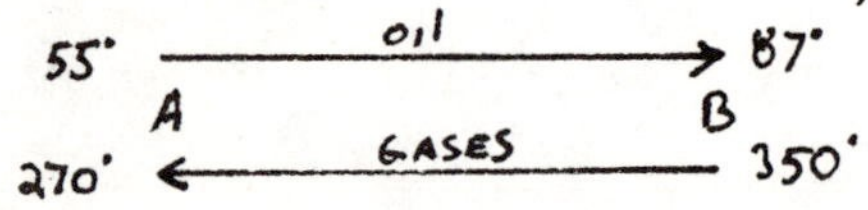

$\Delta T_A = 270 - 55 = 215$

$\Delta T_B = 350 - 87 = 263$

FROM EQUATION 10.66

$$\Delta T_M = \frac{215-263}{\ln\left(\frac{215}{263}\right)} = 238.19^\circ F$$

6 IF THE STEAM IS 87% WET, THE QUALITY IS $x = .13$. FROM EQUATION 6.34

$$v = .01727 + .13(8.515 - .01727)$$

$$= 1.122\ \text{FT}^3/\text{LBM}$$

$$\rho = 1/v = \frac{1}{1.122} = .891\ \frac{\text{LBM}}{\text{FT}^3}$$

7 FROM PAGE 10-23

$$\mu = (.458\ \text{EE-3})\frac{\text{LBM}}{\text{FT-SEC}}(3600)\frac{\text{SEC}}{\text{HR}}$$

$$= 1.6488\ \frac{\text{LBM}}{\text{FT-HR}}$$

8 WE CANNOT EVALUATE F_e SINCE THE EMISSIVITIES OF THE INTERIOR AND EXTERIOR ARE NOT KNOWN.

ASSUMING THAT THE WALLS ARE NON-CONDUCTING, RERADIATING, AND VARY IN TEMPERATURE FROM 2200°F AT THE INSIDE TO 70°F AT THE OUTSIDE, FIGURE 10.5 {USING X = 3"/6"} GIVES $F_A = .38$.

$$q = AE$$

$$= \left[\frac{3\ \text{IN}}{12\ \frac{\text{IN}}{\text{FT}}}\right]^2 (.1713\ \text{EE-8})\frac{\text{BTU}}{\text{HR-FT}^2\text{-}^\circ\text{R}}(F_e)(.38) \times$$

$$\times\left[(2200+460)^4 - (70+460)^4\right] {}^\circ F^4$$

$$= 2033.6\ F_e\ \frac{\text{BTU}}{\text{HR}}$$

9 $$U = \frac{1}{\sum\left(\frac{L_i}{K_i}\right) + \sum\frac{1}{h_j}}$$

$L = \frac{4\ \text{IN}}{12\ \text{IN/FT}} = .333\ \text{FT}$

$K = 13.9\ \text{BTU-FT/FT}^2\text{-HR-}^\circ\text{F}$

$h_{INSIDE} \approx 1.65$ FROM TABLE 10.2

$h_{OUTSIDE}$ IS MORE DIFFICULT. ASSUMING LINEARITY BETWEEN 0 AND 15 MPH,

$$h_{OUTSIDE} \approx 1.65 + \frac{10}{15}(6.00 - 1.65) = 4.55$$

$$U = \frac{1}{\left(\frac{.333}{13.9}\right) + \frac{1}{1.65} + \frac{1}{4.55}} = 1.18\ \frac{\text{BTU}}{\text{FT}^2\text{-HR-}^\circ\text{F}}$$

10 FROM EQUATION 3.62

$$N_{Re} = \frac{DV}{\nu}$$

$$D = \frac{.6\ IN}{12\ \frac{IN}{FT}} = .05\ FT$$

$$V = 2\ FT/SEC$$

ASSUMING SAE 10W AT 100°F
ν = 45 CENTISTOKES {PAGE 4-27, 15TH EDITION, CAMERON HYDRAULIC DATA}

FROM TABLE 3.2,

$$(45)\ CENTISTOKES \left(\frac{1}{100}\right)\frac{STOKES}{CENTISTOKE}\left(\frac{1}{929}\right)\frac{FT^2}{SEC\text{-}STOKE}$$

$$= 4.84\ EE\text{-}4\ FT^2/SEC$$

$$N_{Re} = \frac{(.05)\ FT\ (2)\ FT/SEC}{(4.84\ EE\text{-}4)\ FT^2/SEC} = 206.6$$

CONCENTRATES

1 SINCE THE WALL TEMPERATURES ARE GIVEN {NOT ENVIRONMENT TEMPERATURES} IT IS NOT NECESSARY TO CONSIDER FILMS. USING EQUATION 10.6 ON A 1 SQ. FT BASIS:

$$q = \frac{(1)\ FT^2(1000-200)°F}{\frac{\left(\frac{3}{12}\right)}{.06} + \frac{\left(\frac{5}{12}\right)}{.5} + \frac{\left(\frac{6}{12}\right)}{.8}} = 142.2\ BTU/HR$$

NOW, SOLVING FOR ΔT FROM EQN. 10.6 UP TO THE FIRST INTERFACE

$$\Delta T = \frac{(142.2)\frac{BTU}{HR}\left(\frac{(3/12)\ FT}{.06\ BTU\text{-}FT/FT^2\text{-}HR\text{-}°F}\right)}{(1)\ FT^2} = 592.5°F$$

THE TEMPERATURE AT THE FIRST INTERFACE IS

$$T = 1000 - 592.5 = 407.5°F$$

SIMILARLY FOR THE OTHER INTERFACE,

$$\Delta T = \frac{(142.)\ \frac{(6/12)}{.8}}{(1)} = 88.9$$

$$T = 200 + 88.9 = 288.9°F$$

2 INITIALLY ASSUME h_o = 1.65 BTU/HR-FT²-°F. AND, SINCE THE STEAM IS WET, IT MUST BE CONDENSING. SO, $h_i \approx 2000$, ASSUME STEEL PIPE, K = 25.4 {INTERPOLATED FROM PAGE 10-21}

$$r_a = \frac{(3.5)\ IN}{(2)(12)\ IN/FT} = .1458\ FT$$

$$r_b = \frac{4.0}{(2)(12)} = .1667\ FT$$

$$r_c = \frac{2}{12} + .1667 = .3334\ FT$$

THEN FROM EQUATION 10.11

$$q = \frac{2\pi(100)(350-50)}{\frac{1}{(.1458)(2000)} + \frac{\ln\left(\frac{.1667}{.1458}\right)}{25.4} + \frac{\ln\left(\frac{.3334}{.1667}\right)}{.05} + \frac{1}{(.3334)(1.65)}}$$

$$= \frac{188495.5}{.00343 + .00527 + 13.863 + 1.818}$$

$$= 12013.7\ BTU/HR$$

THIS IS THE FIRST ITERATION, ADDITIONAL ACCURACY MAY BE OBTAINED BY EVALUATING THE FILMS. TO FIND h_o, FIND THE OUTSIDE INSULATION TEMPERATURE BY SOLVING EQN 10.11 FOR ΔT UP TO THE OUTER FILM

$$\Delta T = \frac{12013.7}{(2\pi)(100)}\left[.00343 + .00527 + 13.863\right]$$

$$= 265.2$$

$$\text{SO } T_{INSULATION} = 350 - 265.2 = 84.8$$

THE OUTSIDE FILM SHOULD BE EVALUATED AT $\frac{1}{2}(84.8+50) = 67.4°F$. FROM PAGE 10-24 INTERPOLATING FOR 67.4°F AIR

$$N_{Pr} = .72$$

$$\frac{g\beta\rho^2}{\mu^2} = 2.43\ EE6$$

SO $N_{Gr} = L^3 \Delta T\left(\frac{g\beta\rho^2}{\mu^2}\right)$ FROM EQN 10.57

WHERE L IS THE OUTSIDE DIAMETER IN FEET

$$= (.6667)^3(84.8-50)(2.43\ EE6)$$

$$= 2.51\ EE7$$

THEN

$$N_{Pr}N_{Gr} = (.72)(2.51\ EE7) = 1.87\ EE7$$

FROM TABLE 10.7 WITHIN THIS RANGE

$$h_o = .27\left(\frac{\Delta T}{L}\right)^{.25} = .27\left(\frac{84.8-50}{.6667}\right)^{.25} = .73$$

{MORE}

CONCENTRATE #2 CONTINUED

THE PROCEDURE ON PAGE 10-13 MUST BE USED TO FIND h_i

ASSUME $(T_{sv} - T_s) = 20°F$

SO, THE PIPE WALL IS AT 350-20 = 330°F

THE FILM IS TO BE EVALUATED AT

$$\tfrac{1}{2}(330+350) = 340°F$$

FROM PAGE 6-29

$$\rho_\ell = 1/v_\ell = 1/.01787 = 55.96 \text{ LBM/FT}^3$$

$$\rho_v = 1/3.788 = .26 \text{ LBM/FT}^3$$

$$h_{fg} = 879$$

FROM PAGE 10-23

$$K = .392$$

$$\mu_\ell = (.109 \text{ EE-3}) \frac{\text{LBM}}{\text{FT-SEC}} (3600) \frac{\text{SEC}}{\text{HR}} = .392 \frac{\text{LBM}}{\text{FT-HR}}$$

THEN FROM EQN. 10.59

$$h_i = .725\left[\frac{(55.96)(55.96-.26)(4.17\text{EE8})(879)(.392)^3}{\left(\frac{3.5}{12}\right)(.392)(350-330)}\right]^{.25} = 1698$$

THEN, RETURNING TO EQN. 10.11

$$q = \frac{(2\pi)(100)(350-50)}{\frac{1}{(.1458)(1698)} + .00527 + 13.863 + \frac{1}{(.3334)(.73)}} = 10483 \text{ BTU/HR}$$

3 SINCE NO INFORMATION WAS GIVEN ABOUT THE PIPE TYPE, MATERIAL, OR THICKNESS, IT CAN BE ASSUMED THAT THE PIPE RESISTANCE IS NEGLIGIBLE,

AND SO $T_{PIPE} = T_{SAT}$.

FOR 300 PSIA STEAM, $T_{SAT} = 417.33°F$

THE OUTSIDE FILM SHOULD BE EVALUATED AT $\frac{1}{2}(417.33+70) = 243.7°F$

INTERPOLATING FROM PAGE 10-24

$$N_{Pr} = .715$$

$$\frac{g\beta\rho^2}{\mu^2} = .673 \text{ EE}6$$

$$N_{Gr} = \frac{L^3 g\beta\rho^2 \Delta T}{\mu^2} = \left(\frac{4}{12}\right)^3(.673\text{EE}6)(417.33-70) = 8.66 \text{ EE}6$$

SO

$$N_{Pr}N_{Gr} = (.715)(8.66 \text{ EE}6) = 6.2 \text{ EE}6$$

FROM TABLE 10.7

$$h_o \approx .27\left(\frac{\Delta T}{L}\right)^{.25} = .27\left(\frac{417.33-70}{\frac{4}{12}}\right)^{.25} = 1.53$$

FROM EQN. 10.11

$$q = \frac{2\pi(50)(417.33-70)}{\frac{1}{\left(\frac{2}{12}\right)(1.53)}} = 2.782 \text{ EE}4 \text{ BTU/HR}$$

THE ENTHALPY DECREASE PER POUND IS

$$\frac{2.782 \text{ EE}4 \text{ BTU/HR}}{5000 \text{ LBM/HR}} = 5.56 \text{ BTU/LBM}$$

THIS IS A QUALITY LOSS OF

$$\Delta x = \frac{\Delta h}{h_{fg}} = \frac{5.56}{809} = .007$$

OR .7%

4 THE HEAT LOSS IS

$$\frac{(2)\frac{\text{WATTS}}{\text{FT LENGTH}}(60)\frac{\text{MIN}}{\text{HR}}}{(17.57)\frac{\text{WATT-MIN}}{\text{BTU}}} = 6.83 \frac{\text{BTU}}{\text{HR-FT LENGTH}}$$

FROM PAGE 10-24 FOR 100°F FILM

$$N_{Pr} = .72$$

$$\frac{g\beta\rho^2}{\mu^2} = 1.76 \text{ EE}6$$

FROM EQN 10.57

$$N_{Gr} = L^3 \Delta T\left(\frac{g\beta\rho^2}{\mu^2}\right)$$

$$\Delta T = (T_{WIRE} - 60°F)$$

BUT T_{WIRE} IS UNKNOWN, SO ASSUME 150°F

$$N_{Gr} = \left(\frac{.6}{12}\right)^3(150-60)(1.76 \text{ EE}6) = 1.98 \text{ EE}4$$

$$N_{Pr}N_{Gr} = (.72)(1.98 \text{ EE}4) = 1.42 \text{ EE}4$$

FROM TABLE 10.7

$$h_o \approx .27\left(\frac{\Delta T}{L}\right)^{.25} = .27\left(\frac{150-60}{\frac{.6}{12}}\right)^{.25} = 1.76$$

THEN FROM EQN. 10.38

$$T_{WIRE} = \frac{q}{hA} + T_\infty = \frac{6.83}{(1.76)\left(\frac{.6}{12}\right)(\pi)(1)} + 60 = 84.7$$

THIS CANNOT BE SINCE T_{FILM} CANNOT EXCEED T_{WIRE}. THE LOWEST T_{WIRE} CAN BE IS 100°F.

(MORE)

CONCENTRATES #4, CONTINUED

SO, TRY $T_{WIRE} = 100°F$.

$$N_{gr} = \left(\frac{.6}{12}\right)^3 (100-60)(1.76\ EE6) = 8.8\ EE3$$

$$N_{Pr} N_{gr} = (.72)(8.8\ EE3) = 6.34\ EE3$$

FROM TABLE 10.7

$$h_o \approx .27\left(\frac{\Delta T}{L}\right)^{.25} = .27\left(\frac{100-60}{\frac{.6}{12}}\right)^{.25} = 1.44$$

FROM EQN 10.38

$$T_{WIRE} = \frac{6.83}{(1.44)(.6/12)(\pi)(1)} + 60 = 90.2$$

SINCE THE WIRE TEMPERATURE IS STILL LESS THAN THE FILM TEMPERATURE, THE PROBLEM STATEMENT MUST CONTAIN INCONSISTENT DATA. THIS CONCLUSION IS VERIFIED BY THE OBSCURE METHOD PRESENTED IN 1948 BY W. ELENBAAS IN THE JOURNAL OF APPLIED PHYSICS.

---

5 THE BULK TEMPERATURE OF THE WATER IS $\frac{1}{2}(70°+190°) = 130°$

FROM PAGE 10-23 FOR 130° WATER

$C_p = .999$
$\nu = .582\ EE{-5}$
$N_{Pr} = 3.45$
$K = .376$

THE REQUIRED HEAT IS

$$q = \dot{m} C_p \Delta T = (2940)(.999)(190-70)$$
$$= 352{,}447\ BTU/HR$$

FROM EQUATION 10.40

$$N_{Re} = \frac{VD}{\nu} = \frac{(3)\frac{FT}{SEC}\left(\frac{.9}{12}\right)FT}{.582\ EE{-5}} = 3.87\ EE4$$

THEN FROM EQUATION 10.42

$$h_i = \frac{(.376)(.0225)(3.87EE4)^{.8}(3.45)^{.4}}{\left(\frac{.9}{12}\right)}$$

$$= 866$$

THE PROCEDURE ON PAGE 10-13 MUST BE USED TO FIND h_o. 134 PSIA STEAM HAS $T_{SAT} = 350°F$. ASSUME $T_{SV} - T_S = 20$.

SO, THE WALL IS AT $350-20 = 330°F$ AND THE FILM SHOULD BE EVALUATED AT $\frac{1}{2}(330+350) = 340$

FROM PAGE 6-29

$$\rho_l = 1/v_l = 1/.01787 = 55.96\ LBM/FT^3$$

$$\rho_v = 1/3.788 = .26\ LBM/FT^3$$

$$h_{fg} = 879$$

FROM PAGE 10-23

$K = .392$

$$\mu_l = (.109\ EE{-3})\frac{LBM}{FT\text{-}SEC}(3600)\frac{SEC}{HR}$$
$$= .392\ \frac{LBM}{FT\text{-}HR}$$

THEN FROM EQUATION 10.59

$$h_o = .725\left[\frac{55.96(55.96-.26)(4.17EE8)(879)(.392)^3}{\left(\frac{1}{12}\right)(.392)(350-330)}\right]^{.25}$$

$$= 2323$$

USING EQN 10.71,

$$r_o = \frac{1}{(2)(12)} = .0417$$

$$r_i = \frac{.9}{(2)(12)} = .0375$$

$K_{COPPER} = 216$ {FROM PAGE 10-21, ASSUMING PURE COPPER PIPES}

$$U_o = \frac{1}{\frac{1}{2323} + \frac{.0417}{216}\ln\left(\frac{.0417}{.0375}\right) + \frac{.0417}{(.0375)(866)}}$$

$$= 576.4$$

FOR CROSS-FLOW OPERATION

CONCENTRATE #5 CONTINUED

$\Delta T_A = 350 - 70 = 280$

$\Delta T_B = 350 - 190 = 160$

$$\Delta T_M = \frac{280 - 160}{\ln\left(\frac{280}{160}\right)} = 214.4$$

FROM EQN. 10.67

$$A_o = \frac{q}{U_o \Delta T_M} = \frac{352{,}447}{(576.4)(214.4)} = 2.85 \text{ FT}^2$$

6 THIS IS A TRANSIENT PROBLEM

$C_e = 100{,}000$ BTU/°F {EQN. 10.21}

$R_e = 1/6500 = .0001538$ BTU/HR-°F

FROM EQUATION 10.24

$$T_{8\,HRS} = 40 + (70-40)\, \text{EXP}\left[\frac{-8}{(100{,}000)(.0001538)}\right] = 57.8\,°\text{F}$$

7 SINCE WE KNOW THE DUCT OUTSIDE TEMPERATURE, WE DON'T NEED K_{DUCT} AND h_i. EVALUATE h_o AT $\frac{1}{2}(80+200) = 140°$F

FROM PAGE 10-24

$N_{Pr} = .72$

$\frac{g\beta\rho^2}{\mu^2} = 1.396$ EE6

$$N_{Gr} = L^3 \Delta T \frac{g\beta\rho^2}{\mu^2} = \left(\frac{9}{12}\right)^3 (200-80)(1.396 \text{ EE6}) = 7.07 \text{ EE7}$$

SO $N_{Pr}N_{Gr} = (.72)(7.07 \text{ EE7}) = 5.09 \text{EE7}$

FROM TABLE 10.7

$$h = .27\left(\frac{\Delta T}{L}\right)^{.25} = .27\left(\frac{200-80}{\frac{9}{12}}\right)^{.25} = .96$$

THE DUCT AREA PER FOOT OF LENGTH IS

$$A = pL = \left(\frac{9}{12}\right)\pi(1) = 2.356 \text{ FT}^2$$

THE CONVECTION LOSSES ARE

$$q = hA\Delta T = (.96)(2.356)(200-80) = 271.4 \frac{\text{BTU}}{\text{HR-FT}}$$

THE RADIATION LOSSES ARE GIVEN BY EQN. 10.84. ASSUME $\epsilon_{DUCT} = .95$ {FROM PAGE 10-26}. THEN $F_e = \epsilon_{DUCT} = .95$, $F_A = 1$ SINCE THE DUCT IS ENCLOSED.

$$q = (2.356)(.1713 \text{ EE-8})(.95)\left[(200+460)^4 - (70+460)^4\right] = 424.9 \text{ BTU/HR-FT}$$

$$q_{total} = 271.4 + 424.9 = 696.3 \frac{\text{BTU}}{\text{HR-FT LENGTH}}$$

8 1 FOOT OF ROD HAS THE VOLUME

$$V = \left(\frac{.4}{12}\right)^2 \frac{\pi}{4}(1) = 8.727 \text{ EE-4 FT}^3/\text{FT}$$

THE HEAT OUTPUT PER FOOT OF ROD IS

$$q = (8.727 \text{ EE-4}) \frac{\text{FT}^3}{\text{FT}} (4 \text{ EE7}) \frac{\text{BTU}}{\text{HR-FT}^3} = 3.491 \text{ EE4} \frac{\text{BTU}}{\text{HR-FT}}$$

THE SURFACE TEMPERATURE OF THE CLADDING IS FOUND FROM EQN 10.38

$$T_s = \frac{q}{hA} + T_\infty$$

$$A = \pi\left(\frac{.44}{12}\right)(1) = .1152$$

$$T_s = \frac{3.491 \text{ EE4}}{(10000)(.1152)} + 500 = 530.3$$

FOR THE CLADDING,

$$r_o = \frac{.2+.02}{12} = .01833 \text{ FT}$$

$$r_i = \frac{.2}{12} = .01667$$

$K \approx 10.9$

FROM EQN. 10.71, THE CLADDING COEFFICIENT IS

$$U_{o,cladding} = \frac{1}{\left(\frac{.01833}{10.9}\right)\ln\left(\frac{.01833}{.01667}\right)} = 6264.2$$

THEN, THE TEMPERATURE DIFFERENCE ACROSS THE CLADDING IS

$$\Delta T = \frac{q}{U_o A_o} = \frac{3.491 \text{ EE4}}{(6264.2)\pi\left(\frac{.44}{12}\right)(1)} = 48.4\,°\text{F}$$

$$T_{\text{INSIDE CLADDING}} = T_{\text{OUTSIDE FUEL ROD}} = 530.3 + 48.4\,°\text{F} = 578.7$$

FROM EQN. 10.29

$$T_{CENTER} = T_o + \frac{r_o^2 q^*}{4K} = 578.7 + \frac{\left(\frac{.2}{12}\right)^2 (4 \text{ EE7})}{(4)(1.1)} = 3104$$

9 CONSIDER THIS AN INFINITE CYLINDRICAL FIN WITH

$T_s = 450°F$ $\quad T_\infty = 80°F$ $\quad h = 3$

$$P = \frac{\pi\left(\frac{1}{16}\right)}{12} = .01636$$

$K \approx 215$

$$A = \frac{\pi}{4}\left(\frac{1}{(16)(12)}\right) = 2.131 \text{ EE-5}$$

THEN FROM EQN 10.33

$$q = 2\sqrt{(3)(.01636)(215)(2.131 \text{ EE-5})}\,(450-80)$$

$= 11.1$ BTU/HR THIS IGNORES RADIATION

10 THIS IS A SINGLE TUBE IN CROSS FLOW. REFER TO PAGE 10-10

$$\frac{hD}{K} = C\left(N_{Re}\right)^n$$

THE FILM IS EVALUATED AT

$$\tfrac{1}{2}(100+150) = 125$$

FOR AIR AT 125°F

$K = .0159$

$\nu = .195$ EE-3

$$N_{Re} = \frac{VD}{\nu} = \frac{(100)(.35/12)}{.195 \text{ EE-3}} = 14957$$

BASED ON N_{Re}, C AND n CAN BE FOUND

$C = .174$ AND $n = .618$

$$h = \frac{(.0159)(.174)(14957)^{.618}}{\left(\frac{.35}{12}\right)} = 36.07 \text{ BTU/HR-FT}^2\text{-°F}$$

TIMED

1 ASSUME THAT THE EXPOSED WALL TEMPERATURES ARE EQUAL TO THE RESPECTIVE AMBIENT TEMPERATURES.

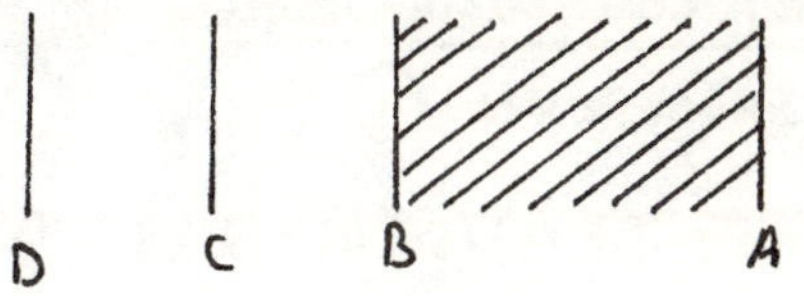

$T_A = 80°F = 540°R$

$$q_{A-B} = \frac{(540-T_B)}{\left(\frac{.4}{12}\right)/.025} = 40.5 - .075\,T_B \quad (1)$$

SINCE THE AIR SPACES ARE EVACUATED, ONLY RADIATION SHOULD BE CONSIDERED FOR B-C AND C-D

$$E_{B-C} = \frac{q_{B-C}}{A} = \sigma F_e F_A\left[T_B^4 - T_C^4\right]$$

$F_A = 1$ SINCE THE FREEZER IS ASSUMED LARGE

$$F_e = \frac{(.5)(.1)}{(.1)+(1-.1)(.5)} = .0909 \quad \text{\{FROM TABLE 10.10\}}$$

$$E_{BC} = (.1713 \text{ EE-8})(.0909)(1)\left[T_B^4 - T_C^4\right]$$

$$= (1.56 \text{ EE-10})\,T_B^4 - (1.56 \text{ EE-10})\,T_C^4 \quad (2)$$

SIMILARLY

$$E_{C-D} = (.1713 \text{ EE-8})(.0909)(1)\left[T_C^4 - (460-60)^4\right]$$

$$= (1.56 \text{ EE-10})\,T_C^4 - 3.99 \quad (3)$$

BUT (2) = (3)

$$(1.56 \text{ EE-10})\,T_B^4 - (1.56 \text{ EE-10})\,T_C^4 = (1.56 \text{ EE-10})\,T_C^4 - 3.99$$

$$T_B^4 - T_C^4 = T_C^4 - (2.56 \text{ EE10})$$

$$T_B^4 = 2T_C^4 - (2.56 \text{ EE10})$$

$$T_C^4 = \tfrac{1}{2}T_B^4 + (1.28 \text{ EE10}) \quad (4)$$

AND (1) = (2)

$$40.5 - .075\,T_B = (1.56 \text{ EE-10})\,T_B^4 - (1.56 \text{ EE-10})\,T_C^4$$

$$T_C^4 = T_B^4 + (4.81 \text{ EE8})\,T_B - 2.6 \text{ EE11} \quad (5)$$

BUT (4) = (5)

$$\tfrac{1}{2}T_B^4 + 1.28 \text{ EE10} = T_B^4 + (4.81 \text{ EE8})\,T_B - 2.6 \text{ EE11}$$

OR $T_B^4 + (9.62 \text{ EE8})\,T_B = 5.456 \text{ EE11}$

BY TRIAL + ERROR, $T_B = 501.4°R$

TIMED #1 CONTINUED

THEN FROM EQN (1)

$$\frac{q}{A} = 40.5-(.075)(501.4) = 2.895 \text{ BTU/FT}^2\text{-HR}$$

FROM EQN. (4)

$$T_c = \sqrt[4]{\left(\tfrac{1}{2}\right)(501.4)^4 + (1.28 \text{ EE}10)} = 459°R$$

NOW, CHECK THE INITIAL ASSUMPTION ABOUT WALL TEMPERATURE. ASSUME A FILM EXISTS OF, SAY, 2 BTU/FT²-HR-°F.

USING EQN 10.38

$$(T_s - T_\infty) = \frac{q}{Ah} = \frac{2.9}{(1)2} = 1.5°F$$

SINCE THIS IS SMALL, OUR ASSUMPTION IS VALID.

2 THE EXPOSED DUCT AREA IS

$$\frac{(18+18+12+12) \text{ IN}}{12 \text{ IN/FT}}(50) \text{ FT} = 250 \text{ FT}^2$$

THE INSIDE FILM COEFFICIENT CANNOT BE EVALUATED WITHOUT KNOWING THE EQUIVALENT DIAMETER OF THE DUCT. FROM TABLE 3.6

$$D_e = \frac{(2)(18)(12)}{(18+12)\, 12 \text{ IN/FT}} = 1.2 \text{ FT}$$

CHECK THE REYNOLDS NUMBER. FROM PAGE 10-24 FOR 100°F AIR,

$\nu = .180 \text{ EE-3}$

$$N_{Re} = \frac{vD}{\nu} = \frac{\frac{800 \text{ FPM}}{60 \text{ SEC/MIN}}(1.2) \text{ FT}}{.180 \text{ EE-3 } \frac{\text{FT}^2}{\text{SEC}}} = 8.9 \text{ EE } 4$$

SINCE THIS IS TURBULENT, EQN 10.43 MAY BE USED. AT 100°F, $\rho = .071$ LBM/FT³

$$G = v\rho = (800)(.071)/60 = .947 \text{ LBM/SEC-FT}^2$$

$$C = .00351 + .000001583(100) = .003668$$

$$h = \frac{(.003668)[(3600)(.947)]^{.8}}{(1.2)^{.2}} = 2.37$$

IGNORING THE DUCT THERMAL RESISTANCE, THE TOTAL HEAT TRANSFER COEFFICIENT IS

$$\frac{1}{U} = \frac{1}{h_i} + \frac{1}{h_o} = \frac{1}{2.37} + \frac{1}{2} = .922$$

$$U = 1/.922 = 1.08$$

THEN $q = UA\Delta T$

$$= (1.08)(250)(100-70) = 8100 \text{ BTU/HR}$$

b) THE FLOW PER HOUR IS

$$3600\, GA = (3600)(.947)\left(\frac{(12)(18)}{144}\right)$$

$$= 5113.8 \text{ LBM/HR}$$

$C_p = .240$

$$\Delta T = \frac{q}{\dot{m} C_p} = \frac{8100}{(5113.8)(.240)} = 6.6°F$$

c) USE PAGE 5-7 FOR CLEAN GALVANIZED DUCTWORK WITH $\epsilon = .0005$ AND ABOUT 40 JOINTS PER 100 FEET.

THE EQUIVALENT DIAMETER FOR THIS CHART IS GIVEN BY EQN 5.30

$$D_e = (1.3)\frac{[(12)(18)]^{.625}}{[(12)+(18)]^{.25}} = 16 \text{ INCHES}$$

THEN, $\Delta p = .057 \frac{\text{IN. WATER}}{100 \text{ FEET}}$

SO, THIS DUCT SYSTEM HAS A LOSS OF

$$\left(\frac{50}{100}\right)(.057) = .029'' \text{ W.G.}$$

3 THIS IS A TRANSIENT PROBLEM (SEE PAGE 10-5). CHECK THE BIOT MODULUS TO SEE IF THE LUMPED PARAMETER METHOD CAN BE USED.

$$L = \frac{V}{A_s} = \frac{\frac{4}{3}\pi r^3}{4\pi r^2} = \frac{r}{3}$$

FOR THE LARGEST BALL,

$$L = \frac{\frac{1.5}{(2)(12)}}{3} = .0208 \text{ FT}$$

EVALUATE K FOR STEEL AT

$$\tfrac{1}{2}(1800+250) \approx 1000°F$$

FROM PAGE 10-21 FOR MILD STEEL, $K \approx 22$. (ACTUALLY K FOR A HIGH ALLOY STEEL IS ABOUT 12-15)

$$N_{Bi} = \frac{hL}{K} = \frac{(56)(.0208)}{22} = .053$$

FOR SMALLER BALLS, N_{Bi} WILL BE EVEN SMALLER.

(MORE)

TIMED #3 CONTINUED

SINCE $N_{Bi} < .10$, THE LUMPED PARAMETER METHOD CAN BE USED. THE ASSUMPTIONS ARE:

- HOMOGENEOUS BODY TEMPERATURE
- MINIMAL RADIATION LOSSES
- OIL BATH REMAINS AT 110°F
- CONSTANT h (NO VAPORIZATION OF OIL)

FROM EQN. 10.21 AND 10.22

$$R_eC_e = \frac{c_p \rho V}{h A_s} = \frac{c_p \rho r}{3h}$$

USE $\rho = 490$ LBM/FT³ AND $C = .11$ EVEN THOSE ARE FOR 32°F. (ACTUALLY, $C \approx .16$ AT 900°F)

$$R_eC_e = \frac{(.11)(490)r}{(3)(56)} = .3208\, r_{FEET}$$

$$= .01337\, D_{INCHES}$$

TAKING ln OF EQN 10.24

$$\ln(T_t - T_\infty) = \ln(\Delta T e^{-t/R_eC_e})$$

$$\ln(T_t - T_\infty) = \ln \Delta T + \ln(e^{-t/R_eC_e})$$

$$\ln(T_t - T_\infty) = \ln \Delta T - \frac{t}{R_eC_e}$$

BUT $T_t = 250$ AND $T_\infty = 110$

AND $\Delta T = 1800 - 110 = 1690$

$$\ln(250 - 110) = \ln(1690) - \frac{t}{R_eC_e}$$

$$4.942 = 7.432 - \frac{t}{R_eC_e}$$

$$t = R_eC_e(2.49)$$

$$= (.01337)D(2.49)$$

$$= .0333\, D_{INCHES}$$

b) FROM EQUATION 10.23, THE TIME CONSTANT IS

$$\frac{c_p \rho L}{h} = \frac{(.11)(490)\frac{D}{(2)(3)(12)}}{56} = \frac{D}{74.81} \text{ (hr)}$$

4

USING STEAM TABLES,

$h_1 = 167.99$ BTU/LBM
$h_2 = 364.17$
$h_3 = 1201.0$
$h_4 = 374.97$

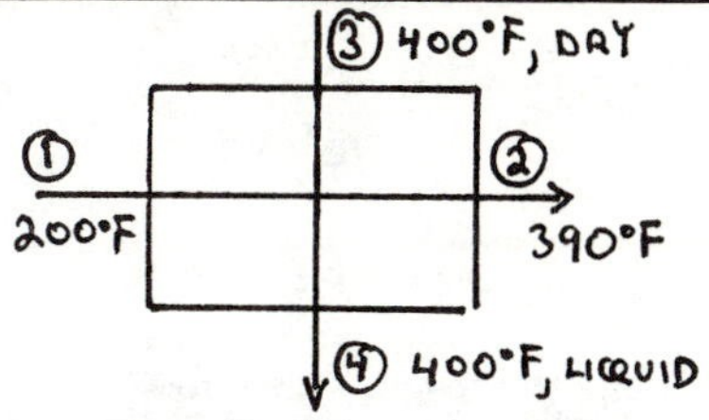

SO, THE HEAT TRANSFER IS

$$q = m \Delta h = (500{,}000)(364.17 - 167.99)$$
$$= 9.809 \text{ EE7 BTUH}$$

THE FLOW AREA PER PIPE IS

$$A = \frac{\pi}{4} D^2 = \frac{\left(\frac{\pi}{4}\right)(.875 - 2 \times .0625)^2}{144} = .003068 \text{ FT}^2$$

AT 390°F THE WATER VOLUME IS LARGEST. FROM THE STEAM TABLES,

$$\rho_2 = \frac{1}{.01850} = 54.05$$

THE WATER VOLUME IS

$$Q = \frac{m}{\rho} = \frac{500{,}000}{54.05} = 9250.7 \text{ FT}^3/\text{HR}$$

THE REQUIRED NUMBER OF TUBES IS

$$\# = \frac{Q}{A_{tube} v} = \frac{9250.7}{(.003068)(5)(3600)}$$
$$= 167.5 \quad (\text{SAY } 168)$$

(NORMALLY, 20% WOULD BE ADDED)

NOW, USE EQUATION 10.68

$$q = F_c U A \Delta T_m$$

CALCULATE ΔT_m AS IF IN COUNTERFLOW.

$$\Delta T_A = 400 - 390 = 10$$
$$\Delta T_B = 400 - 200 = 200$$
$$\Delta T_m = \frac{10 - 200}{\ln\left(\frac{10}{200}\right)} = 63.4 \text{ °F}$$

F_c IS NOT REQUIRED EVEN THOUGH THIS IS A 2-PASS HEAT EXCHANGER BECAUSE ONE OF THE FLUIDS HAS A CONSTANT TEMPERATURE.

THE OUTSIDE AREA OF ONE FOOT OF 143 TUBES IS

$$A = \frac{(168)(1)(\pi)(.875)}{12} = 38.48 \text{ FT}^2$$

THE REQUIRED AREA IS

$$A = \frac{q}{U \Delta T_m} = \frac{9.809 \text{ EE7}}{(700)(63.4)} = 2210 \text{ FT}^2$$

THE EFFECTIVE TUBE LENGTH IS

$$L = \frac{2210}{38.48} = 57.4 \text{ FT}$$

5

$$h_{probe} = \frac{(.024)(3480)^{.8}}{\left(\frac{.5}{12}\right)^{.4}} = 58.3$$

$T_{walls} = 600°F = 1060°R$

THE THERMOCOUPLE GAINS HEAT BY RADIATION FROM THE WALLS. IT ALSO LOSES HEAT BY CONVECTION TO THE GAS. NEGLECT CONDUCTION AND THE INSIGNIFICANT KINETIC ENERGY.

$$E_{LOSS,\,CONVECTION} = E_{GAIN,\,RADIATION}$$

$$E = \frac{q}{A_{probe}} = h(T_{GAS} - T_{probe}) = \sigma\epsilon\left(T_{walls}^4 - T_{probe}^4\right)$$

(a) 300°F = 760°R, ASSUME $T_{GAS} = T_{probe}$ TO GET STARTED.

$$58.3(760 - T_{probe}) = (.1713\ EE\text{-}8)(.8)\left[(1060)^4 - (T_{probe})^4\right]$$

$$4438 - 58.3\,T_{probe} = 1730 - (1.37\ EE\text{-}9)\,T_{probe}^4$$

$$42578 = 58.3\,T_{probe} - (1.37\ EE\text{-}9)\,T_{probe}^4$$

$$730.3 = T_{probe} - 2.35\ EE\text{-}11\,T_{probe}^4$$

BY TRIAL AND ERROR, $T_{probe} = 737°R$

AT 737°R,

$$E = (.1713\ EE\text{-}8)(.8)\left[(1060)^4 - (737)^4\right] = 1325$$

THEN, FROM $E = h(T_{GAS} - T_{probe})$

$$T_{GAS} = T_{probe} + \frac{E}{h} = 737 + \frac{1325}{58.3} = \boxed{760°R}$$

(b) 300°F = 760°R

ASSUME E IS APPROXIMATELY THE SAME

$$T_{probe} = T_{GAS} - \frac{E}{h} = 760 - \frac{1325}{58.3} = \boxed{737.3\ °R}$$

6 THE OUTSIDE DUCT AREA IS

$$A = (\pi)(1)(50) = 157.1\ FT^2$$

ASSUME 14.7 PSIA INSIDE THE AIR DUCT. IGNORE ANY MOISTURE CONTENT.

$$\rho_{AIR} = \frac{P}{RT} = \frac{(14.7)(144)}{(53.3)(460+45)} = .0786\ \frac{LBM}{FT^3}$$

SO, THE MASS FLOW IS

$$\dot{m} = (500)(.0786)(60) = 2358\ LBM/HR$$

SINCE THE ROOM AND DUCT "SEE" EACH OTHER COMPLETELY, $F_A = 1$. THE HEAT EXCHANGE DUE TO RADIATION IS

$$q = (.1713\ EE\text{-}8)(.28)(157.1)\left[(460+80)^4 - (460+45)^4\right] = 1508\ BTU/HR$$

FINDING THE CONVECTIVE HEAT TRANSFER IS MORE DIFFICULT SINCE NO FILM COEFFICIENTS WERE GIVEN.

FIRST, FIND h_o

THIS IS NATURAL CONVECTION. EVALUATE THE FILM AT THE DUCT MID-SECTION. AT THAT POINT THE INSIDE AIR TEMPERATURE IS APPROXIMATELY 50°F, WHICH WILL BE TAKEN AS THE OUTSIDE SURFACE TEMPERATURE OF THE DUCT.

$$T_{film} = \tfrac{1}{2}(50 + 80) = 65°F$$

AT 65°F, $N_{Pr} = .72$

$$N_{Gr} = \left(\frac{12}{12}\right)^3(2.48\ EE6)(80-50) = 7.44\ EE7$$

SO, $N_{Gr}N_{Pr} = 5.36\ EE7$

THE APPROXIMATE OUTSIDE FILM COEFFICIENT FOR A 1-FOOT DIAMETER CYLINDER IS TAKEN FROM TABLE 3.8.

$$h_o = .27\left(\frac{80-50}{1}\right)^{.25} = .63$$

NOW, FIND h_i

THIS IS FORCED CONVECTION. USE EQUATION 3.47 BECAUSE IT'S EASIER. AGAIN WORK WITH 50° MOVING AIR.

$$C = .00351 + .000001583(50) = .00359$$

CROSS SECTIONAL DUCT AREA: $\frac{\pi(1)^2}{4} = .7854\ FT^2$

$$G = \frac{2358}{(3600)(.7854)} = .834\ LBM/SEC\text{-}FT^2$$

$$h_i = \frac{.00359\,[(3600)(.834)]^{.8}}{(1)^{.2}} = 2.17$$

THE OVERALL HEAT TRANSFER COEFFICIENT IGNORES THE DUCT THERMAL RESISTANCE

$$\frac{1}{U} = \frac{1}{2.17} + \frac{1}{.63} = 2.048$$

$$U = .488$$

THE HEAT TRANSFER DUE TO CONVECTION IS

$$q = (.488)(157.1)(80-50) = 2300\ BTU/HR$$

AT 50°F, THE SPECIFIC HEAT OF AIR IS .240. THE TOTAL INCREASE IN TEMPERATURE IS

$$\Delta T = \frac{q}{\dot{m}c_p} = \frac{2300 + 1508}{(2358)(.240)} = 6.73°F$$

$$T_{FINAL} = 45° + 6.73° = \boxed{51.7°\ F}$$

7 (a) $D_o = \frac{4.25}{12} = .354$

$A_{PIPE} = (.354)(\pi)(35') = 38.9\ FT^2$

THE ENTERING DENSITY IS

$$\rho = \frac{(25)(144)}{(53.3)(460+500)} = .0704$$

$$\dot{m} = (200)(.0704)(60) = 844.8\ LBM/HR$$

AT LOW PRESSURES, THE AIR ENTHALPY IS FOUND FROM PAGE 6-35

$h_1 = 231.06\ BTU/LBM$ AT $960°R$

$h_2 = 194.25\ BTU/LBM$ AT $810°R$

$$q_{LOSS} = \dot{m}\Delta h = 844.8(231.06 - 194.25)$$
$$= 31097\ BTU/HR$$

ASSUMING THE MID-POINT PIPE SURFACE IS AT

$$\tfrac{1}{2}(500+350) = 425°$$

THEN,

$$\bar{h} = \frac{q}{A\Delta T} = \frac{31097}{(38.9)(425-70)}$$

$$= \boxed{2.25\ \frac{BTU}{FT^2\text{-}HR\text{-}°F}}$$

(b) IGNORE THE PIPE THERMAL RESISTANCE, IGNORE THE INSIDE FILM WHICH WILL NOT BE A LARGE FACTOR COMPARED TO THE OUTSIDE FILM AND RADIATION.

WORK WITH THE MIDPOINT PIPE TEMPERATURE OF 425°F.

RADIATION

ASSUME $F_a = 1$

$F_e = e \approx .9$ FOR 500°F ENAMEL PAINT OF ANY COLOR.

$$E = (.1713\ EE\text{-}8)(.9)\left[(425+460)^4 - (70-460)^4\right]$$
$$= 824.1\ BTU/HR\text{-}FT^2$$

NOTICE THE OMISSION OF A WHICH WOULD BE DIVIDED OUT IN THE NEXT STEP.

THE RADIANT HEAT TRANSFER COEFFICIENT IS

$$h_r = \frac{E}{\Delta T} = \frac{824.1}{425-70} = 2.32\ \frac{BTU}{HR\text{-}FT^2\text{-}°F}$$

OUTSIDE COEFFICIENT

EVALUATE THE FILM AT THE MIDPOINT, SO THE FILM TEMPERATURE IS

$$\tfrac{1}{2}(425+70) = 247.5 \quad (\text{SAY } 250°F)$$

$N_{Pr} = .72$

$N_{Gr} = (.354)^3(425-70)(.647\ EE6) = 1.02\ EE7$

SINCE $N_{Pr}N_{Gr} < EE9$, THE SIMPLIFIED COEFFICIENT IS

$$\bar{h} = (.27)\left(\frac{425-70}{.354}\right)^{.25} = 1.52$$

THE OVERALL FILM COEFFICIENT IS

$$h_t = h_r + \bar{h} = 2.25 + 1.52 = \boxed{3.77}$$

THIS IS CONSIDERABLY MORE THAN ACTUAL.

(c) - LOWER EMISSIVITY DUE TO DIRTY OUTSIDE OF DUCT.

- h_r AND $\bar{h}$ ARE NOT REALLY ADDITIVE.
- MID-POINT CALCULATIONS SHOULD BE REPLACED WITH INTEGRATION ALONG THE LENGTH
- THE INTERNAL FILM RESISTANCE WAS IGNORED

8 $100\ gpm = (100)(.1337)(60)(62.4) = 50057\ \frac{LBM}{HR}$

$$q_1 = MC_p\Delta T = (50057)(1)(140-70)$$
$$= 350.4\ EE4\ \frac{BTU}{HR}$$

$\Delta T_A = 230 - 70 = 160$

$\Delta T_B = 230 - 140 = 90$

$$\Delta T_M = \frac{160-90}{\ln\left(\frac{160}{90}\right)} = 121.66°$$

$$U_1 = \frac{q}{A\Delta T_M} = \frac{350.4\ EE4}{(50)(121.66)} = 576\ \frac{BTU}{HR\text{-}FT^2\text{-}°F}$$

$$q_2 = (50057)(1)(122-70) = 260.3\ EE4$$

$$U_2 = \frac{260.3\ EE4}{(50)(121.66)} = 427.9\ \frac{BTU}{HR\text{-}FT^2\text{-}°F}$$

FROM EQUATION 10.73

$$\frac{1}{U_2} = \frac{1}{U_1} + R_f$$

$$R_f = \frac{1}{427.9} - \frac{1}{576}$$

$$= \boxed{.000601\ \frac{HR\text{-}FT^2\text{-}°F}{BTU}}$$

RESERVED FOR FUTURE USE

HVAC

WARM-UPS

1 THE TOTAL PERIMETER INCLUDING THE CENTER-LINE IS

$3(4) + 10 = 22$ FT

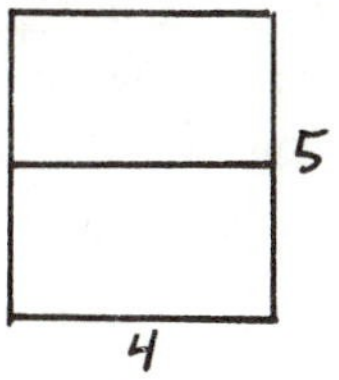

ASSUME NO WEATHER-STRIPPING. FROM TABLE 11.4,

$B = 80 \frac{FT^3}{HR\text{-}FT}$

SO $Q = BL = (80)\frac{FT^3}{HR\text{-}FT}(22)\ FT = 1760\ FT^3/HR$

2 THE OFFICE VOLUME IS

$60 \times 95 \times 10 = 57000\ FT^3$

FROM TABLE 11.5 , CHOOSE 6 AS THE NUMBER OF AIR CHANGES PER HOUR. THEN,

$$Q = \frac{(57000)\ FT^3\ (6)\ 1/HR}{(60)\ \frac{MIN}{HR}} = 5700\ CFM$$

b) FROM PAGE , THE PREFERRED VENTILATION RATE IS

$$Q = \tfrac{1}{2}(45)(15) + \tfrac{1}{2}(45)(40) = 1238\ CFM$$

3 USE EQN. 11.18 AND TABLE 11.8

FOR THE METHANOL:

FROM STEP 2,

ADD 2 BECAUSE TLV = 200
ADD 1 ASSUMING UNIFORM EVOLUTION
ADD 3 ASSUMING POOR BOOTH VENTILATION
K = 6

$$CFH = \frac{(4\ EE8)(.792)(2)(6)}{(32.04)(200)} = 6.0\ EE5$$

FOR THE METHYLENE CHLORIDE

$$CFH = \frac{(4\ EE8)(1.336)(2)(6)}{(84.94)(500)} = 1.5\ EE5$$

TOTAL $= (6.0 + 1.5)\ EE5 = 7.5\ EE5$

4 A PSYCHROMETRIC CHART IS NEEDED FOR THIS.

$\omega = 79$ GRAINS/LBM DRY AIR

$$= \frac{79\ GRAINS}{7000\ \frac{GRAINS}{LBM}} = .01129\ \frac{LBM}{LBM\ DRY\ AIR}$$

$h = 31.5$ BTU/LBM DRY AIR

FROM TABLE 6.5 $\bar{c}_p$ IS GRAVIMETRICALLY WEIGHTED. ASSUME $c_p = .241$ BTU/LBM-°F FOR AIR.

$$G_{AIR} = \frac{1}{1 + .01129} = .989$$

$$G_{WATER} = \frac{.01129}{1 + .01129} = .011$$

SO

$$\bar{c}_p = (.989)(.241) + (.011)(1) = .249\ \frac{BTU}{LBM\text{-}°F}$$

5
$$q = \frac{(12000)\ W\ (3413)\ \frac{BTU}{KW\text{-}HR}\ (1.2)}{(1000)\ \frac{W}{KW}} + \frac{(12)(.8)(10)\ HP\ (2545)\ \frac{BTU}{HP\text{-}HR}}{.9}$$

$= 3.206\ EE5$ BTU/HR

6 THE TOTAL DEGREE DAYS PER YEAR CANNOT BE USED BECAUSE SOME HEATING OCCURS IN THE OTHER MONTHS. THE NUMBER OF DEGREE DAYS IN THE CENTRAL PARK AREA OF NEW YORK CITY FOR (ALL OF) OCTOBER AND (ALL OF) MAY IS 4772. (ASHRAE 1966 GUIDE AND DATA BOOK, PAGE 249.)

FROM PAGE 23-6, BUNKER C OIL (#6)

$$HV = \tfrac{1}{2}(151{,}300 + 155{,}900) = 153{,}600\ BTU/GAL$$

FROM TABLE 9.6 ASSUMING $T_i = 70°$, $\eta = .70$

$$\text{FUEL CONSUMPTION} = \frac{(24)\frac{HR}{DAY}(3.5\ EE6)\frac{BTU}{HR}(4772)\ °F\text{-DAYS}}{(70-0)\ °F\ (.70)(153{,}600)\ \frac{BTU}{GAL}}$$

$= 53260$ GAL

COST $= (53260)\ GAL\ (.15)\ \$/GAL = \$\ 7989$

7

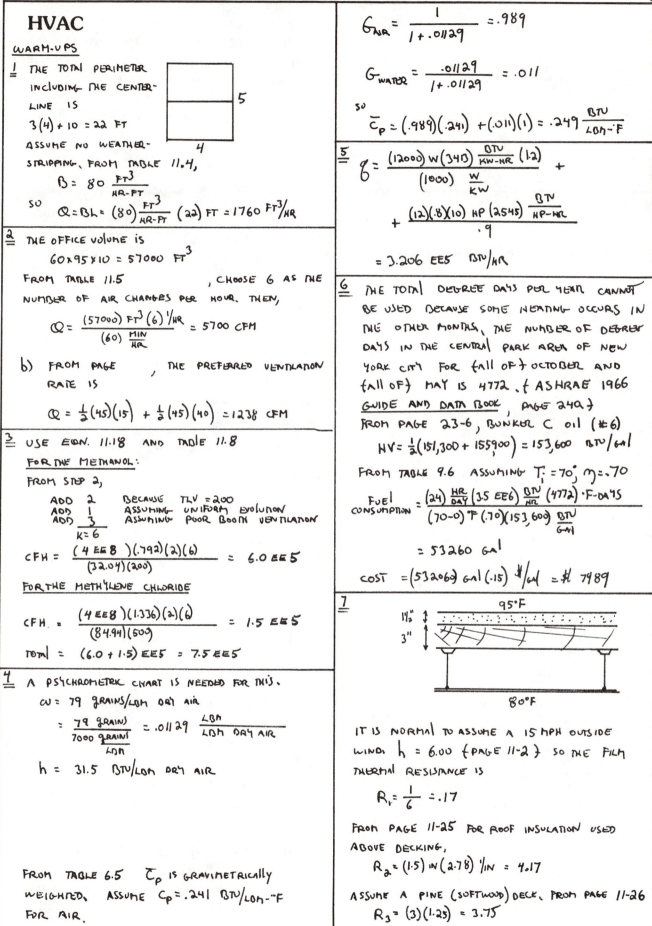

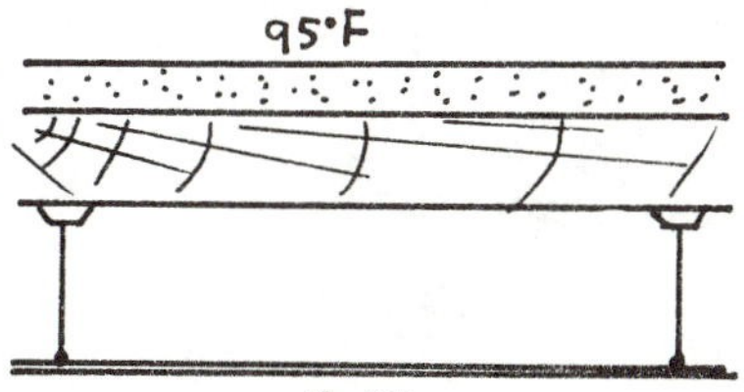

IT IS NORMAL TO ASSUME A 15 MPH OUTSIDE WIND. $h = 6.00$ (PAGE 11-2) SO THE FILM THERMAL RESISTANCE IS

$$R_1 = \frac{1}{6} = .17$$

FROM PAGE 11-25 FOR ROOF INSULATION USED ABOVE DECKING,

$$R_2 = (1.5)\ IN\ (2.78)\ 1/IN = 4.17$$

ASSUME A PINE (SOFTWOOD) DECK. FROM PAGE 11-26

$$R_3 = (3)(1.25) = 3.75$$

WARM-UP PROBLEM #7 CONTINUED

FROM PAGE 10-26

$\epsilon_{wood} = .93$

$\epsilon_{paper} = .95$ (USED FOR THE ACCOUSTICAL TILE)

FROM EQN 11.2

$$E = \frac{1}{\frac{1}{.93} + \frac{1}{.95} - 1} = .89$$

FROM TABLE 11.2 (ASSUMING 4" SEPARATION)

$C \approx .81$

SO $R_4 = 1/.81 = 1.23$

FOR $\frac{1}{2}$" ACCOUSTICAL TILE (FROM PAGE 11-25)

$R_5 = 1.78$

FOR THE INSIDE FILM, $h = 1.65$ (PAGE 11-2)

$$R_6 = \frac{1}{1.65} = .61$$

THE TOTAL RESISTANCE IS

$$R_t = R_1 + R_2 + R_3 + R_4 + R_5 + R_6$$
$$= .17 + 4.17 + 3.75 + 1.23 + 1.78 + .61$$
$$= 11.71$$

$$U = 1/R_t = .0854 \frac{BTU}{HR\text{-}FT^2\text{-}°F}$$

8 $BF = \left(\frac{1}{3}\right)^4 = .0123$

9 USE THE SLAB EDGE METHOD.

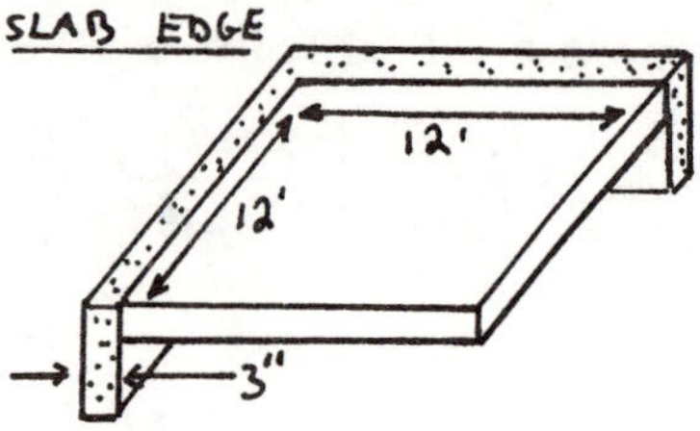

$P = 12 + 12 = 24$ FT.

CHOOSE $F = .55$ (PAGE 11-3)

ASSUME $T_i = 70°F$

FROM EQN 10.7

$$q = (24)\,FT\,(.55)\frac{BTU}{HR\text{-}FT\text{-}°F}(70-(-10))$$
$$= 1056\ BTU/HR$$

10 a) THE TOTAL VENTILATION IS

$$Q = (60)\,CFM/PERSON\,(4500)\,PEOPLE\,(60)\frac{MIN}{HR}$$
$$= 1.62\ EE7\ CFH$$

BECAUSE $T_0 = 0°F$ (LESS THAN FREEZING) THERE WILL BE NO MOISTURE IN THE AIR, THE OUTSIDE DENSITY IS

$$\rho = \frac{P}{RT} = \frac{(14.6)\,PSI\,(144)}{(53.3)\,(460+0°)} = .08575 \frac{LBM}{FT^3}$$

SO, $\dot{W} = Q\rho = (1.62\ EE7)\ CFH\,(.08575)\frac{LBM}{FT^3}$

$$= 1.389\ EE6\ LBM/HR$$

ASSUME $C_p = .241$ BTU/LBM-°F

THE INCREASE IN AIR TEMPERATURE ONCE IT ENTERS THE AUDITORIUM IS DUE TO THE BODY HEAT LESS ANY CONDUCTIVE LOSSES THROUGH THE WALLS, ETC. FROM TABLE 11.13 THE OCCUPANT LOAD IS 225 BTU/HR-PERSON.

SO,

$$q_{IN\ FROM\ PEOPLE} = (225)(4500) = 1.01\ EE6\ BTU/HR$$

SINCE THE AIR LEAVES AT 70°F, FROM

$$q = WC_p\Delta T$$

$$(1.01\ EE6) = (1.389\ EE6)(.241)(70 - T_{IN})$$

OR $T_{IN} = 66.98°F$

b) THE COIL HEAT IS

$$q = WC_p\Delta T$$
$$= (1.389\ EE6)(.241)(66.98 - 0)$$
$$= 2.24\ EE7\ BTU/HR$$

CONCENTRATES

1

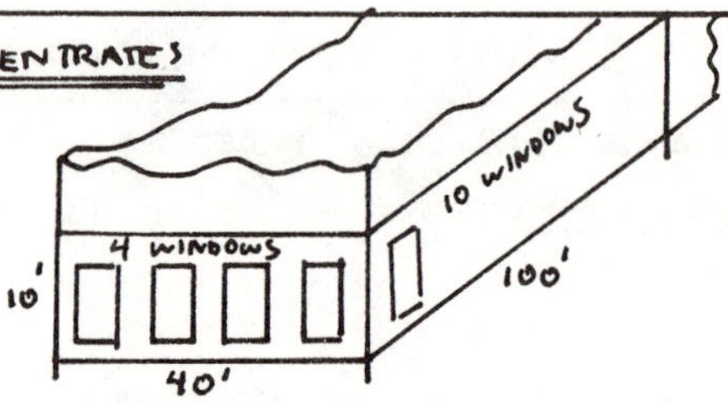

THE TOTAL EXPOSED (GLASS AND WALL) AREA IS

$$10(100 + 100 + 40) = 2400\ FT^2$$

THE WINDOW AREA IS

$$(10 + 10 + 4)\ WINDOWS\,(4 \times 6)\,FT^2 = 576\ FT^2$$

THE INSIDE AND OUTSIDE FILM COEFFICIENTS ARE 1.65 AND 6.00 $\frac{BTU}{HR\text{-}FT^2\text{-}°F}$ RESPECTIVELY (PAGE 11-2)

$$U_{wall} = \frac{1}{\frac{1}{1.65} + \frac{1}{.2} + \frac{1}{6.00}} = .173$$

SO, THE HEAT LOSS FOR THE WALLS IS

$$q_1 = UA\Delta T = (.173)(2400 - 576)(75 - (-10))$$
$$= 26822\ BTU/HR$$

FROM PAGE 11-26 FOR THE WINDOWS ASSUMING A $\frac{1}{4}$" AIR SPACE,

$$U_{windows} = 1/1.63 = .61$$

SO, THE HEAT LOSS FOR THE WINDOWS IS

$$q_2 = (.61)(576)(75 - (-10))$$
$$= 29866\ BTU/HR$$

FOR THE SAKE OF COMPLETENESS, CALCULATE THE AIR INFILTRATION. ASSUME DOUBLE-HUNG, UNLOCKED

(MORE)

CONCENTRATE #1 CONTINUED

FROM TABLE 11.4, $B = 32 \frac{CFH}{FT}$

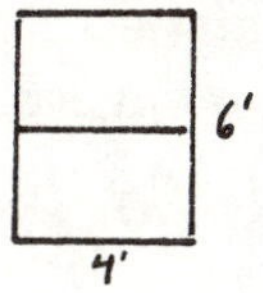

$L = 4+4+4+6+6 = 24$ FT

SO, $Q = (14)_{\text{windows}}(32)(24)$

$= 10752$ CFH

{USE WINDOWS ON 2 SIDES FOR 3 EXPOSED walls}

FROM EQN 11.5

$q_3 = (.018)(10752)(75-(-10)) = 16451$

SO, $q_{total} = q_1 + q_2 + q_3$

$= 26,822 + 29,866 + 16,451$

$= 73,139$ BTU/HR

2 SEE PAGE 11-11 (PARAGRAPH C). LOCATE POINTS A + B ON THE PSYCHROMETRIC CHART.

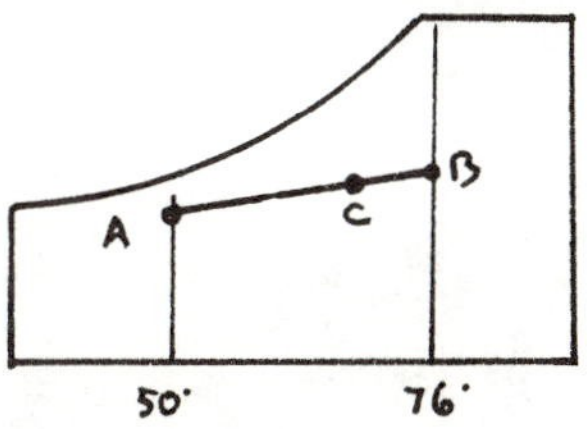

THE LENGTH OF THE LINE (A-B) ON MY CHART IS 6.6 CM. READING FROM THE CHART,

$v_A = 13.0$ FT³/LBM

$v_B = 13.69$ FT³/LBM

SO $\rho_A = 1/13.0 = .0769$ LBM/FT³

$\rho_B = 1/13.69 = .073$

$M_A = \rho Q_A = (.0769)(1000) = 76.9$

$M_B = (.073)(1500) = 109.5$

THE GRAVIMETRIC FRACTION OF FLOW A IS

$$\frac{76.9}{76.9 + 109.5} = .41$$

SO, POINT C IS LOCATED $(.41)(6.6) = 2.7$ CM FROM POINT B. THIS DETERMINES

$T_{dry\ bulb} = 65.5$ °F

$\omega = 57$ grains/LBM

$T_{dp} = 51$ °F

3 a) $BF = 1 - .70 = .30$

b) FROM EQN 11.59

$$.70 = \frac{60 - T_{out}}{60 - 45}$$

$T_{out} = 49.5$ °F

4 FROM THE PSYCHROMETRIC CHART

AT 1 $\omega_1 = 99$ GRAINS/LBM

$h_1 = 38.3$ BTU/LBM

$v_1 = 14.29$ FT³/LBM

AT 2 $\omega_2 = 165.8$

$h_2 = 46.4$

SO, THE MOISTURE ADDED IS

$$\frac{(165.8 - 99)\ \frac{\text{GRAINS}}{\text{LBM AIR}}}{(7000)\ \frac{\text{GRAINS}}{\text{LBM WATER}}\ (14.29)\ \frac{\text{FT}^3}{\text{LBM AIR}}} = 6.68\text{ EE-4 } \frac{\text{LBM}}{\text{FT}^3}$$

THE ENTHALPY CHANGE IS

$$\frac{(46.4 - 38.3)}{14.29} = .567\ \text{BTU/FT}^3\ \text{AIR}$$

5 Follow THE PROCEDURE STARTING ON PAGE 11-15

$$SHR = \frac{200,000}{200,000 + 50,000} = .8$$

DRAW THE CONDITION LINE:

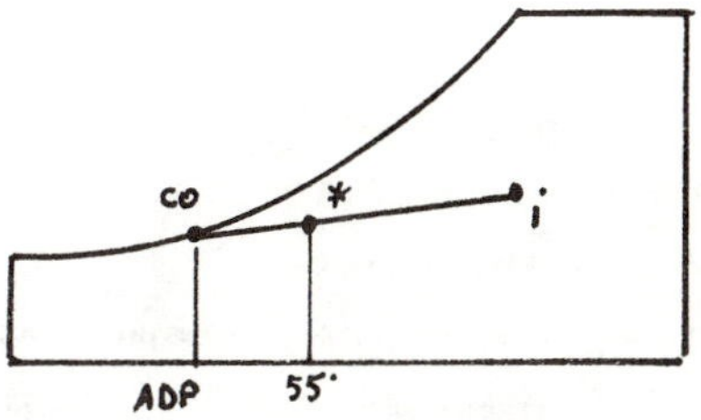

ADP = 50.8 °F. ASSUME BF = 0, SO

$T_{co} =$ ADP {FROM EQN 10.83}.

CHOOSE 20° AS A TEMPERATURE GAIN,

$T_{db,IN} = (75 - 20) = 55$ °F

FROM EQN 11.66 (IN CFM NOT CFH)

$$CFM_{IN} = \frac{200,000}{(1.1)(75-55)} = 9091 \text{ CFM}$$

THIS IS A MIXING PROBLEM

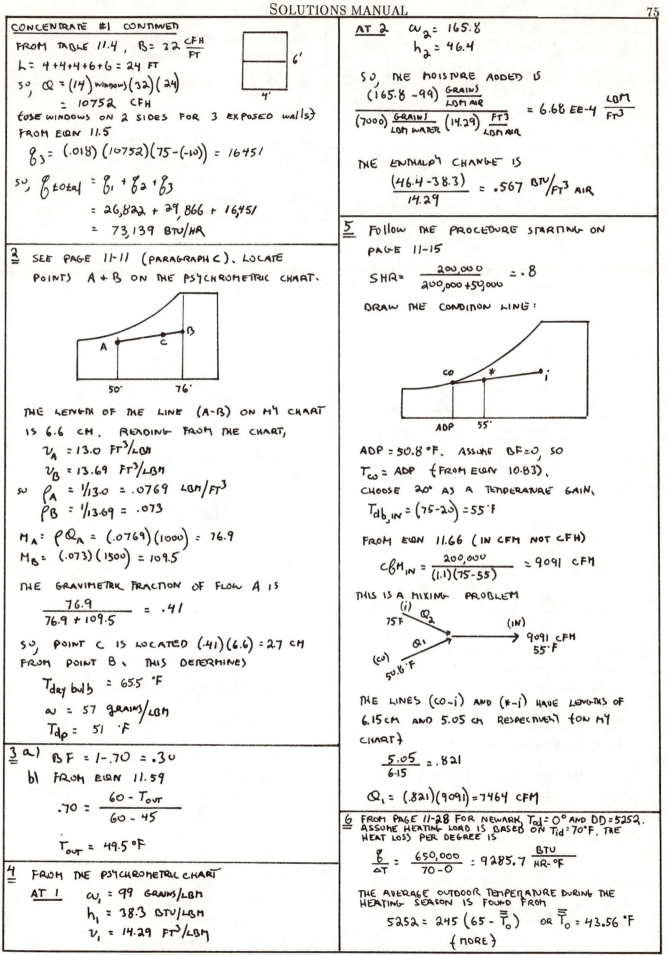

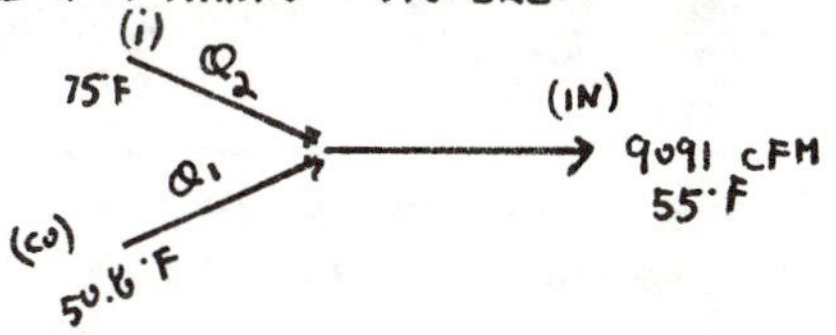

THE LINES (CO-i) AND (*-i) HAVE LENGTHS OF 6.15 CM AND 5.05 CM RESPECTIVELY {ON MY CHART}

$$\frac{5.05}{6.15} = .821$$

$Q_1 = (.821)(9091) = 7464$ CFM

6 FROM PAGE 11-28 FOR NEWARK, $T_{od} = 0°$ AND DD = 5252. ASSUME HEATING LOAD IS BASED ON $T_{id} = 70$°F. THE HEAT LOSS PER DEGREE IS

$$\frac{q}{\Delta T} = \frac{650,000}{70 - 0} = 9285.7\ \frac{\text{BTU}}{\text{HR-°F}}$$

THE AVERAGE OUTDOOR TEMPERATURE DURING THE HEATING SEASON IS FOUND FROM

$5252 = 245(65 - \bar{\bar{T}}_o)$ OR $\bar{\bar{T}}_o = 43.56$ °F

{MORE}

CONCENTRATES # 6 CONTINUED

8:30 A.M. TO 5:30 P.M. IS 5 HOURS. THE TOTAL WINTER HEAT LOSS IS

$$(245)(9285.7)\left[(9)(70-43.56) + (15)(50-43.56)\right]$$

$$= 7.61 \text{ EE8 BTU}$$

$$\text{FUEL CONSUMPTION} = \frac{7.61 \text{ EE8}}{(13{,}000)(.70)} = 83{,}626 \text{ LB/YR}$$

7 THIS IS A MIXING PROBLEM. THE PROCESS CANNOT BE REPRESENTED BY A CONSTANT ENTHALPY LINE BECAUSE $T_{WATER} < T_{wb,\ INCOMING}$

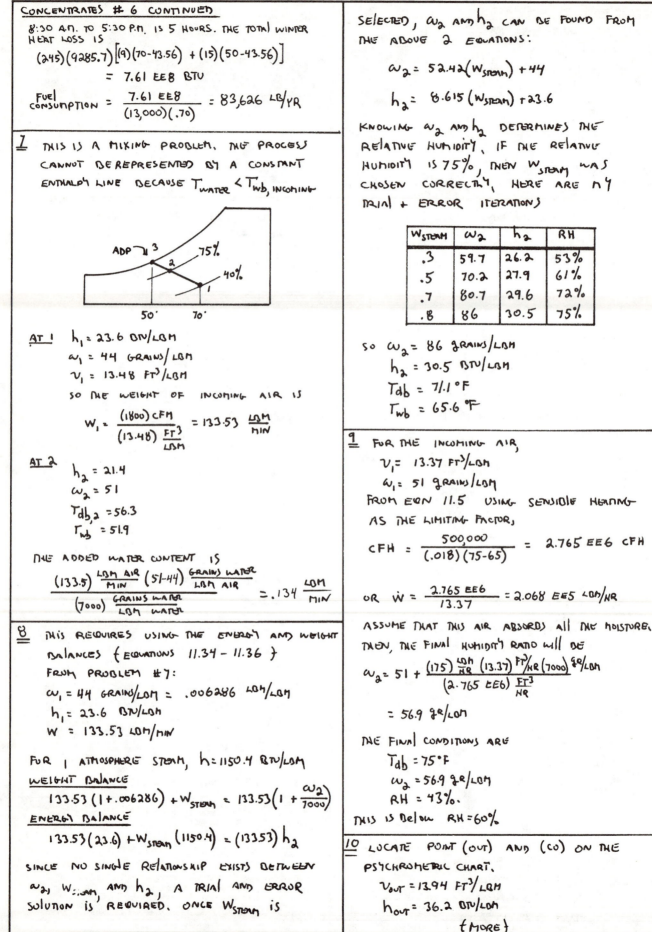

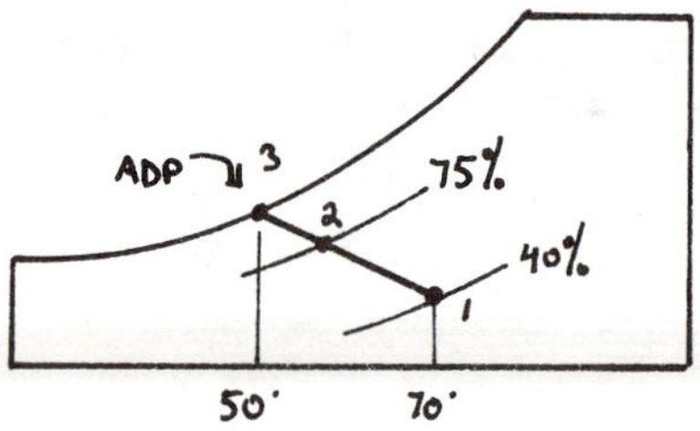

AT 1 $h_1 = 23.6$ BTU/LBM
$\omega_1 = 44$ GRAINS/LBM
$v_1 = 13.48$ FT³/LBM

SO THE WEIGHT OF INCOMING AIR IS

$$W_1 = \frac{(1800)\text{ CFM}}{(13.48)\ \frac{\text{FT}^3}{\text{LBM}}} = 133.53\ \frac{\text{LBM}}{\text{MIN}}$$

AT 2 $h_2 = 21.4$
$\omega_2 = 51$
$T_{db,2} = 56.3$
$T_{wb} = 51.9$

THE ADDED WATER CONTENT IS

$$\frac{(133.5)\ \frac{\text{LBM AIR}}{\text{MIN}}\ (51-44)\ \frac{\text{GRAINS WATER}}{\text{LBM AIR}}}{(7000)\ \frac{\text{GRAINS WATER}}{\text{LBM WATER}}} = .134\ \frac{\text{LBM}}{\text{MIN}}$$

8 THIS REQUIRES USING THE ENERGY AND WEIGHT BALANCES {EQUATIONS 11.34 – 11.36}

FROM PROBLEM #7:

$\omega_1 = 44$ GRAINS/LBM $= .006286$ LBM/LBM
$h_1 = 23.6$ BTU/LBM
$W = 133.53$ LBM/MIN

FOR 1 ATMOSPHERE STEAM, $h = 1150.4$ BTU/LBM

WEIGHT BALANCE

$$133.53(1+.006286) + W_{STEAM} = 133.53\left(1 + \frac{\omega_2}{7000}\right)$$

ENERGY BALANCE

$$133.53(23.6) + W_{STEAM}(1150.4) = (133.53)\,h_2$$

SINCE NO SINGLE RELATIONSHIP EXISTS BETWEEN ω_2, W_{STEAM}, AND h_2, A TRIAL AND ERROR SOLUTION IS REQUIRED. ONCE W_{STEAM} IS SELECTED, ω_2 AND h_2 CAN BE FOUND FROM THE ABOVE 2 EQUATIONS:

$$\omega_2 = 52.42(W_{STEAM}) + 44$$

$$h_2 = 8.615(W_{STEAM}) + 23.6$$

KNOWING ω_2 AND h_2 DETERMINES THE RELATIVE HUMIDITY. IF THE RELATIVE HUMIDITY IS 75%, THEN W_{STEAM} WAS CHOSEN CORRECTLY. HERE ARE MY TRIAL + ERROR ITERATIONS

W_{STEAM}	ω_2	h_2	RH
.3	59.7	26.2	53%
.5	70.2	27.9	61%
.7	80.7	29.6	72%
.8	86	30.5	75%

SO $\omega_2 = 86$ GRAINS/LBM
$h_2 = 30.5$ BTU/LBM
$T_{db} = 71.1$ °F
$T_{wb} = 65.6$ °F

9 FOR THE INCOMING AIR,
$v_1 = 13.37$ FT³/LBM
$\omega_1 = 51$ GRAINS/LBM

FROM EQN 11.5 USING SENSIBLE HEATING AS THE LIMITING FACTOR,

$$\text{CFH} = \frac{500{,}000}{(.018)(75-65)} = 2.765 \text{ EE6 CFH}$$

$$\text{OR } \dot{W} = \frac{2.765 \text{ EE6}}{13.37} = 2.068 \text{ EE5 LBM/HR}$$

ASSUME THAT THIS AIR ABSORBS ALL THE MOISTURE. THEN, THE FINAL HUMIDITY RATIO WILL BE

$$\omega_2 = 51 + \frac{(175)\ \frac{\text{LBM}}{\text{HR}}\ (13.37)\ \frac{\text{FT}^3}{\text{HR}}\ (7000)\ \text{GR/LBM}}{(2.765 \text{ EE6})\ \frac{\text{FT}^3}{\text{HR}}}$$

$$= 56.9 \text{ GR/LBM}$$

THE FINAL CONDITIONS ARE
$T_{db} = 75$°F
$\omega_2 = 56.9$ GR/LBM
RH = 43%.

THIS IS BELOW RH = 60%

10 LOCATE POINT (OUT) AND (CO) ON THE PSYCHROMETRIC CHART.
$v_{OUT} = 13.94$ FT³/LBM
$h_{OUT} = 36.2$ BTU/LBM

{MORE}

CONCENTRATES #10 CONTINUED

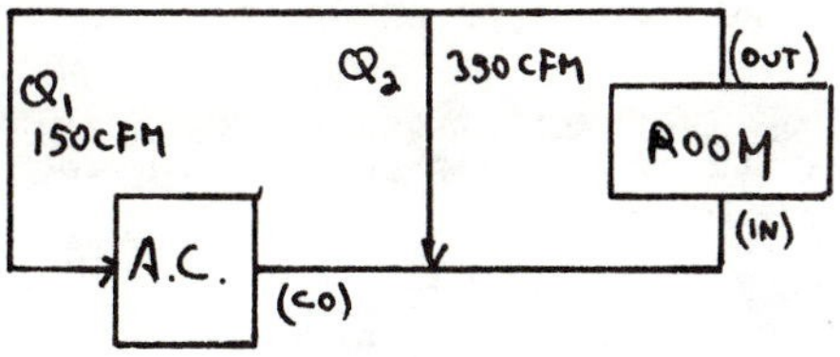

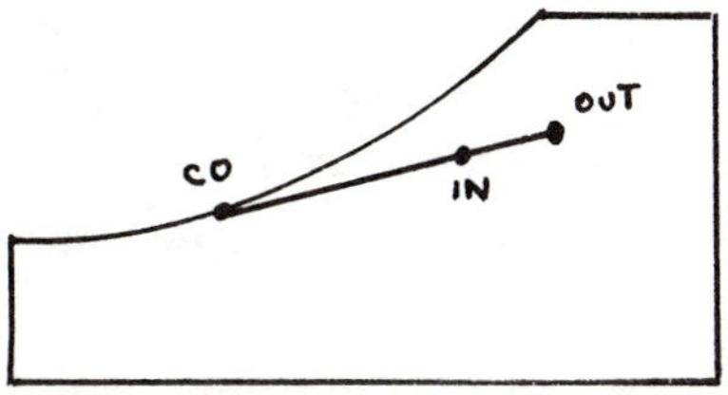

$h_{co} = 20.2$

SO $\dot{W}_1 = \dfrac{(150)\ \text{FT}^3/\text{MIN}}{(13.94)\ \frac{\text{FT}^3}{\text{LBM}}} = 10.76\ \text{LBM/MIN}$

$\dot{W}_2 = \dfrac{350}{13.94} = 25.11$

THE % BYPASS IS

$x = \dfrac{25.11}{25.11 + 10.76} = .70$

ON MY CHART, THE LENGTH OF LINE (CO–OUT) IS 8.5 CM, POINT (IN) IS LOCATED $(.70)(8.5) = 6$ CM FROM POINT CO.

a) AT THAT POINT

$T_{db,IN} = 71.2\ °F$

$\omega_{IN} = 92\ \text{GR/LBM}$

$RH_{,IN} = 80\%$

b) THE CONDITIONER CAPACITY IS

$$\frac{(10.76)\ \frac{\text{LBM}}{\text{MIN}}\ (36.2 - 20.2)\ \frac{\text{BTU}}{\text{LBM}}}{(200)\ \frac{\text{BTU}}{\text{MIN-TON}}} = .86\ \text{TON}$$

TIMED

1 AT 1

$h_1 = 50.7\ \text{BTU/LBM}$

$v_1 = 14.55\ \text{FT}^3/\text{LBM}$

$\omega_1 = 177\ \text{GR/LBM}$

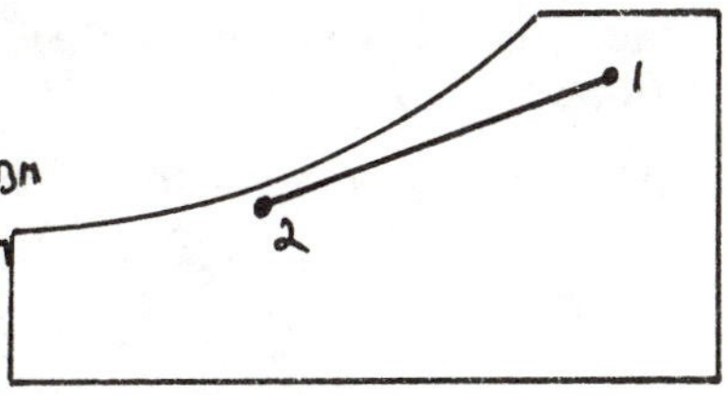

AT 2

$h_2 = 25.8$

$\omega_2 = 73$

b) THE AIR WEIGHT IS

$$\frac{(5000)\ \frac{\text{FT}^3}{\text{MIN}}}{(14.55)\ \frac{\text{FT}^3}{\text{LBM}}} = 343.6\ \text{LBM/MIN}$$

THE WATER DECREASE IS

$$\frac{(343.6)\ \frac{\text{LBM AIR}}{\text{MIN}}\ (177 - 73)\ \frac{\text{GR}}{\text{LBM-AIR}}}{(7000)\ \frac{\text{GR}}{\text{LBM WATER}}} = 5.1\ \frac{\text{LBM}}{\text{MIN}}$$

c) $(343.6)\ \frac{\text{LBM}}{\text{MIN}}\ (50.7 - 25.8)\ \frac{\text{BTU}}{\text{LBM}} = 8555.6\ \frac{\text{BTU}}{\text{MIN}}$

d) IT'S NOT CLEAR FROM THE PROBLEM WHETHER THIS IS A WET OR DRY CYCLE. ASSUME WET SINCE THE PROBLEM SAYS 'SATURATED' AT 100°F

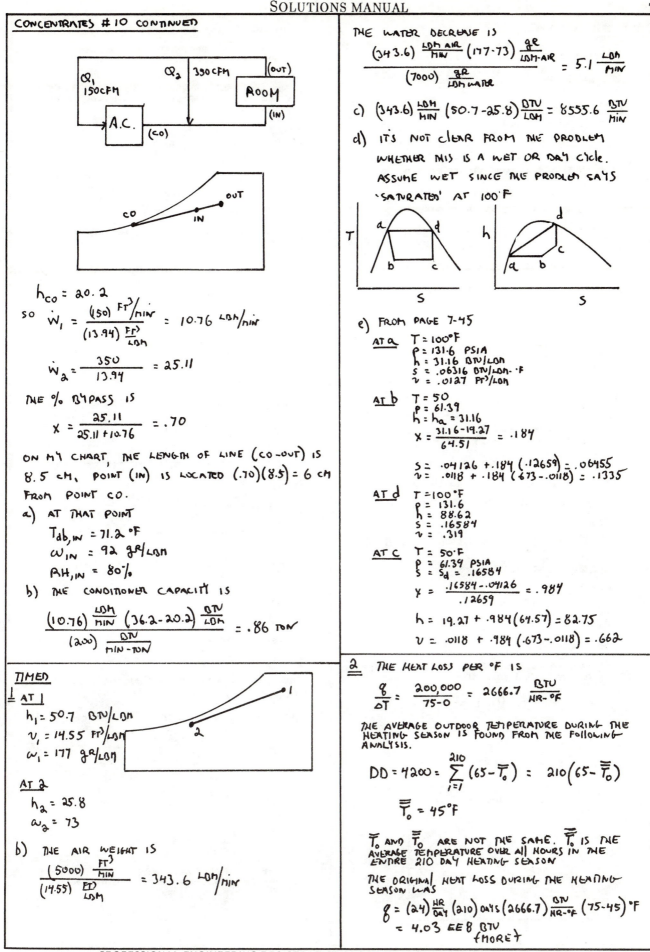

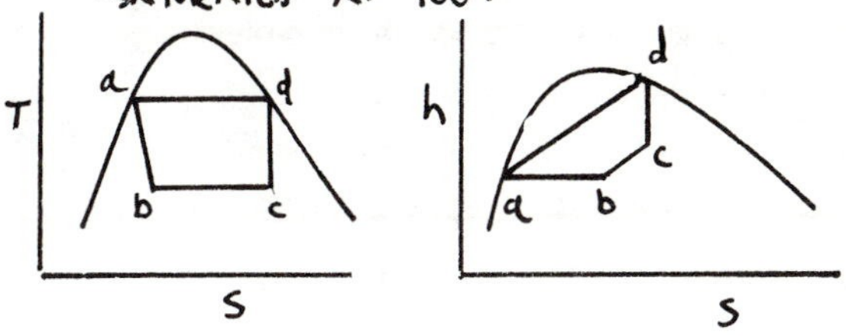

e) FROM PAGE 7-45

AT a
$T = 100°F$
$p = 131.6$ PSIA
$h = 31.16$ BTU/LBM
$s = .06316$ BTU/LBM-°F
$v = .0127$ FT³/LBM

AT b
$T = 50$
$p = 61.39$
$h = h_a = 31.16$
$x = \dfrac{31.16 - 19.27}{64.51} = .184$

$s = .04126 + .184\,(.12659) = .06455$
$v = .0118 + .184\,(.673 - .0118) = .1335$

AT d
$T = 100°F$
$p = 131.6$
$h = 88.62$
$s = .16584$
$v = .319$

AT c
$T = 50°F$
$p = 61.39$ PSIA
$s = s_d = .16584$
$x = \dfrac{.16584 - .04126}{.12659} = .984$

$h = 19.27 + .984\,(64.57) = 82.75$

$v = .0118 + .984\,(.673 - .0118) = .662$

2 THE HEAT LOSS PER °F IS

$$\frac{q}{\Delta T} = \frac{200{,}000}{75 - 0} = 2666.7\ \frac{\text{BTU}}{\text{HR-°F}}$$

THE AVERAGE OUTDOOR TEMPERATURE DURING THE HEATING SEASON IS FOUND FROM THE FOLLOWING ANALYSIS.

$$DD = 4200 = \sum_{i=1}^{210} (65 - \bar{T}_o) = 210(65 - \bar{\bar{T}}_o)$$

$$\bar{\bar{T}}_o = 45°F$$

$\bar{T}_o$ AND $\bar{\bar{T}}_o$ ARE NOT THE SAME. $\bar{\bar{T}}_o$ IS THE AVERAGE TEMPERATURE OVER ALL HOURS IN THE ENTIRE 210 DAY HEATING SEASON

THE ORIGINAL HEAT LOSS DURING THE HEATING SEASON WAS

$$q = (24)\ \frac{\text{HR}}{\text{DAY}}\ (210)\ \text{DAYS}\ (2666.7)\ \frac{\text{BTU}}{\text{HR-°F}}\ (75 - 45)\ °F$$
$$= 4.03\ \text{EE}8\ \text{BTU}$$

(MORE)

TIMED #2, CONTINUED

THE REDUCED HEAT LOSS IS

$$q = (24)(210)(2666.7)(68-45) = 3.09 \text{ EE8 BTU}$$

THE REDUCTION IS

$$\frac{4.03-3.09}{4.03} = .233 \quad \boxed{23.3\%}$$

3 THIS IS JUST A STANDARD BYPASS PROBLEM

THE INDOOR CONDITIONS ARE GIVEN:

$$T_i = 75°F \qquad \phi_i = 50\%$$

THE OUTDOOR CONDITIONS ARE GIVEN:

$$T_o = 90°F \text{ db} \qquad T_{o,wb} = 76°F$$

THE VENTILATION IS GIVEN AS 2000 CFM

THE LOADS ARE GIVEN

$$q_s = 200{,}000 \text{ BTUH}$$

$$q_l = 450{,}000 \text{ gr/HR}$$

WE WANT TO FIND THE SENSIBLE HEAT RATIO. BUT, q_l MUST BE EXPRESSED IN BTUH.

$\omega \approx .0095$ FOR THE ROOM CONDITIONS. (ACTUALLY, IT SHOULD BE A LITTLE LESS SINCE MOISTURE IS REMOVED BETWEEN T_{IN} AND T_i)

ASSUME P = 14.7 PSIA, THEN, FROM EQN 11.26,

$$.0095 = \frac{.622\,P_W}{14.7 - P_W} \approx \frac{.622\,P_W}{14.7}$$

SO, $P_W = .22$ psiA

FOR P = .22 psiA, FROM THE STEAM TABLES, $h_{fg} \approx 1060$ BTU/LBM

EACH POUND OF WATER EVAPORATED REQUIRES APPROXIMATELY 1060 BTU

$$q_l = \frac{(450{,}000)\frac{gr}{HR}(1060)\frac{BTU}{LB}}{(7000)\frac{gr}{LB}} = 68143 \text{ BTU/HR}$$

FROM EQUATION 11.64

$$SHR = \frac{200{,}000}{200{,}000 + 68143} = .75$$

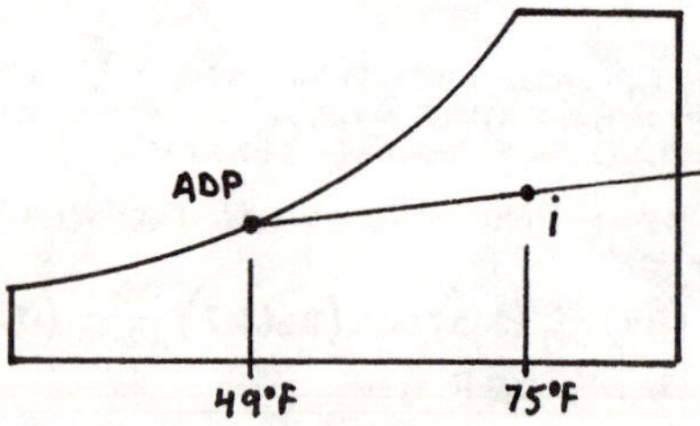

USING T_i AND ϕ_i, LOCATE POINT i ON THE PSYCH CHART AND DRAW THE CONDITION LINE WITH SLOPE = .75 THROUGH IT.

THE LEFT-HAND INTERSECTION SHOWS ADP = 49°F. SINCE THE AIR LEAVES THE CONDITIONER SATURATED, WE KNOW

$$BF_{coil} = 0$$

$$\boxed{T_{CO} = 49°F}$$

NOW CALCULATE THE AIRFLOW THROUGH THE ROOM.

$$(CFM)_{IN} = \frac{200{,}000}{1.1\,(75-58)} = \boxed{10{,}700}$$

$$Q_{IN} = 60(10{,}700) = 642{,}000 \text{ CFH}$$

SINCE $T_{IN} = 58°F$ IS GIVEN, LOCATE THIS DRY BULB TEMPERATURE ON THE CONDITION LINE

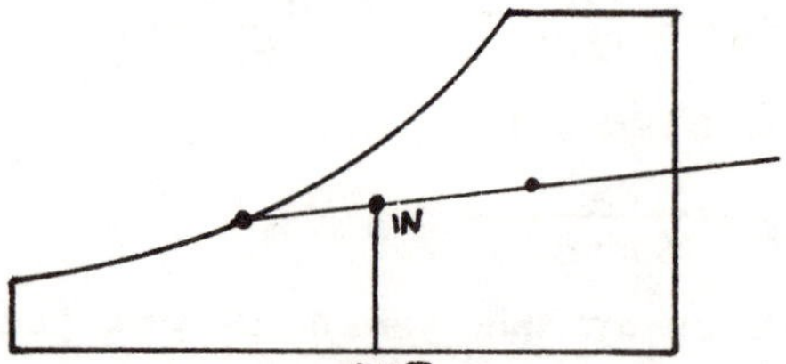

READING FROM THE CHART,

$$\boxed{\omega_{IN} = 56.5 \text{ gr/pound}}$$

4 ALTHOUGH SOME HIGH-TEMPERATURE PSYCH CHARTS EXIST, IT IS NOT NECESSARY TO USE THEM.

SEE EQUATION 11.27 AT $T_i = 200°F$ AND 100% RELATIVE HUMIDITY,

$P_W = 11.526$ (FROM PAGE 6-29)

SINCE $P_t = 25$ psiA,

$$P_A = 25 - 11.526 = 13.474 \text{ psiA}$$

THE SPECIFIC GAS CONSTANT OF WATER IS 85.8 {TABLE 6.4}

THE MASS OF THE WATER VAPOR IS CONSTANT

$$M_W = \frac{PV}{RT} = \frac{(11.526)(144)(1500)}{(85.8)(200+460)} = 43.96 \text{ LBM}$$

THE MASS OF THE AIR IS ALSO CONSTANT

$$M_A = \frac{(13.474)(144)(1500)}{(53.3)(200+460)} = 82.73 \text{ LBM}$$

$$\omega = \frac{43.96}{82.73} = .531$$

THE MASSES OF AIR AND WATER DO NOT CHANGE, SO THE MOLE FRACTIONS AND PARTIAL PRESSURE RATIOS ALSO DO NOT CHANGE.

{MORE}

TIMED #4, CONTINUED

$P_{w,2} = 11.526$

AT $T_2 = 400°F$, $P_{w,SAT} = 247.31$ psi (PAGE 6-29)

$$\phi_2 = \frac{11.526}{247.31} = \boxed{.0466}$$

ω IS UNCHANGED.

$$\omega_2 = \boxed{.531}$$

THE ENERGY REQUIRED MUST BE OBTAINED IN TWO PARTS

AIR: FROM PAGE 6.35

$h_1 = 157.92$ BTU/LBM (AT 660°R)

$h_2 = 206.46$ BTU/LBM (AT 860 °R)

$q_1 = 206.46 - 157.92$

$= 48.54$ BTU/LBM-DRY AIR

WATER: A MOLLIER DIAGRAM, LOCATE POINT 1 (T = 200, P = 11.526)

$h_1 = 1146$ (APPEARS TO BE SATURATED)

NOW, FOLLOW A CONSTANT PRESSURE CURVE UP TO 400°F.

$h_2 = 1240$ BTU/LBM

$$q_2 = \frac{(43.96)(1240-1146)}{82.73}$$

$= 49.95$ BTU/LBM DRY AIR

$$q_{total} = 48.54 + 49.95 = \boxed{98.49 \ \frac{\text{BTU}}{\text{LBM DRY AIR}}}$$

THE DEW POINT IS THE TEMPERATURE AT WHICH THE WATER STARTS TO CONDENSE OUT IN A CONSTANT-PRESSURE PROCESS. FOLLOWING THE CONSTANT PRESSURE LINE BACK TO THE SATURATION LINE,

$$\boxed{T_{dp} = 200°F}$$

RESERVED FOR FUTURE USE

Statics

WARM-UPS

1 $T = F \cdot r = \dfrac{(3)\text{ LBF}(3)\text{ IN}}{(12)\left(\frac{\text{IN}}{\text{FT}}\right)} = .75 \text{ FT-LBF}$

2 FROM TABLE 14.1,

$E_{STEEL} = 30 \text{ EE6 psi}$

$E_{ALUM} = 10 \text{ EE6 psi}$

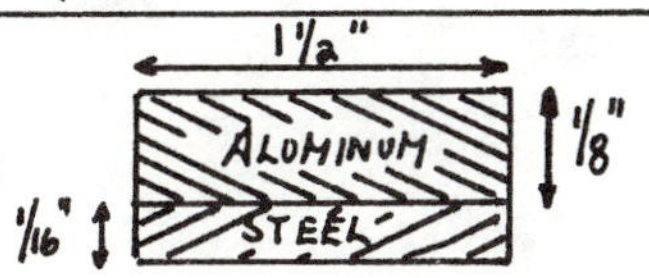

REFER TO THE PROCEDURE ON PAGE 14-7

$N = \dfrac{30 \text{ EE6}}{10 \text{ EE6}} = 3$

SO, THE EQUIVALENT, ALL ALUMINUM CROSS SECTION IS

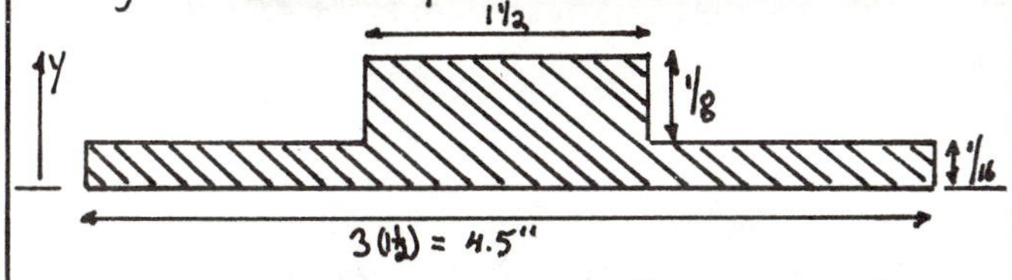

THE CENTROID OF THIS COMPOSITE STRUCTURE IS FOUND FROM EQN 12.73

$A_1 = (1/16)(4.5) = .28125$

$A_2 = (1/8)(1.5) = .1875$

$y_1 = (1/16)/2 = .03125$

$y_2 = \frac{1}{16} + (1/8)/2 = .125$

$$\bar{y} = \frac{(.28125)(.03125) + (.1875)(.125)}{.28125 + .1875} = .06875$$

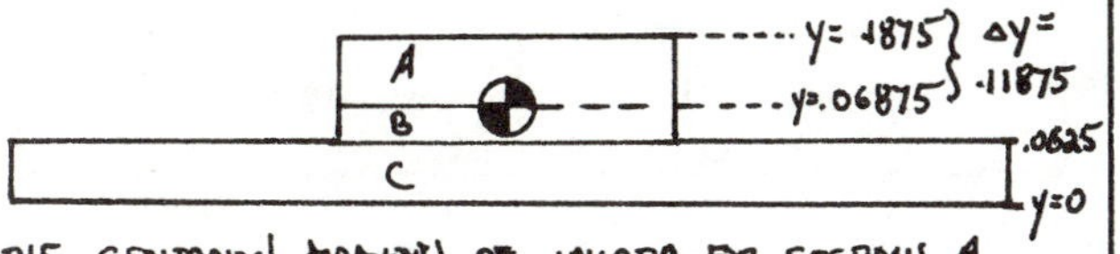

THE CENTROIDAL MOMENTS OF INERTIA FOR SECTIONS A AND B ARE

$I_A = \dfrac{bh^3}{3} = \dfrac{(1.5)(.11875)^3}{3} = 8.373 \text{ EE-4}$

$I_B = \dfrac{(1.5)(.00625)^3}{3} = 1.221 \text{ EE-7}$

THE MOMENT OF INERTIA OF SECTION C ABOUT ITS OWN CENTROID IS

$I_{ox} = \dfrac{bh^3}{12} = \dfrac{(4.5)(.0625)^3}{12} = 9.155 \text{ EE-5}$

THE DISTANCE FROM C'S CENTROID TO THE COMPOSITE CENTROID IS

$\bar{y} = \left(\dfrac{.0625}{2}\right) + .00625 = .0375$

FROM THE PARALLEL AXIS THEOREM (EQN 12.80)

$I_C = I_{ox} + A(\bar{y})^2$

$= (9.155 \text{ EE-5}) + (.28125)(.0375)^2$

$= 4.871 \text{ EE-4}$

THE TOTAL MOMENT OF INERTIA IS

$I_t = I_A + I_B + I_C = 1.324 \text{ EE-3 IN}^4$

3 FROM TABLE 14.1

$\rho_{CAST\ IRON} = .256 \dfrac{\text{LBM}}{\text{IN}^3}$

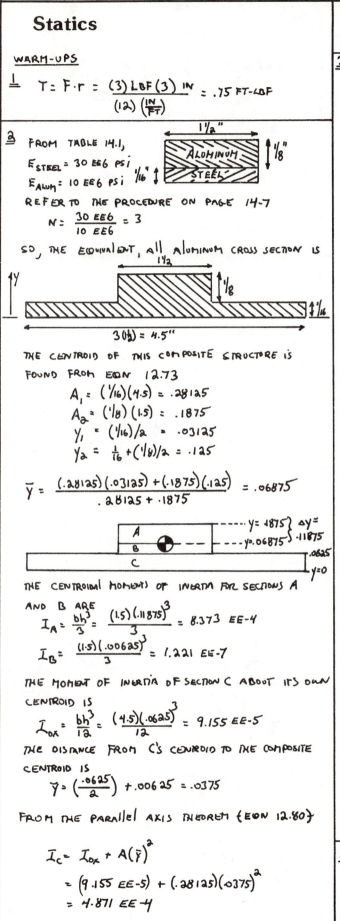

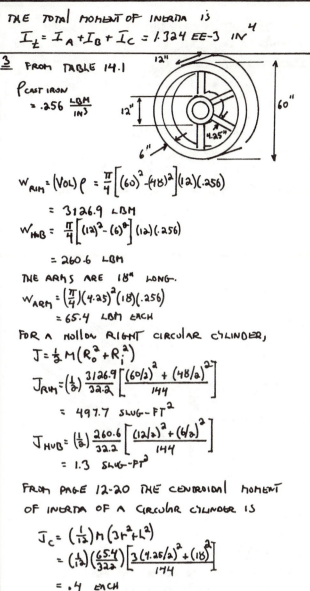

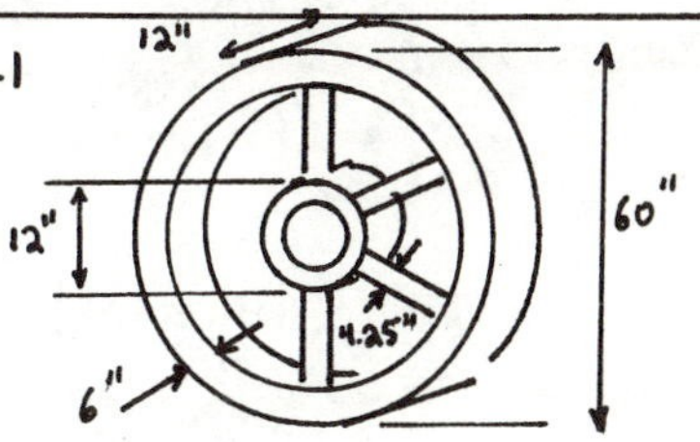

$W_{RIM} = (\text{VOL})\rho = \frac{\pi}{4}\left[(60)^2 - (48)^2\right](12)(.256)$

$= 3126.9 \text{ LBM}$

$W_{HUB} = \frac{\pi}{4}\left[(12)^2 - (6)^2\right](12)(.256)$

$= 260.6 \text{ LBM}$

THE ARMS ARE 18" LONG.

$W_{ARM} = \left(\frac{\pi}{4}\right)(4.25)^2(18)(.256)$

$= 65.4 \text{ LBM EACH}$

FOR A HOLLOW RIGHT CIRCULAR CYLINDER,

$J = \frac{1}{2}M(R_o^2 + R_i^2)$

$J_{RIM} = \left(\frac{1}{2}\right)\dfrac{3126.9}{32.2}\left[\dfrac{(60/2)^2 + (48/2)^2}{144}\right]$

$= 497.7 \text{ SLUG-FT}^2$

$J_{HUB} = \left(\frac{1}{2}\right)\dfrac{260.6}{32.2}\left[\dfrac{(12/2)^2 + (6/2)^2}{144}\right]$

$= 1.3 \text{ SLUG-FT}^2$

FROM PAGE 12-20 THE CENTROIDAL MOMENT OF INERTIA OF A CIRCULAR CYLINDER IS

$J_c = \left(\frac{1}{12}\right)M(3r^2 + L^2)$

$= \left(\frac{1}{12}\right)\left(\frac{65.4}{32.2}\right)\left[\dfrac{3(4.25/2)^2 + (18)^2}{144}\right]$

$= .4 \text{ EACH}$

ANALOGOUS TO EQN 12.80, THE MOMENT OF INERTIA ABOUT THE ROTATIONAL AXIS (15" AWAY FROM THE ARM'S CENTROID) IS

$J_{ARM} = J_c + Mr^2$

$= .4 + \left(\frac{65.4}{32.2}\right)\left(15/12\right)^2$

$= 3.57 \text{ SLUG-FT}^2 \text{ EACH}$

FOR 6 ARMS,

$J_{ARMS} = 6(3.57) = 21.4 \text{ SLUG-FT}^2$

THEN,

$J_{total} = J_{RIM} + J_{HUB} + J_{ARMS}$

$= 520.4 \text{ SLUG-FT}^2$

4 REFER TO PAGE 12-10. FROM FIGURE 12.12,

$w = 2 \text{ LBM/FT}$

$S = 10$

$a = 50$

(MORE)

WARM-UPS # 4 CONTINUED

SOLVING EQN 12.51 BY TRIAL AND ERROR GIVES THE CONSTANT C:

$$S = c\left[\cosh\left(\frac{a}{c}\right) - 1\right]$$

$$c = 126.6$$

THE MID-POINT {HORIZONTAL} TENSION IS GIVEN BY EQN 12.53

$$H = wc = (2)(126.6) = 253.2 \text{ LBF}$$

THE END-POINT TENSION IS GIVEN BY EQN

$$T = wy = w(c+S)$$
$$= 2(126.6+10) = 273.2 \text{ LBF}$$

CONCENTRATES

1

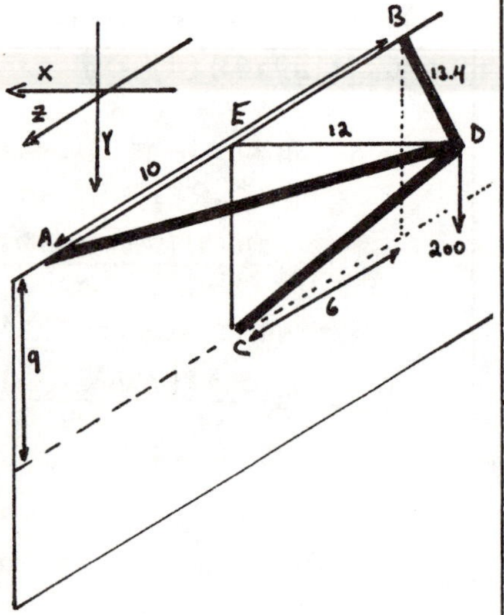

FOR TRIANGLE EBD

LENGTH BE =

$$\sqrt{(13.4)^2 - (12)^2} = 5.96$$

{SAY 6}

SO LENGTH AE IS $(10-6) = 4$.

USING POINT D AS THE ORIGIN,

A IS AT (12,0,4)

B IS AT (12,0,-6)

C IS AT (12,9,0)

D IS AT (0,0,0)

FOLLOWING THE PROCEDURE ON PAGE 12-11, THE DIRECTION COSINES OF THE APPLIED LOAD ARE

$\cos\theta_x = 0$ $\quad \cos\theta_y = 1$ $\quad \cos\theta_z = 0$

SO $F_x = 0$ $\quad F_y = 200$ $\quad F_z = 0$

THE LENGTHS OF THE LEGS ARE

$$AD = \sqrt{(12-0)^2 + (0-0)^2 + (4-0)^2} = 12.65$$

BD = 13.4 (GIVEN)

$$CD = \sqrt{(12)^2 + (9)^2 + (0)^2} = 15.00$$

THE LEG DIRECTION COSINES ARE

FOR AD

$$\cos\theta_{XA} = \frac{12}{12.65} = .949$$

$$\cos\theta_{YA} = \frac{0}{12.65} = 0$$

$$\cos\theta_{ZA} = \frac{4}{12.65} = .316$$

FOR BD

$$\cos\theta_{XB} = \frac{12}{13.4} = .896$$

$$\cos\theta_{YB} = \frac{0}{13.4} = 0$$

$$\cos\theta_{ZB} = \frac{-6}{13.4} = -.448$$

FOR CD

$$\cos\theta_{XC} = \frac{12}{15.00} = .8$$

$$\cos\theta_{YC} = \frac{9}{15.00} = .6$$

$$\cos\theta_{ZC} = \frac{0}{15.00} = 0$$

USING EQUATIONS 12.65 – 12.67

$$\left.\begin{array}{l} .949\,F_A + .896F_B + .8\,F_C + 0 = 0 \\ \qquad\qquad\qquad\qquad .6\,F_C + 200 = 0 \\ .316\,F_A - .448\,F_B \qquad\quad + 0 = 0 \end{array}\right\}$$

THE SIMULTANEOUS SOLUTION TO THESE 3 EQUATIONS IS

F_C = -333.3 LBF (COMPRESSION)

F_B = 118.9 (TENSION)

F_A = 168.6 (TENSION)

2 $\quad \Sigma M_C$ (+): $D_y(6) - 8000(6) + 1600(16) = 0$

$$D_y = 3733$$

SINCE DE IS THE ONLY VERTICAL MEMBER LEAVING POINT D,

DE = 3733 {COMPRESSION}

3 DIVIDE INTO 3 AREAS

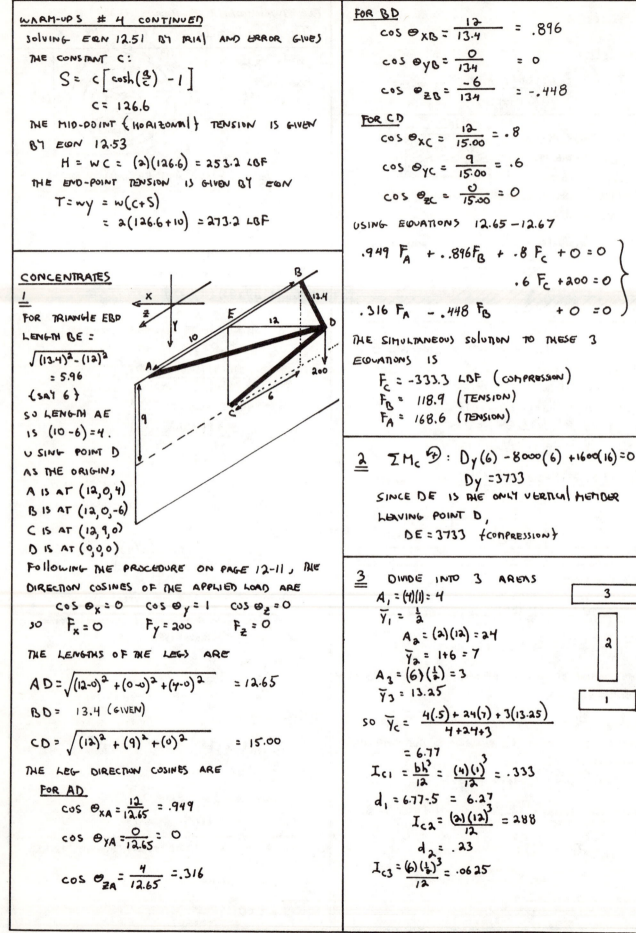

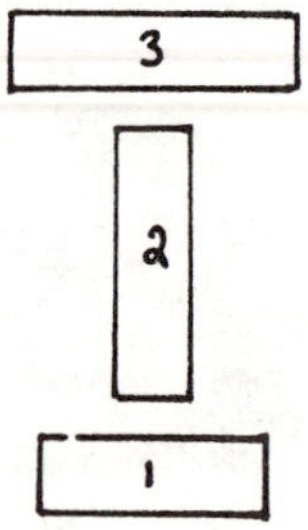

$A_1 = (4)(1) = 4$

$\bar{Y}_1 = \frac{1}{2}$

$A_2 = (2)(12) = 24$

$\bar{Y}_2 = 1+6 = 7$

$A_3 = (6)(\frac{1}{2}) = 3$

$\bar{Y}_3 = 13.25$

$$\text{SO } \bar{Y}_c = \frac{4(.5) + 24(7) + 3(13.25)}{4+24+3} = 6.77$$

$$I_{c1} = \frac{bh^3}{12} = \frac{(4)(1)^3}{12} = .333$$

$d_1 = 6.77 - .5 = 6.27$

$$I_{c2} = \frac{(2)(12)^3}{12} = 288$$

$d_2 = .23$

$$I_{c3} = \frac{(6)(\frac{1}{2})^3}{12} = .0625$$

$d_3 = 6.48$

USING THE PARALLEL AXIS THEOREM {EQN 12.80}

$$I_{total} = .333 + 4(6.27)^2 + 288 + 24(.23)^2 + .0625 + 3(6.48)^2$$
$$= 572.88$$

4 USE THE PROCEDURE ON PAGE 12-11. BUT FIRST, MOVE THE ORIGIN TO THE APEX OF THE TRIPOD SO THAT THE COORDINATES OF THE TRIPOD BASES BECOME

$A = (5, -12, 0)$

$B = (0, -8, -8)$

$C = (-4, -7, 6)$

BY INSPECTION,

$F_x = 1200 \quad F_y = 0 \quad F_z = 0$

$$L_A = \sqrt{(5)^2 + (-12)^2 + (0)^2} = 13$$

$$\cos\theta_{xA} = \frac{5}{13} = .385$$

$$\cos\theta_{yA} = \frac{-12}{13} = -.923$$

$$\cos\theta_{zA} = \frac{0}{13} = 0$$

$$L_B = \sqrt{(0)^2 + (-8)^2 + (-8)^2} = 11.31$$

$$\cos\theta_{xB} = \frac{0}{11.31} = 0$$

$$\cos\theta_{yB} = \frac{-8}{11.31} = -.707$$

$$\cos\theta_{zB} = \frac{-8}{11.31} = -.707$$

$$L_C = \sqrt{(-4)^2 + (-7)^2 + (6)^2} = 10.05$$

$$\cos\theta_{xC} = \frac{-4}{10.05} = -.398$$

$$\cos\theta_{yC} = \frac{-7}{10.05} = -.697$$

$$\cos\theta_{zC} = \frac{6}{10.05} = .597$$

FROM EQUATIONS 12.65-12.67

$$.385 F_A \qquad\qquad -.398 F_C = -1200$$
$$-.923 F_A - .707 F_B - .697 F_C = 0$$
$$-.707 F_B + .597 F_C = 0$$

SOLVING THESE SIMULTANEOUSLY YIELDS

$F_A = -1793$ {COMPRESSION}

$F_B = 1080$ {TENSION}

$F_C = 1279$ {TENSION}

5 BY SYMMETRY, $A_y = L_y = 160$ KIPS

FOR DE

CUT AS SHOWN AND SUM VERTICAL FORCES.

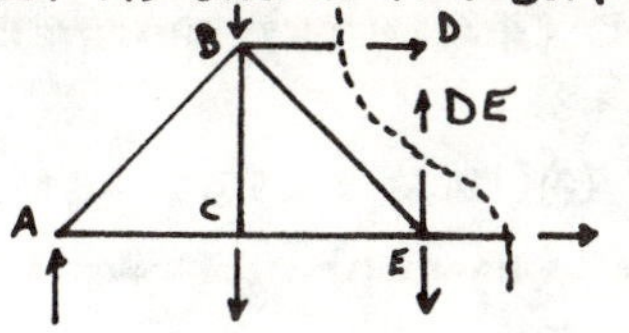

$\Sigma F_y + \uparrow$: $160 + DE - 60 - 60 - 4 = 0$

$DE = -36$ KIPS {COMPRESSION}

FOR HJ

CUT AS SHOWN AND TAKE MOMENTS ABOUT I

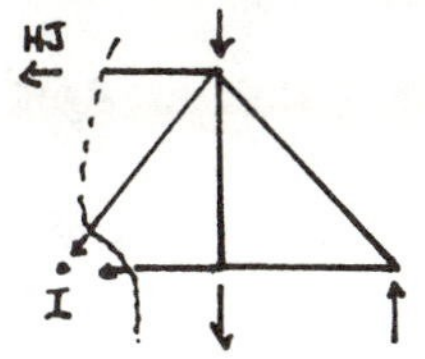

ΣM_I ⟲: $160(60) - 60(30) - 4(30) + HJ(20) = 0$

$HJ = -384$ {COMPRESSION}

6 IN THE X-Y PLANE

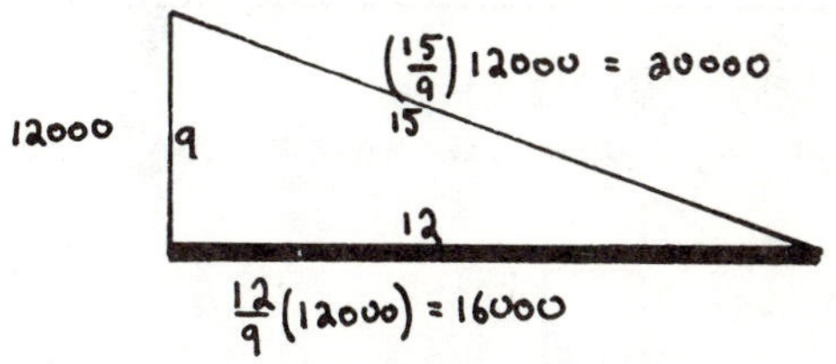

IN THE X-Z PLANE

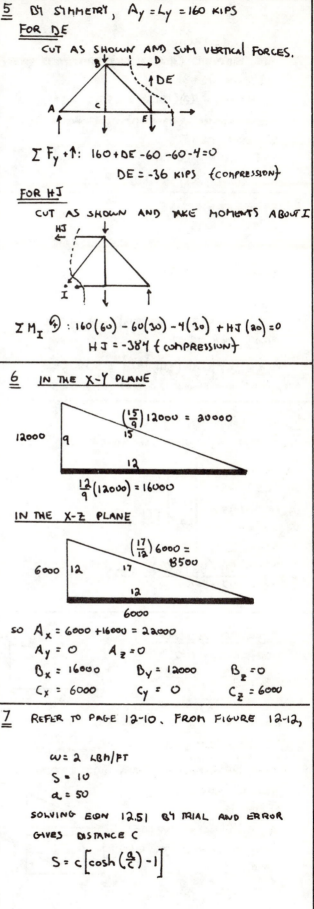

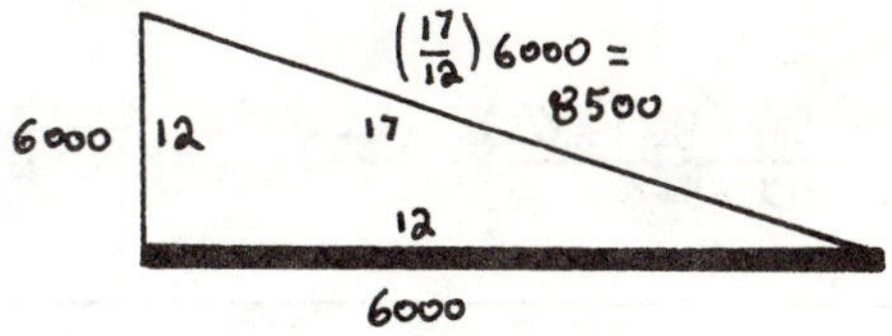

SO $A_x = 6000 + 16000 = 22000$

$A_y = 0 \quad A_z = 0$

$B_x = 16000 \quad B_y = 12000 \quad B_z = 0$

$C_x = 6000 \quad C_y = 0 \quad C_z = 6000$

7 REFER TO PAGE 12-10. FROM FIGURE 12-12,

$w = 2$ LBM/FT

$S = 10$

$a = 50$

SOLVING EQN 12.51 BY TRIAL AND ERROR GIVES DISTANCE C

$$S = c\left[\cosh\left(\frac{a}{c}\right) - 1\right]$$

$c = 126.6$

THE MIDPOINT (HORIZONTAL) TENSION IS GIVEN BY EQN 12.53

$H = wc = (2)(126.6) = 253.2$ LBF

THE ENDPOINT (MAXIMUM) TENSION IS GIVEN BY EQN 12.55

$T = wy = w(c+s)$

$= 2(126.6 + 10) = 273.2$

---

8 FROM EQN 12.55

$T = wy$ OR

$y = \frac{T}{w} = \frac{500}{2} = 250$ AT RIGHT SUPPORT

FROM EQN 12.48

$250 = c\left(\cosh\left(\frac{50}{c}\right)\right)$

$c = 245$ BY TRIAL AND ERROR

FROM EQN 12.51

$s = 245\left[\cosh\left(\frac{50}{245}\right) - 1\right] = 5.12$ ft

(OR, NOTICE FROM FIGURE 12.12 THAT $s = y - c = 5$)

---

9 BY INSPECTION, $\bar{y} = 0$

TO FIND $\bar{x}$, DIVIDE THE OBJECT INTO 3 PARTS

$A_1 = 8(4) = 32$

$\bar{x}_1 = 2$

$A_2 = A_3 = 2(4) = 8$

$\bar{x}_2 = \bar{x}_3 = 6$

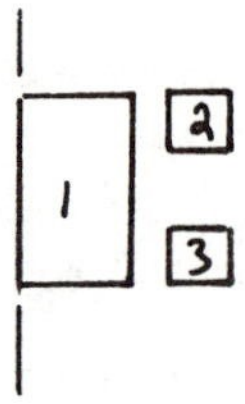

$$\bar{x}_c = \frac{32(2) + 8(6) + 8(6)}{32 + 8 + 8} = 3.333$$

---

10 FOR THE PARABOLA

FROM PAGE 1-2

$A = \frac{2bh}{3} = \frac{2(300)(3)}{3}$

$= 600$ LBF

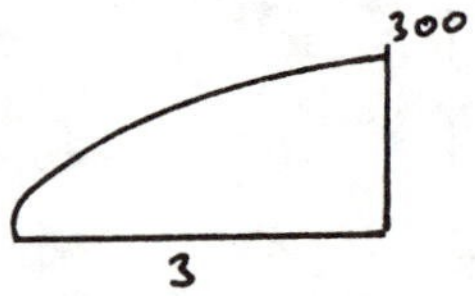

FROM PAGE 12-19 CENTROID IS

$\frac{3h}{5} = \frac{3(3)}{5} = 1.8'$ FROM TIP

DIVIDE THE REMAINING AREA INTO A TRIANGLE AND A RECTANGLE

RECTANGLE

$A_R = (8)(300) = 2400$ LBF

CENTROID IS 4' FROM RIGHT END

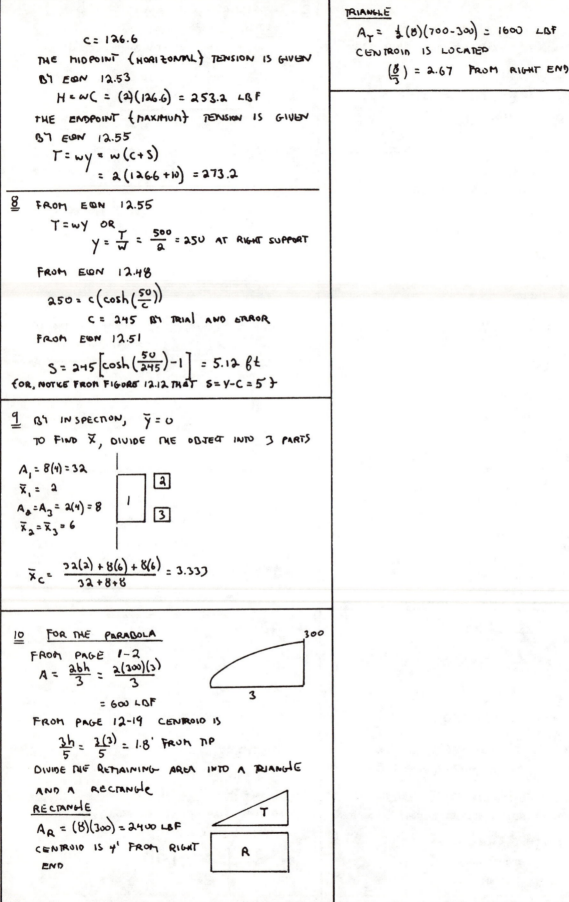

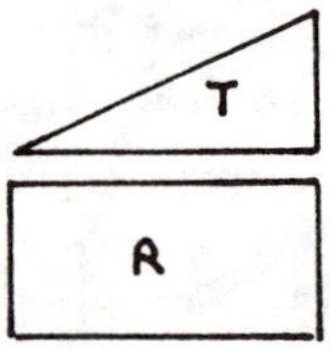

TRIANGLE

$A_T = \frac{1}{2}(8)(700-300) = 1600$ LBF

CENTROID IS LOCATED

$\left(\frac{8}{3}\right) = 2.67$ FROM RIGHT END

RESERVED FOR FUTURE USE

Materials Science

1 a)

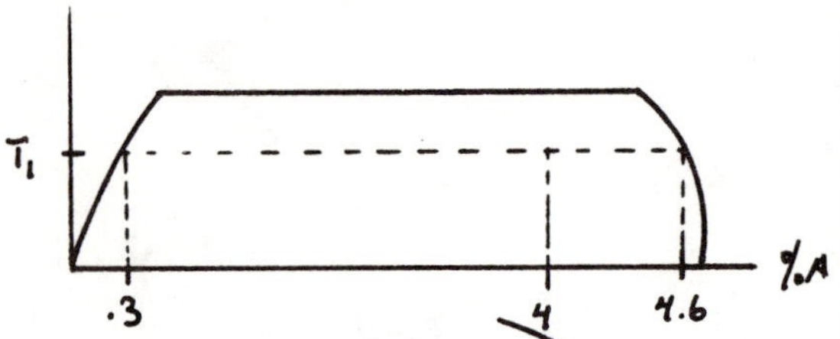

α IS .3% A, 99.7% B
β IS 4.6% A, 95.4% B

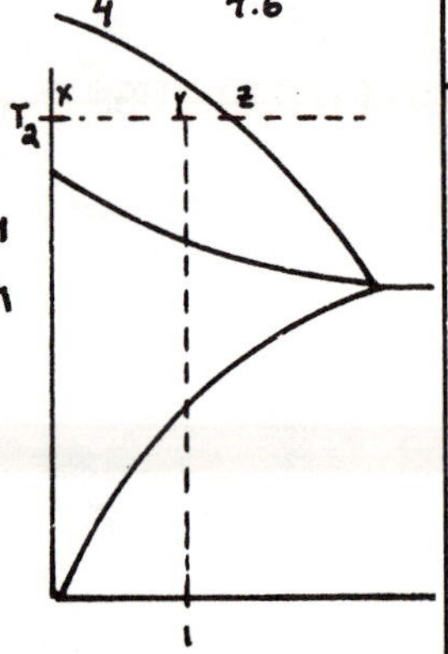

b) LINE SEGMENT XY = 13 MM
LINE SEGMENT XZ = 15 MM

% LIQUID = $\frac{13}{15}$ = 87%

% SOLID = 13%

2 % CHANGE IN DIAMETER = (.3)(.020) = .006

$$\sigma_{true} = \frac{P}{A_o(1-.006)^2} = 20242 \text{ psi}$$

$$\epsilon_{true} = \ln\left(\frac{L}{L_o}\right) = \ln(1+.02) = .0198 \text{ in/in}$$

3 a)

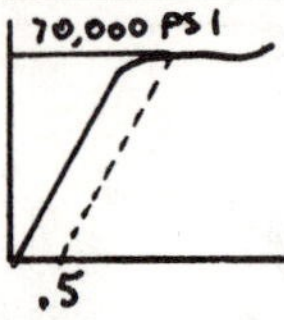

b)

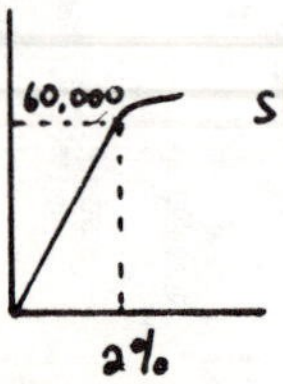

$$\text{slope} = \frac{60{,}000}{.02} = 3 \text{ EE6 PSI}$$

c) THE HIGHEST STRESS REACHED IS APPROXIMATELY 80 KSI

d) THE STRESS AT FRACTURE IS APPROXIMATELY 70 KSI

e) STRAIN AT FRACTURE ≈ 8%

4 FROM EQN 14.77

$$G = \frac{E}{2(1+\mu)} = \frac{3 \text{ EE6}}{2(1+.3)} = 1.15 \text{ EE6 psi}$$

5 DUCTILITY $= \frac{4-3.42}{4} = .145$ REDUCTION IN AREA

6 TOUGHNESS IS THE AREA UNDER THE CURVE: DIVIDE THE AREA INTO (20 KSI × 1%) SQUARES. THERE ARE ABOUT 25 SQUARES

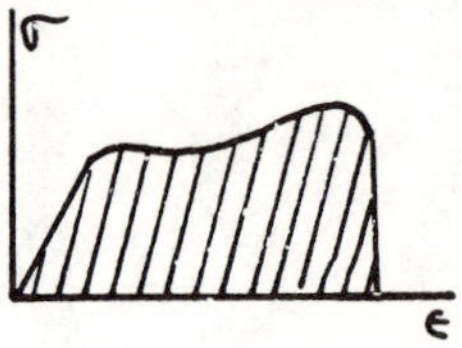

$$(25)(20{,}000)\frac{\text{LBF}}{\text{IN}^2}(.01)\frac{\text{IN}}{\text{IN}} = 5000 \frac{\text{IN-LBF}}{\text{IN}^3}$$

7 PLOT THE DATA AND DRAW A STRAIGHT LINE

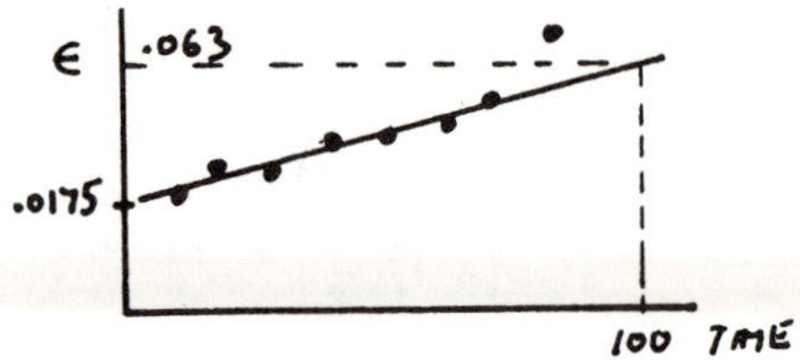

$$\text{slope} = \frac{.063-.0175}{100} = .000455 \text{ 1/HR}$$

8 THE VINYL CHLORIDE MER IS

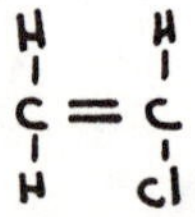

THE MOLECULAR WEIGHT IS

$$2(12) + 3(1) + 1(35.5) = 62.5$$

WITH 20% EFFICIENCY, WE WILL NEED 5 MOLECULES OF HCl PER PVC MOLECULE TO SUPPLY THE END -Cl ATOM. THIS IS THE SAME AS 5 MOLES HCl PER MOLE PVC, OR,

$$\frac{(5)(6.023 \text{ EE } 23)}{7000} = (4.3 \text{ EE } 20) \frac{\text{molecules HCl}}{\text{gram PVC}}$$

9 THE ETHYLENE MOLECULE IS

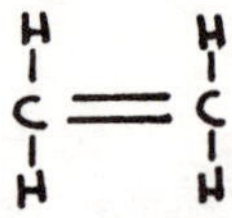

EACH POLYETHYLENE MOLECULE HAS THE FORM

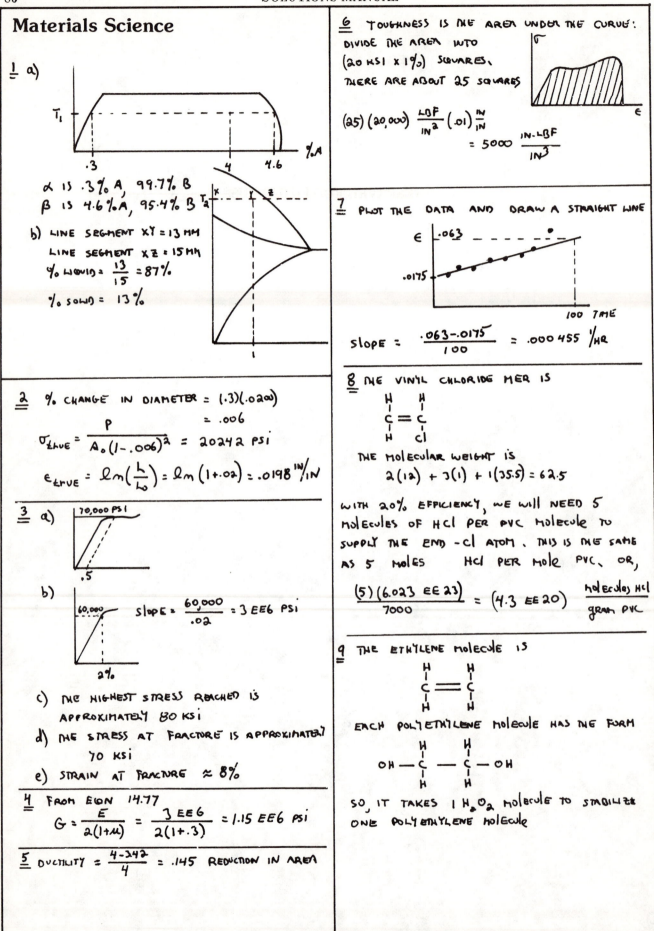

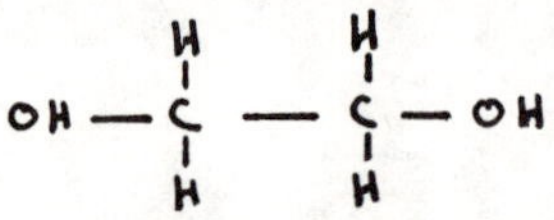

SO, IT TAKES 1 H_2O_2 MOLECULE TO STABILIZE ONE POLYETHYLENE MOLECULE

RESERVED FOR FUTURE USE

Mechanics of Materials

1 FIRST, FIND THE REACTIONS L AND R.

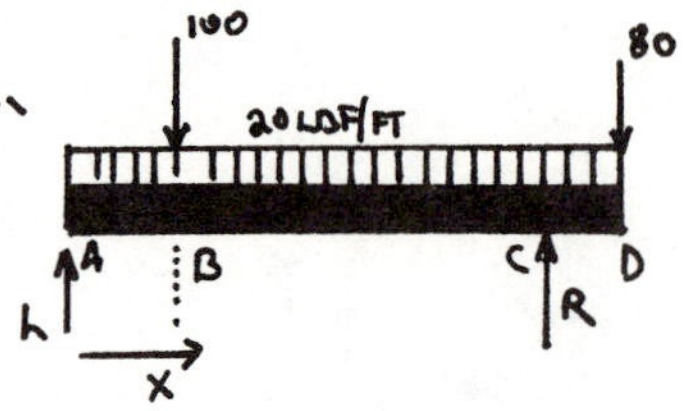

$\sum M_L \circlearrowright : 100(2) + (20)(12)(6) + (80)(12) - R(10) = 0$

$R = 260$

$\sum F_y + \uparrow : L + 260 - 100 - 80 - 20(12) = 0$

$L = 160$

SHEAR

AT A^+: $V = 160$

AT B^-: $V = 160 - 2(20) = 120$

AT B^+: $V = 120 - 100 = 20$

AT C^-: $V = 20 - 20(8) = -140$

AT C^+: $V = -140 + 260 = 120$

AT D^-: $V = 120 - 20(2) = 80$

AT D^+: $V = 80 - 80 = 0$

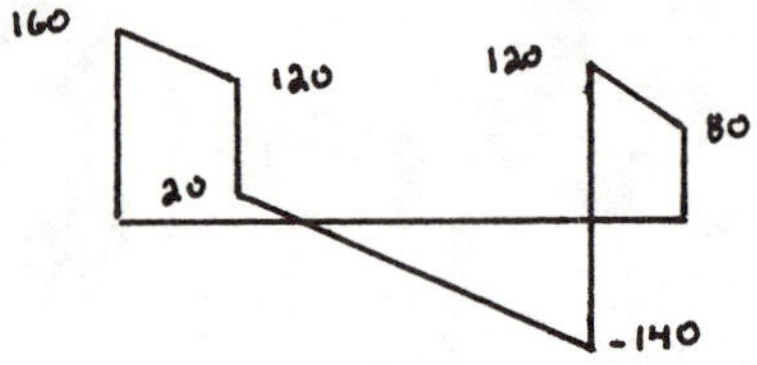

SO $V_{MAX} = 160$ LBF AT LEFT END

MOMENT

M_{MAX} OCCURS WHEN $V = 0$. THIS OCCURS AT

$x = 2 + \left(\frac{20}{160}\right)(8) = 3.00$

$M_{MAX} = 160(3) - 100(1.00) - (20)(3)\left(\frac{3}{2}\right)$

$= 290$ ft-lbf

2 $I = \frac{bh^3}{12}$

$= \frac{6(4)^3}{12} = 32 \text{ IN}^4$

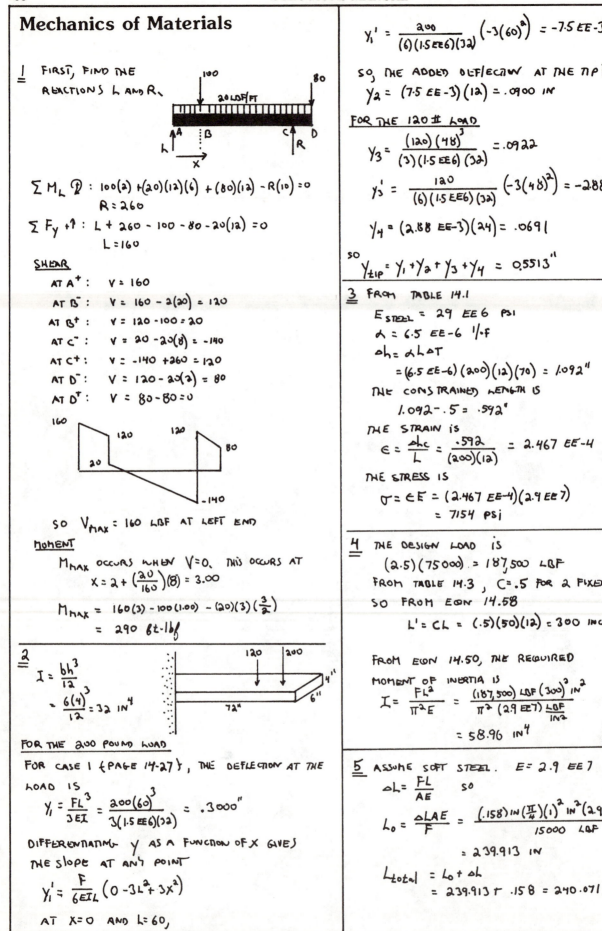

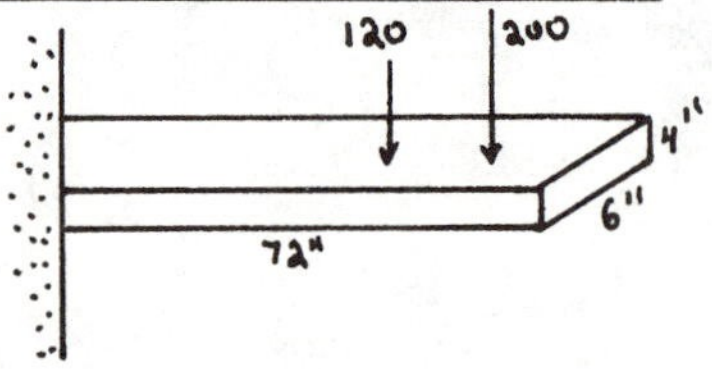

FOR THE 200 POUND LOAD

FOR CASE 1 {PAGE 14-27}, THE DEFLECTION AT THE LOAD IS

$y_1 = \frac{FL^3}{3EI} = \frac{200(60)^3}{3(1.5 EE6)(32)} = .3000''$

DIFFERENTIATING y AS A FUNCTION OF x GIVES THE SLOPE AT ANY POINT

$y_1' = \frac{F}{6EIL}\left(0 - 3L^2 + 3x^2\right)$

AT $x = 0$ AND $L = 60$,

$y_1' = \frac{200}{(6)(1.5EE6)(32)}\left(-3(60)^2\right) = -7.5 EE\text{-}3$ IN/IN

SO, THE ADDED DEFLECTION AT THE TIP IS

$y_2 = (7.5 EE\text{-}3)(12) = .0900$ IN

FOR THE 120 # LOAD

$y_3 = \frac{(120)(48)^3}{(3)(1.5 EE6)(32)} = .0922$

$y_3' = \frac{120}{(6)(1.5 EE6)(32)}\left(-3(48)^2\right) = -2.88 EE\text{-}3$ IN/IN

$y_4 = (2.88 EE\text{-}3)(24) = .0691$

SO

$y_{tip} = y_1 + y_2 + y_3 + y_4 = 0.5513''$

3 FROM TABLE 14.1

$E_{STEEL} = 29 EE6$ PSI

$\alpha = 6.5 EE\text{-}6$ 1/°F

$\Delta L = \alpha L \Delta T$

$= (6.5 EE\text{-}6)(200)(12)(70) = 1.092''$

THE CONSTRAINED LENGTH IS

$1.092 - .5 = .592''$

THE STRAIN IS

$\epsilon = \frac{\Delta L_c}{L} = \frac{.592}{(200)(12)} = 2.467 EE\text{-}4$

THE STRESS IS

$\sigma = \epsilon E = (2.467 EE\text{-}4)(2.9 EE7)$

$= 7154$ psi

4 THE DESIGN LOAD IS

$(2.5)(75000) = 187{,}500$ LBF

FROM TABLE 14.3, $C = .5$ FOR 2 FIXED ENDS,

SO FROM EQN 14.58

$L' = CL = (.5)(50)(12) = 300$ INCHES

FROM EQN 14.50, THE REQUIRED MOMENT OF INERTIA IS

$I = \frac{FL^2}{\pi^2 E} = \frac{(187{,}500) \text{ LBF} (300)^2 \text{ IN}^2}{\pi^2 (2.9 EE7) \frac{\text{LBF}}{\text{IN}^2}}$

$= 58.96 \text{ IN}^4$

5 ASSUME SOFT STEEL. $E = 2.9 EE7$ PSI

$\Delta L = \frac{FL}{AE}$ SO

$L_o = \frac{\Delta L A E}{F} = \frac{(.158) \text{IN} \left(\frac{\pi}{4}\right)(1)^2 \text{ IN}^2 (2.9 EE7) \frac{\text{LBF}}{\text{IN}^2}}{15000 \text{ LBF}}$

$= 239.913$ IN

$L_{total} = L_o + \Delta L$

$= 239.913 + .158 = 240.071$

6 ASSUMING ASTM A36 STEEL, THE YIELD STRENGTH IS APPROXIMATELY 36000 psi

$$FS = \frac{36000}{8240} = 4.37$$

CONCENTRATES

1

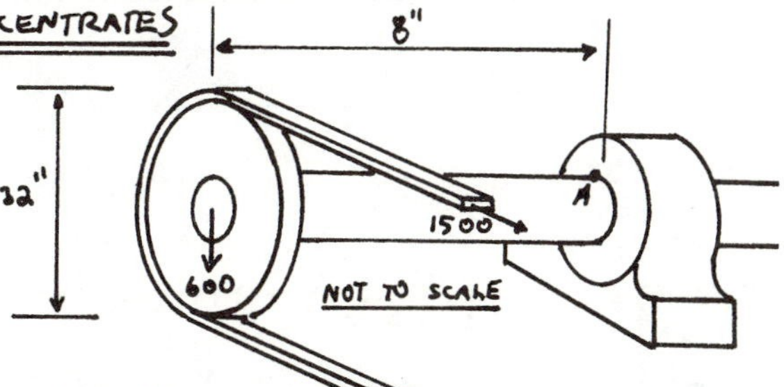

FOR THE SHAFT

$$A = \frac{\pi}{4}d^2 = \frac{\pi}{4}(3)^2 = 7.07 \text{ in}^2$$

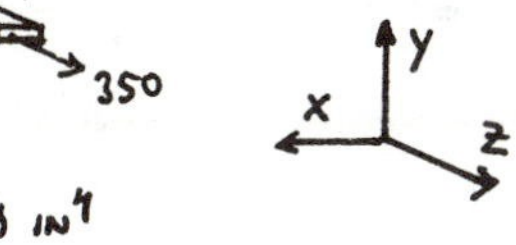

$$I = \frac{\pi}{4}r^4 = \frac{\pi}{4}(1.5)^4 = 3.98 \text{ in}^4$$

$$J = \frac{\pi}{2}r^4 = \frac{\pi}{2}(1.5)^4 = 7.98 \text{ in}^4$$

THE MOMENT AT THE BEARING FACE DUE TO THE PULLEY WEIGHT IS

$$M_y = (600)\text{ LBF }(8)\text{ IN} = 4800 \text{ IN-LBF}$$

THE STRESS IS

$$\sigma_y = \frac{Mc}{I} = \frac{(4800)(1.5)}{3.98} = 1809 \text{ PSI}$$

THE MOMENT AT THE BEARING FACE DUE TO THE BELT TENSIONS IS

$$M_z = (1500 + 350)(8) = 14800 \text{ IN-LBF}$$

THE STRESS IS

$$\sigma_z = \frac{(14800)(1.5)}{3.98} = 5578 \text{ PSI}$$

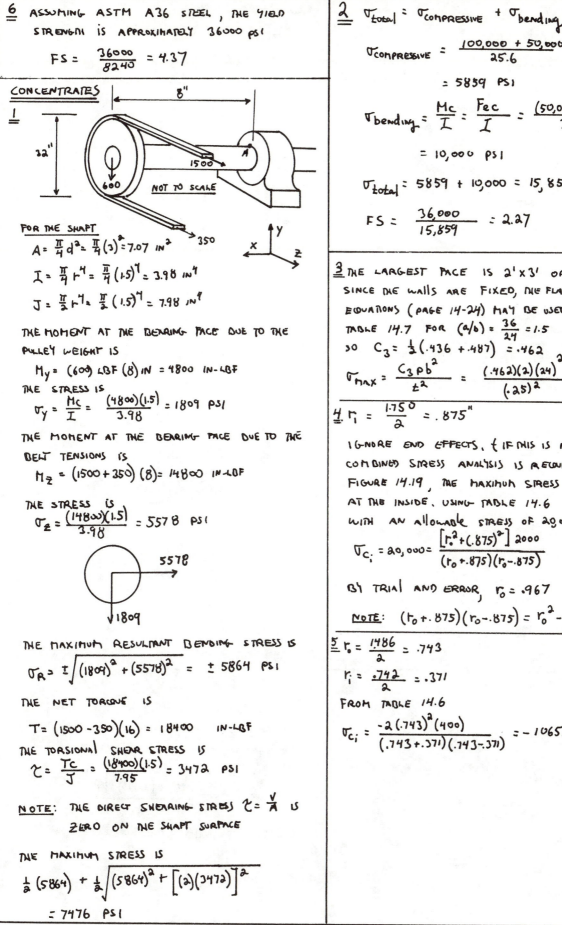

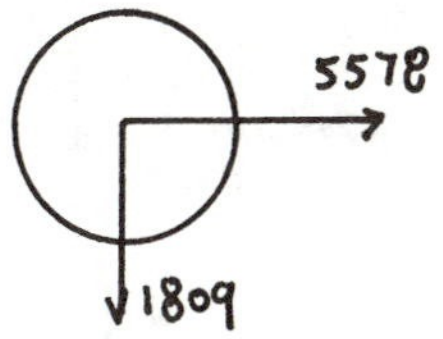

THE MAXIMUM RESULTANT BENDING STRESS IS

$$\sigma_R = \pm\sqrt{(1809)^2 + (5578)^2} = \pm 5864 \text{ PSI}$$

THE NET TORQUE IS

$$T = (1500 - 350)(16) = 18400 \text{ IN-LBF}$$

THE TORSIONAL SHEAR STRESS IS

$$\tau = \frac{Tc}{J} = \frac{(18400)(1.5)}{7.95} = 3472 \text{ PSI}$$

NOTE: THE DIRECT SHEARING STRESS $\tau = \frac{V}{A}$ IS ZERO ON THE SHAFT SURFACE

THE MAXIMUM STRESS IS

$$\frac{1}{2}(5864) + \frac{1}{2}\sqrt{(5864)^2 + [(2)(3472)]^2}$$

$$= 7476 \text{ PSI}$$

2 $\sigma_{total} = \sigma_{compressive} + \sigma_{bending}$

$$\sigma_{compressive} = \frac{100{,}000 + 50{,}000}{25.6}$$

$$= 5859 \text{ PSI}$$

$$\sigma_{bending} = \frac{Mc}{I} = \frac{Fec}{I} = \frac{(50{,}000)(10)(7)}{350}$$

$$= 10{,}000 \text{ PSI}$$

$$\sigma_{total} = 5859 + 10{,}000 = 15{,}859 \text{ PSI}$$

$$FS = \frac{36{,}000}{15{,}859} = 2.27$$

3 THE LARGEST FACE IS 2' x 3' OR 24" x 36". SINCE THE WALLS ARE FIXED, THE FLAT PLATE EQUATIONS (PAGE 14-24) MAY BE USED. FROM TABLE 14.7 FOR $(a/b) = \frac{36}{24} = 1.5$

SO $C_3 = \frac{1}{2}(.436 + .487) = .462$

$$\sigma_{max} = \frac{C_3 p b^2}{t^2} = \frac{(.462)(2)(24)^2}{(.25)^2} = 8515.6 \text{ PSI}$$

4. $r_i = \frac{1.750}{2} = .875"$

IGNORE END EFFECTS. (IF THIS IS NOT TRUE, A COMBINED STRESS ANALYSIS IS REQUIRED). FROM FIGURE 14.19, THE MAXIMUM STRESS WILL OCCUR AT THE INSIDE. USING TABLE 14.6 WITH AN ALLOWABLE STRESS OF 20,000,

$$\sigma_{c_i} = 20{,}000 = \frac{[r_o^2 + (.875)^2]\,2000}{(r_o + .875)(r_o - .875)}$$

BY TRIAL AND ERROR, $r_o = .967$

NOTE: $(r_o + .875)(r_o - .875) = r_o^2 - (.875)^2$

5 $r_o = \frac{1486}{2} = .743$

$$r_i = \frac{.742}{2} = .371$$

FROM TABLE 14.6

$$\sigma_{c_i} = \frac{-2(.743)^2(400)}{(.743 + .371)(.743 - .371)} = -1065.7 \text{ PSI (COMP)}$$

6

$A = \frac{\pi}{4}(d^2) = \frac{\pi}{4}(1)^2 = .7854$

$I = \frac{\pi}{4}(r)^4 = .04909$

$J = \frac{\pi}{2}(r)^4 = .09817$

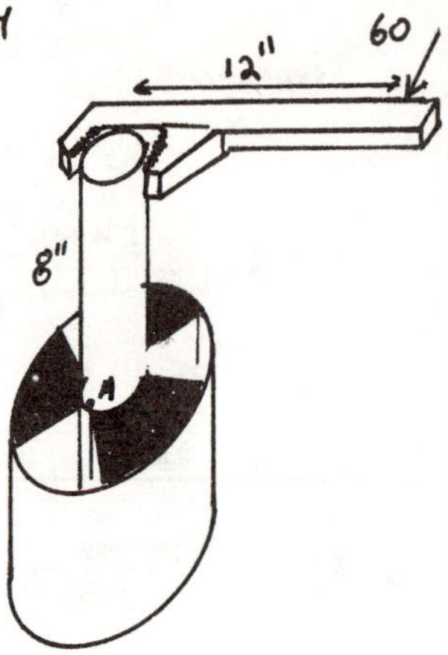

M AT CHUCK IS

(60) LBM (8) IN = 480 IN-LBF

$\sigma = \frac{Mc}{I} = \frac{(480)(.5)}{.04909}$

$= 4889$ PSI

THE APPLIED TORQUE IS

$T = (60)$ LBF (12) IN

$= 720$ IN-LBF

SO, THE SHEAR STRESS IS

$\tau = \frac{Tc}{J} = \frac{(720)(.5)}{.09817} = 3667$ PSI

FROM EQN 14.47

$\tau_{MAX} = \frac{1}{2}\sqrt{(4889)^2 + [(2)(3667)]^2} = 4407$ PSI

FROM EQN 14.46

$\sigma_{MAX} = \frac{1}{2}(4889) + 4407$

$= 6852$ PSI

7 $\sigma_1 = \frac{18,000}{1.5} = -12,000$

NEG. BECAUSE COMPRESSIVE

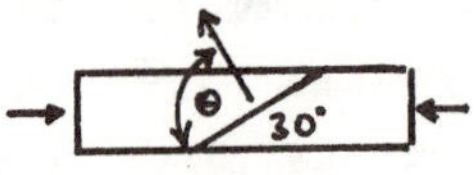

$\sigma_2 = 0$

FOR A 30° INCLINE, $\theta = 60°$, FROM EQN 14.44

$\sigma_\theta = \frac{1}{2}(-12,000) + \frac{1}{2}(-12,000)\cos 120° +$

$(4000) \sin 120° = 464$ PSI

FROM EQN 14.45

$\tau_\theta = -\frac{1}{2}(-12,000)\sin 120 + (4000)\cos 120°$

$= 3196$ PSI

8 THE AREA IN TENSION IS

$\frac{\pi}{4}\left[(16)^2 - (15.8)^2\right] = 5$ IN^2

SO $\sigma_N = \frac{F}{A} = \frac{40,000}{5} = 8000$ PSI

$J = \frac{1}{32}\pi\left[(16)^4 - (15.8)^4\right] = 315.7$

SO $\tau = \frac{Tc}{J} = \frac{(400,000)(8)}{315.7} = 10,136$ PSI

FROM EQN 14.47

$\tau_{MAX} = \frac{1}{2}\sqrt{(8000)^2 + [(2)(10136)]^2}$

$= 10897$

FROM EQN 14.46

$\sigma_N = \frac{1}{2}(8000) \pm 10897$

OR $+14,897$, -6897

TIMED

1

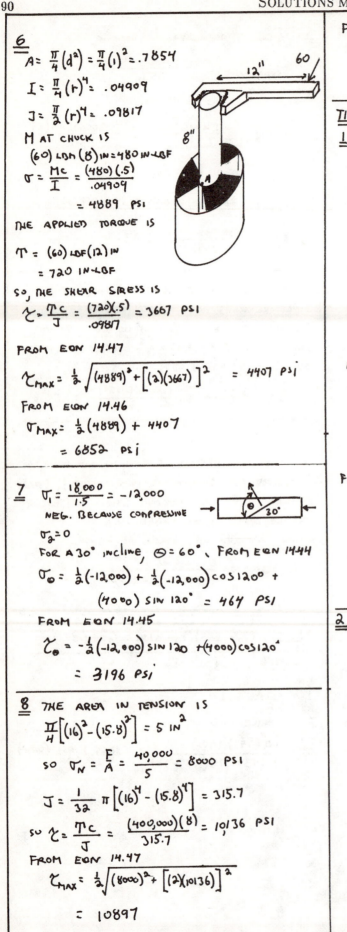

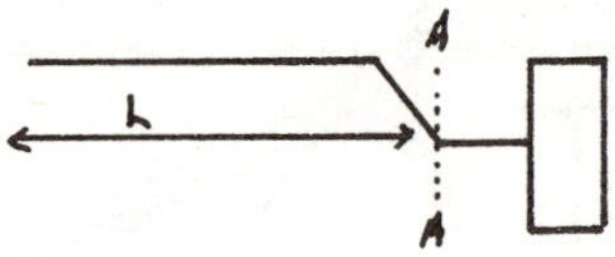

$L = 14 + 3 = 17$

$I = \frac{\pi}{4}(r)^4 = \frac{\pi}{4}\left(\frac{5}{(8)(2)}\right)^4 = .00749$

$J = \frac{\pi}{2}(r)^4 = \frac{\pi}{2}\left(\frac{5}{(8)(2)}\right)^4 = .01498$

THE MOMENT AT A-A IS

$M = FL = (50)(17) = 850$ IN-LBF

THE BENDING STRESS IS

$\sigma = \frac{Mc}{I} = \frac{(850)(5/8)/2}{.00749} = 35464$ PSI

THE TORQUE AT SECTION A-A DUE TO THE ECCENTRICITY IS

$T = (50)(3) = 150$ IN-LBF

THE SHEAR STRESS IS

$\tau = \frac{Tc}{J} = \frac{(150)(5/8)/2}{.01498} = 3129$ PSI

FROM EQN 14.44, THE MAXIMUM SHEAR IS

$\tau_{MAX} = \frac{1}{2}\sqrt{(35464)^2 + [(2)(3129)]^2}$

$= 18006$ PSI

F/A DIRECT SHEAR IS IGNORED

FROM EQN 14.43

$\sigma_1, \sigma_2 = \frac{1}{2}(35464) \pm 18006$

$= 35738, -274$ PSI

2 FOR THE JACKET

$r_o = \frac{12}{2} = 6"$ $\quad r_i = \frac{7.75}{2} = 3.875"$

FOR THE TUBE

$r_o = \frac{7.75}{2} = 3.875"$ $\quad r_i = \frac{4.7}{2} = 2.35$

FOR THE JACKET, THIS IS NO DIFFERENT THAN EXPOSURE TO AN INTERNAL PRESSURE, P. FROM TABLE 14.6, PAGE 14-22, σ_{ci} IS THE HIGHEST STRESS, SO

$\sigma_{ci} = 18000 = \frac{\left[(6)^2 + (3.875)^2\right]P}{(6+3.875)(6-3.875)}$

OR THE EQUIVALENT PRESSURE IS

$P = 7404$ PSI

AND $\sigma_{r_i} = -7404$

FROM EQN 14.88

$$\Delta D = \frac{D}{E}\left[\sigma_c - \mu(\sigma_r - \sigma_L)\right]$$

$$= \frac{7.75}{(29.6\ EE6)}\left[18000 - (.3)(-7404)\right]$$

$$= .00529''$$

OF COURSE, THE TUBE IS ALSO SUBJECTED TO THE PRESSURE. SO

$$\sigma_{ro} = -7404$$

$$\sigma_{co} = \frac{-\left[(3.875)^2 + (2.35)^2\right]}{\left[(3.875)^2 - (2.35)^2\right]}(7404) = -16018$$

$$\Delta D = \frac{7.75}{(29.6\ EE6)}\left[-16018 - (.3)(-7404)\right]$$

$$= -.00361$$

SO, THE TOTAL INTERFERENCE IS

$$.00529 + .00361 = .0089''$$

b) MAXIMUM STRESS OCCURS AT INNER JACKET FACE. $\sigma_{ci} = 18000$ psi

c) MINIMUM STRESS OCCURS AT OUTER TUBE FACE. $\sigma_{ro} = -P = -7404$ psi

$\sigma_{co} = -16018$ psi

3

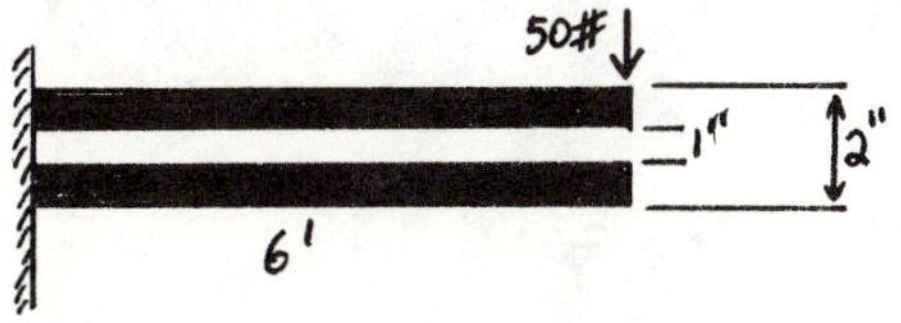

a) YES, THE SUGGESTION IS WITH MERIT. ANYTHING IN THE CENTER VOID WILL HAVE A HIGHER MODULUS OF ELASTICITY (STIFFNESS) THAN AIR.

b) CALCULATE I AS THE DIFFERENCE BETWEEN TWO CIRCLES

$$I_{brass} = \frac{\pi}{4}\left[\left(\frac{2}{2}\right)^4 - \left(\frac{1}{2}\right)^4\right] = .736\ \text{IN}^4$$

$$E_{brass} = 15\ EE6\ \text{psi}$$

$$EI_{brass} = (1.5\ EE7)(.736) = 1.104\ EE7$$

$$I_{steel} = \frac{\pi}{4}\left(\frac{1}{2}\right)^4 = .0491\ \text{IN}^4$$

ASSUME SOFT STEEL, SO

$$E_{steel} = 2.9\ EE7\ \text{psi}$$

$$EI_{steel} = (2.9\ EE7)(.0491) = .1424\ EE7$$

FROM PAGE 14-27, CASE I

$$y_{tip} = \frac{FL^3}{3EI}$$ {INVERSELY PROPORTIONAL TO EI}

SO, THE % CHANGE WOULD BE

$$\% = \frac{y_{old} - y_{new}}{y_{old}} = \frac{\frac{1}{1.104} - \frac{1}{.1424 + 1.104}}{\frac{1}{1.104}}$$

$$= .114\ (11.4\%)$$

4 FROM PAGE 14-30, S_y FOR 6061 T4 ALUMINUM IS 19,000 psi.

THE FOLLOWING STRESSES ARE PRESENT IN THE SPOOL.

- LONG STRESS DUE TO PRESSURE ON END DISKS
- COMPRESSIVE HOOP STRESS
- BENDING STRESS AT DISK AND TUBE JUNCTION {END MOMENT}
- CIRCUMFERENTIAL (TANGENTIAL) STRESS

ASSUMPTIONS

- BENDING STRESS IS IGNORED
- CIRCUMFERENTIAL STRESS IS NEGLIGIBLE DUE TO THIN-WALL CONSTRUCTION
- FACTOR OF SAFETY CALCULATION DOES NOT DEPEND ON OBSCURE COLLAPSING PRESSURE THEORIES, BUT RATHER ON BASIC CONCEPTS.

CALCULATIONS

$$\text{END DISK AREA} = \frac{\pi}{4}\left[(2)^2 - (1)^2\right] = 2.356\ \text{IN}^2$$

$$\text{LONG FORCE} = pA = (500)(2.356) = 1178\ \text{LBF}$$

ANNULUS AREA ABSORBING THE LONG FORCE

$$= (.050)(\pi)(1) = .157$$

$$\sigma_{LONG} = \frac{1178}{.157} = 7500\ \text{PSI (TENSILE)}$$

$$\sigma_{hoop} = \frac{Pr}{t} = \frac{(500)(\frac{1}{2})}{.050} = -5000\ \text{psi (COMP)}$$

THESE ARE THE PRINCIPAL STRESSES. FROM DISTORTION ENERGY THEORY, THE VON MISES STRESS IS

$$\sigma' = \sqrt{\sigma_1^2 - \sigma_1\sigma_2 + \sigma_2^2}$$

$$= \sqrt{(7500)^2 - (7500)(-5000) + (-5000)^2} = 10897\ \text{PSI}$$

$$FS = \frac{19000}{10897} = 1.74$$

ANSWER WILL DEPEND ON THE VALUE OF S_y USED.

5

THE TOP IS BALANCED AGAINST TRANSVERSE MOTION. ONLY BUCKLING NEEDS TO BE CONSIDERED.

DUE TO SYMMETRY THE TWO SIDE GUY WIRES CAN BE COMBINED,

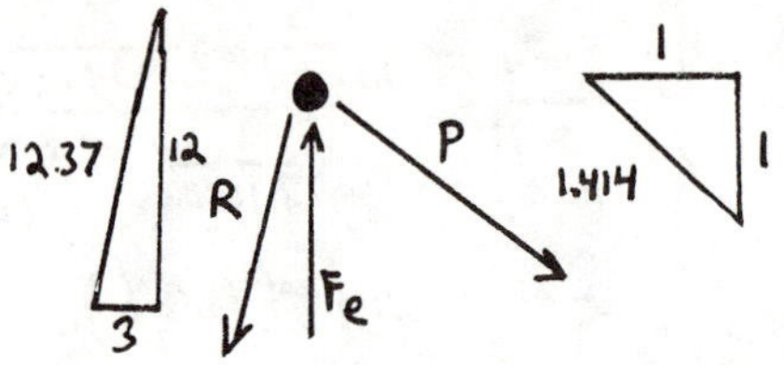

THE HORIZONTAL COMPONENTS MUST BALANCE.

HORIZONTAL FORCES

$$P_x = \left(\frac{1}{1.414}\right)P = .707P$$

$$R_x = \left(\frac{3}{12.37}\right)R = .243R$$

$$.707P = .243R$$

$$R = 2.91P$$

THE IMPOSED VERTICAL LOADS ARE

$$P_y + R_y = \left(\frac{1}{1.414}\right)P + \left(\frac{12}{12.37}\right)R$$

$$= \left(\frac{1}{1.414}\right)P + \left(\frac{12}{12.37}\right)P$$

$$= 3.53P$$

NOW, GET PROPERTIES OF SCHEDULE 40, $1\frac{1}{2}$" PIPE

$E = 2.9$ EE7

$D_o = 1.900$

$D_i = 1.610$

$$I = \frac{\pi}{4}\left[\left(\frac{1.900}{2}\right)^4 - \left(\frac{1.610}{2}\right)^4\right] = .31 \text{ IN}^4$$

THE BASE APPEARS TO BE FIXED. THE TOP IS TRANSLATION FIXED BUT ROTATION FREE. FROM PAGE 14.17, C = .707

$$F_e = \frac{\pi^2 EI}{(CL)^2}$$

$$= \frac{(\pi^2)(2.9\text{EE}7)(.31)}{[(.707)(12)(12)]^2} = 8558 \text{ LBF}$$

SO, EQUATING THE VERTICAL COMPONENTS WITH F_e,

$$8558 = 3.53P \quad \text{OR} \quad P = 2424 \text{ LBF}$$

$$P_{allowable} = \frac{2424}{FS} = \frac{2424}{2} = 1212$$

IMPORTANT CHECKS

- EULER'S EQUATION CAN ONLY BE USED IF THE STRESS IS LESS THAN THE YIELD STRENGTH.

$$A_{ring} = \frac{\pi}{4}\left[(1.900)^2 - (1.610)^2\right] = .800$$

SO, $F_{yield} = (30{,}000)(.8) = 24000$ (OKAY)

- THIS IS OBVIOUSLY A SLENDER COLUMN. BUT, YOU SHOULD CHECK L/k TO BE SURE.

6 PUT ALL LENGTHS IN TERMS OF RADIUS IN METERS

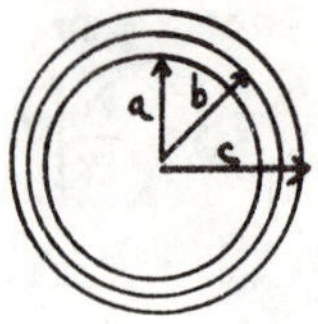

$$a = \left[\frac{120-2-2-3-3}{2}\right]10^{-3}$$

$= 5.5$ EE-2 M

$$b = \left[\frac{120-2-2}{2}\right]\text{EE-3}$$

$= 5.8$ EE-2 M

$$c = \left[\frac{120}{2}\right]\text{EE-3} = 6.0 \text{ EE-2 M}$$

$I = .3$ EE-3 $= 3$ EE-4 M

IN GENERAL, THE INTERFERENCE (I) IS

$$I = |\Delta D_{INNER}| + |\Delta D_{OUTER}|$$

ΔD IS CALCULATED THE SAME FOR BOTH. THERE IS NO LONG STRESS, SO

$$\Delta D = \frac{D}{E}\left[\sigma_c - \mu\sigma_r\right]$$

INNER CYLINDER

- EXTERNAL PRESSURE
- $r_i = a = 5.5$ EE-2
- $r_o = b = 5.8$ EE-2
- $t = b - a = .3$ EE-2
- $D = 2b = .116$ m

$$\sigma_{c,o} = \frac{-(r_o^2 + r_i^2)P}{(r_o + r_i)t} = \frac{-[(5.8)^2 + (5.5)^2](\text{EE-2})^2 P}{(5.8+5.5)(.3)(\text{EE-2})^2}$$

$= -18.85P$

$\sigma_{r,o} = -P$

$$\Delta D = \frac{.116}{207 \text{ EE}9}\left[-18.85P - (.3)(-P)\right]$$

$$= -(1.04 \text{ EE-11})P$$

OUTER CYLINDER

- INTERNAL PRESSURE
- $r_i = b = 5.8$ EE-2 M
- $r_o = c = 6.0$ EE-2 M
- $t = .2$ EE-2 M
- $D = .120$ M

$\sigma_{r,i} = -P$

$$\sigma_{c,i} = \frac{(r_o^2 + r_i^2)P}{(r_o + r_i)t} = \frac{[(6)^2 + (5.8)^2]P}{(6.0+5.8)(.2)}$$

$= 29.51P$

$$\Delta D = \frac{.116}{207 EE9}\left[29.51P - (.3)(-P)\right]$$

$$= +(1.64 EE\text{-}11)\,P$$

SINCE I IS KNOWN TO BE 3 EE-4 M,

$$3 EE\text{-}4 = P(1.04 EE\text{-}11 + 1.64 EE\text{-}11)$$

$$P = 1.12\ EE7\ N/M^2\ (\text{PASCALS})$$

NOW THAT P IS KNOWN, THE CIRCUMFERENTIAL STRESSES CAN BE FOUND:

INNER CYLINDER

$$\sigma_{c,o} = 18.85P = 2.11\ EE8\ N/M^2$$

OUTER CYLINDER

$$\sigma_{c,i} = 29.51P = 3.31\ EE8\ N/M^2$$

RESERVED FOR FUTURE USE

Machine Design

WARM-UPS

1

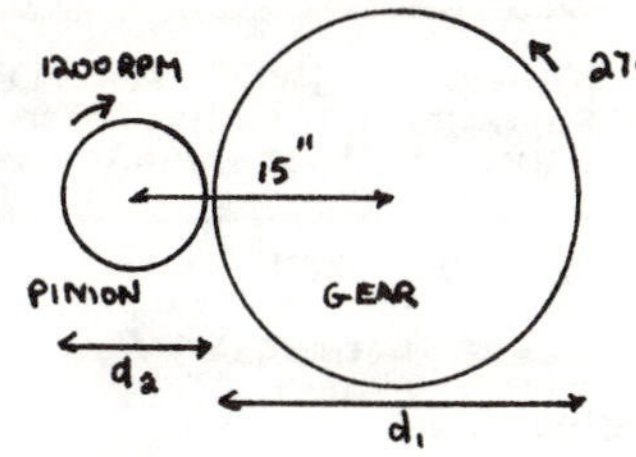

FIRST, FIND THE PITCH DIAMETERS. LET GEAR 2 BE THE PINION, AND GEAR 2 BE THE GEAR. THEN

$$\frac{d_1}{2} + \frac{d_2}{2} = 15$$

$$\frac{d_1}{d_2} = \frac{1200}{270}$$

SOLVING THESE SIMULTANEOUSLY,

$d_1 = 24.49$ AND $d_2 = 5.51$

FROM EQUATION 15.75, THE PITCH LINE VELOCITY IS

$$v = \frac{\pi(\text{RPM})d}{12} = \frac{\pi(270)(24.49)}{12} = 1731.1 \text{ FPM}$$

THIS VELOCITY IS THE SAME FOR BOTH GEARS.

TO GET STARTED, ASSUME FOR THE PINION:
$W = 1.5"$ AND $Y = .3$

THE ENDURANCE LIMIT IS CHOSEN AS THE MAXIMUM STRESS SINCE NO HARDNESS DATA WAS GIVEN. ASSUME FOR THE 1045 STEEL

$$S_{ENDURANCE} = 90 \text{ KSI}$$

USE A FACTOR OF SAFETY OF 3. SO, THE ALLOWABLE STRESS IS

$$\sigma_{ALLOWABLE} = \frac{90}{3} = 30 \text{ KSI}$$

THE BARTH SPEED FACTOR (EQUATION 15.103) IS

$$K_d = \frac{600}{600 + 1731.1} = .257$$

NEGLECTING THE SMALL AMOUNT OF RADIAL STRESS, FROM EQUATIONS 15.98 AND 15.104

$$\frac{(HP)(33{,}000)}{v} = \frac{\sigma W Y}{P K_d}$$

OR

$$P = \frac{(30{,}000)(1.5)(.30)(1731.1)}{(.257)(550)(33{,}000)} = 5.01$$

SO, CHOOSE THE NUMBER OF TEETH PER INCH AS 5. THEN, THE NUMBER OF TEETH ON THE PINION AND GEAR ARE

$$N_{PINION} = (5)(5.51) = 27.55 \text{ (SAY 28)}$$

$$N_{GEAR} = (5)(24.49) = 122.45 \text{ (SAY 122)}$$

NOW THAT THE APPROXIMATE NUMBERS OF TEETH ARE KNOWN, TABLE 15.6 GIVES THE ACTUAL FORM FACTOR.

$Y_{PINION} = .314$ $\quad Y_{GEAR} = .371$

EQUATION 15.98 {AND 15.104} MAY BE USED TO FIND W BASED ON THE NEW INFORMATION.

$$W = \frac{(HP)(33000)(P)(K_d)}{v\sigma Y}$$

$$W_{PINION} = \frac{(550)(33000)(5)(.257)}{(1731.1)(30{,}000)(.314)} = 1.43"$$

ASSUME $\sigma_{MAX} = 50{,}000$ PSI FOR THE CAST STEEL GEAR

$$W_{GEAR} = \frac{(550)(33000)(5)(.257)}{(1731.1)\left(\frac{50{,}000}{3}\right)(.371)} = 2.18$$

THIS IGNORES STRESS CONCENTRATION FACTORS. BOTH GEARS WOULD BE 2.18" OR LARGER.

2 THE MAXIMUM SPEED RATIO IS $\frac{96}{12} = 8$.

WITH 3 STAGES: $(8)^3 = 512$ {TOO SMALL}

WITH 4 STAGES $(8)^4 = 4096$

SO, 4 SETS OF GEARS ARE NEEDED. TRY TO KEEP ALL GEAR SETS AS CLOSE AS POSSIBLE

$$\sqrt[4]{600} \approx 5$$

$$\left(\frac{60}{12}\right)\left(\frac{60}{12}\right)\left(\frac{60}{12}\right)\left(\frac{72}{15}\right) = 600$$

OTHER COMBINATIONS ARE POSSIBLE

3 THIS PROBLEM CANNOT BE SOLVED WITH THE MECHANICAL ENGINEERING REVIEW MANUAL. REFER TO PAGES 8-51 THROUGH 8-54, MARKS' STANDARD HANDBOOK FOR MECHANICAL ENGINEERS, 8TH ED.

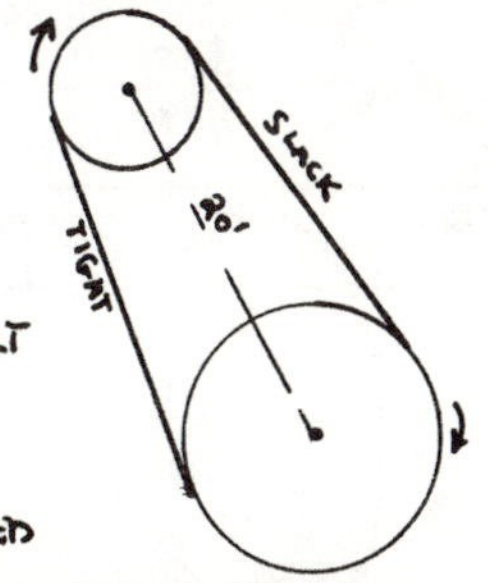

BELT SPEED =

$$\pi\left(\frac{8}{12}\right)\text{FT}(2000)\text{ RPM}$$

$= 4188.8$ FPM

FROM TABLE 49 (MARKS) THE CAPACITY OF A 20/64" BELT IS

11.3 HP/INCH WIDTH

THE FOLLOWING CORRECTION FACTORS ARE ALSO NEEDED

$M = 2.5$ {MOTOR LOAD CORRECTION FOR D.C. MOTOR WITH SLIP RINGS}
$P = .6$ {PULLEY CORRECTION FOR FIBER PULLEY}
$F = 1.2$ {VERTICAL DRIVE CORRECTION}

THEN

$$W = \frac{RMF}{KP} = \frac{(25)\text{ HP}(2.5)(1.2)}{(11.3)\frac{\text{HP}}{\text{IN}}(.6)} = 11.06"$$

SAY 11"

4

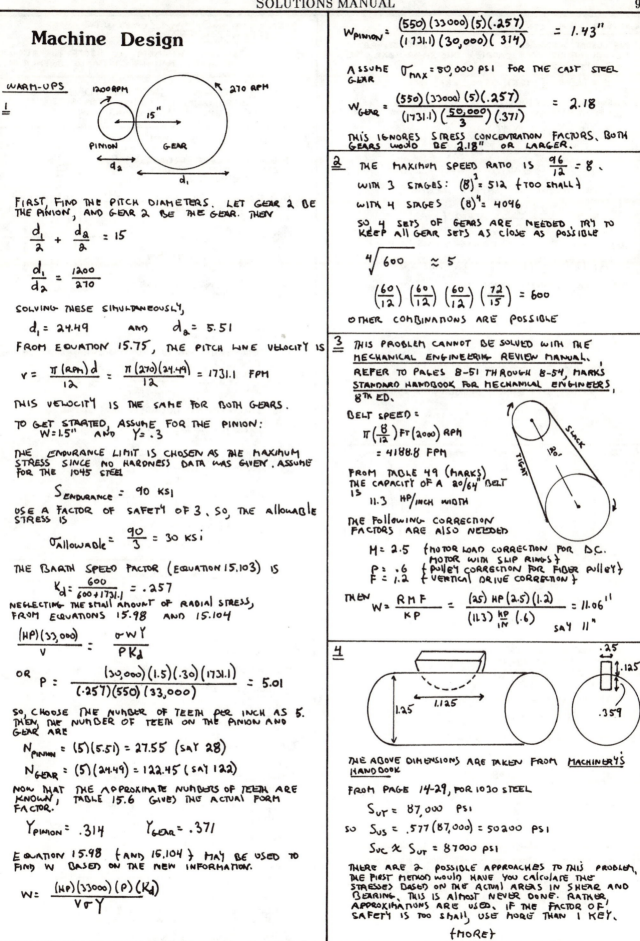

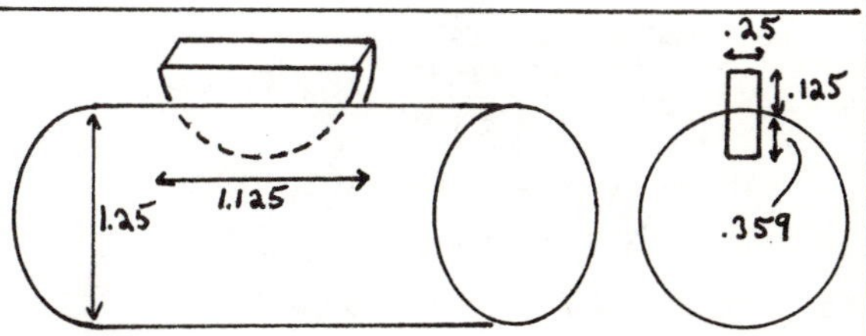

THE ABOVE DIMENSIONS ARE TAKEN FROM MACHINERY'S HANDBOOK

FROM PAGE 14-29, FOR 1030 STEEL

$$S_{UT} = 87{,}000 \text{ PSI}$$

SO $S_{US} = .577(87{,}000) = 50200$ PSI

$$S_{UC} \approx S_{UT} = 87000 \text{ PSI}$$

THERE ARE 2 POSSIBLE APPROACHES TO THIS PROBLEM. THE FIRST METHOD WOULD HAVE YOU CALCULATE THE STRESSES BASED ON THE ACTUAL AREAS IN SHEAR AND BEARING. THIS IS ALMOST NEVER DONE. RATHER, APPROXIMATIONS ARE USED. IF THE FACTOR OF SAFETY IS TOO SMALL, USE MORE THAN 1 KEY.

{MORE}

WARM-UP #4 CONTINUED

THE USUAL APPROACH IS TO ASSUME THE ENTIRE LENGTH OF THE KEY CARRIES THE SHEAR LOAD, THE AREA IN SHEAR IS

$$A = (L)(W) = (1.125)(.25) = .2813 \text{ IN}^2$$

ASSUMING THE TORQUE IS CARRIED AT THE SHAFT SURFACE, THE MAXIMUM TORQUE WOULD BE

$$\tau = F \cdot r = (S_{us})A(r)$$
$$= (50{,}200)(.2813)\left(\frac{1.25}{2}\right)$$
$$= 8825.8 \text{ IN-LBF}$$

SO, THE FACTOR OF SAFETY IN SHEAR

$$FS_s = \frac{8825.8}{4200} = 2.1$$

THE SMALLEST AREA IN BEARING OCCURS IN THE HUB AND IS

$$A = (L)(d) = (1.125)(.125) = .1406 \text{ IN}^2$$

THE ACTUAL APPLIED FORCE IS

$$F = \frac{\tau}{r} = \frac{4200 \text{ IN-LBF}}{\left(\frac{1.25}{2}\right) \text{IN}} = 6720 \text{ LBF}$$

THE BEARING STRESS IS

$$\sigma = \frac{F}{A} = \frac{6720}{.1406} = 47795 \text{ PSI}$$

THE FACTOR OF SAFETY IN BEARING IS

$$FS_b = \frac{87000}{47795} = 1.8$$

5

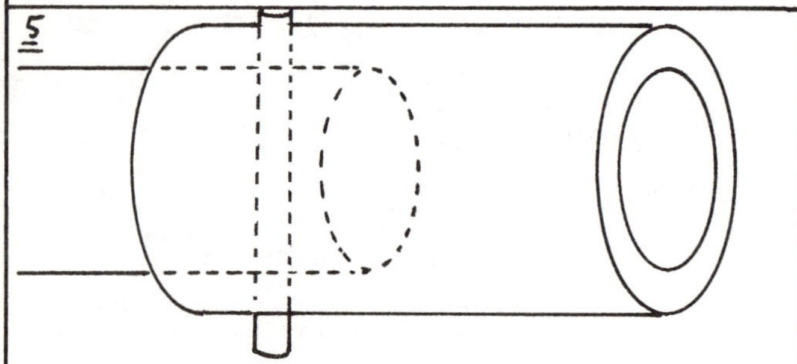

THE AREA OF THE PIN IN SHEAR IS

$$(2)\text{SURFACES}\left(\frac{\pi}{4}\right)d^2 = (1.571\, d^2) \text{ IN}^2$$

THE TOTAL SHEAR IS

$$V = \frac{\tau}{r} = \frac{(400) \text{ IN-LBF}}{\left(\frac{1.125}{2}\right) \text{IN}} = 711.1 \text{ LBF}$$

FROM PAGE 14-29 FOR 1030 STEEL

$$S_{ut} = 87{,}000 \text{ psi}$$

SO $S_{us} = .577(87{,}000) = 50{,}200$ psi

USE A FACTOR OF SAFETY OF 2.5

$$\sigma_{MAX} = \frac{50{,}200}{2.5} = 20080 \text{ psi}$$

SO $20{,}080 = \dfrac{711.1}{1.571 d^2}$

$$d = .149''$$

6 ANNEALING: HEAT TO ABOVE THE RECRYSTALLIZATION TEMPERATURE (600°F) AND AIR COOL TO ROOM TEMPERATURE. THE ALCOA ALUMINUM HANDBOOK RECOMMENDS 775°F HEAT FOR 2-3 HOURS, COOLING AT 50°F/HR DOWN TO 500°F WITH SUBSEQUENT COOLING RATE NOT IMPORTANT.

PRECIPITATION HARDENING (BASED ON THE ALCOA ALUMINUM HANDBOOK)

1. SOLUTION TREATMENT AT 950°F FOR ABOUT 4 HOURS FOLLOWED BY RAPID QUENCH IN COLD WATER
2. REHEAT TO 320°F FOR 12-16 HOURS, FOLLOWED BY COOLING AT A RATE WHICH IS "NOT UNDULY SLOW"

7 ASSUME $E = 2.9$ EE7 PSI

NEGLECT PLATE COMPRESSION
NEGLECT NUT AND HEAD DEFORMATION

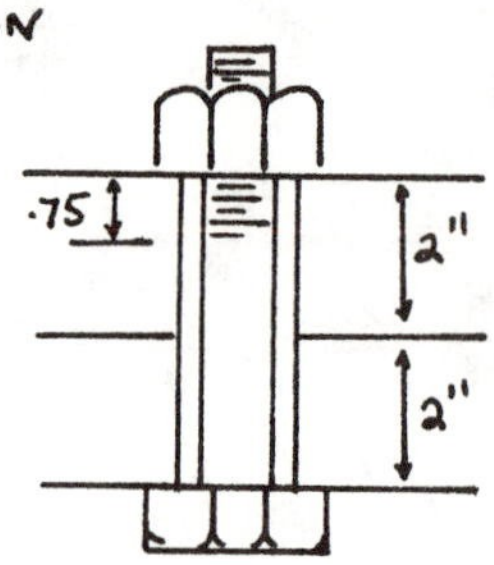

THE BODY OF THE BOLT HAS A DIAMETER AND AREA OF

$$D = .75''$$
$$A = \frac{\pi}{4}(.75)^2 = .4418 \text{ IN}^2$$

THE STRESS AREA IN THE THREAD AREA IS FROM PAGE 14-34

$$A = .3724 \text{ IN}^2$$

THE ELONGATION IN THE UNTHREADED PART OF THE BOLT IS

$$\delta = \frac{FL}{AE} = \sigma\frac{L}{E} = \frac{(40000)\text{ PSI}(3.25)\text{ IN}}{2.9\text{EE7 PSI}} = .00448''$$

THE STRESS IN THE THREADED PART OF THE BOLT IS

$$40{,}000\left(\frac{.4418}{.3724}\right) = 47454 \text{ psi}$$

THE ELONGATION IN THE THREADED PART, INCLUDING 3 THREADS IN THE NUT, IS

$$\delta = \sigma\frac{L}{E} = \frac{(47454)\left[.75 + 3\left(\frac{1}{10}\right)\right]}{2.9\text{EE7}} = .00153$$

THE TOTAL ELONGATION IS

$$\delta_t = .00448 + .00153 = .00601$$

8 $F_{AVE} = \frac{1}{2}(1000 + 8000) = 4500$ LBF

$F_{ALT} = \frac{1}{2}(8000 - 1000) = 3500$

THE THREADED SECTION HAS A MINIMUM AREA OF

$$A_{MIN} = .2018 \text{ IN}^2$$

THE STRESS CONCENTRATION FACTOR FOR ROLLED THREADS IS 2.2 (SHIGLEY, MECHANICAL ENGINEERING DESIGN, 3RD EDITION, PAGE 252)

SO
$$\sigma_{AVE} = \frac{4500}{.2018} = 22300 \text{ psi}$$

$$\sigma_{ALT} = \frac{3500(2.2)}{.2018} = 38160 \text{ PSI}$$

SINCE S_{ut} IS NOT KNOWN, THE SODERBERG DIAGRAM MUST BE USED.

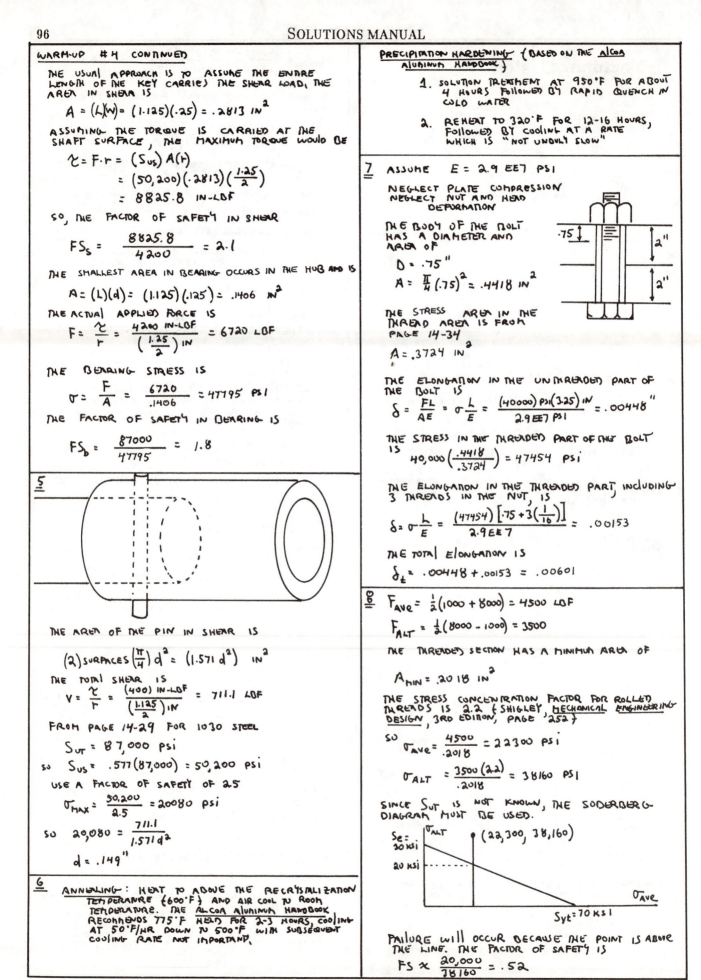

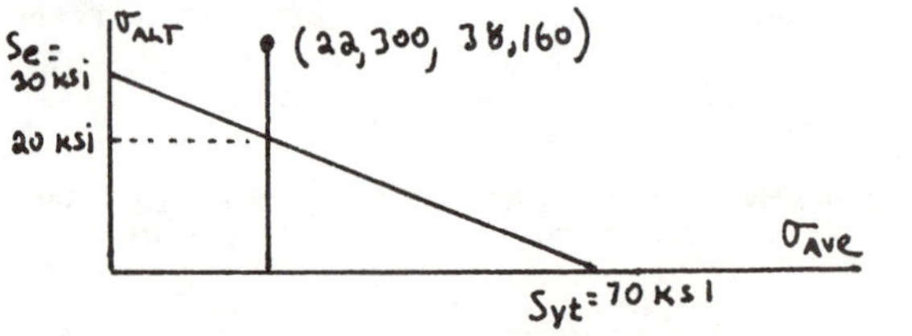

FAILURE WILL OCCUR BECAUSE THE POINT IS ABOVE THE LINE. THE FACTOR OF SAFETY IS

$$FS \approx \frac{20{,}000}{38160} = .52$$

9 FROM PAGE 14-27 {CASE 1}

$$y = \frac{FL^3}{3EI}$$

$$I = \frac{bh^3}{12} = \frac{6(h)^3}{12} = .5h^3$$

$$L = 2(12) = 24 \text{ IN}$$

$$2'' = \frac{800(24)^3}{(3)(29\text{EE}7)(.5)h^3}$$

OR $h = .503''$

A COMPLETE SOLUTION WOULD ALSO CHECK BENDING STRESS.

10

THE TENSION IN THE CABLE DUE TO THE CAR IS

$$\frac{(20{,}000)\text{ LBM }(\sin 30^\circ)}{(2000)\ \frac{\text{LBM}}{\text{TON}}} = 5 \text{ TONS}$$

LET THE CABLE WEIGHT BE W LBM/FT. ASSUME THE COEFFICIENT OF FRICTION IS .3 BETWEEN THE ROPE AND DRAG SURFACE.

THE CABLE WEIGHT IS

$$(2400)\text{ FT }(w)\ \text{LBM/FT} = 2400\,w$$

THE NORMAL FORCE IS

$$(2400\,w)(\cos 30^\circ) = 2078.5\,w$$

THE FRICTIONAL FORCE IS

$$\frac{(2078.5)\,w\,(.3)}{2000\ (\text{LBM/TON})} = .312w \text{ TONS}$$

ADDITIONAL TENSION IS CREATED BY THE ACCELERATION OF THE CART. THE MASS BEING ACCELERATED IS

$$m = \frac{20000 + 2400(w)}{(32.2)(2000)} = (.311 + .0373w)\ \frac{\text{SLUG-TON}}{\text{LBM}}$$

THE VELOCITY PARALLEL TO THE SHAFT IS

$$v = \frac{(100)\ \text{FT/MIN}}{(\sin 30^\circ)(60)\ \text{SEC/MIN}} = 3.33 \text{ FT/SEC}$$

$$a = \frac{3.33 \text{ FT/SEC}}{10} = .333 \text{ FT/SEC}^2$$

$$F = ma = (.333)(.311 + .0373w) = .104 + .0124w \text{ TON}$$

SINCE THE DRUM SIZE IS NOT GIVEN, ALLOW 2 TONS FOR BENDING LOAD. THE TOTAL TENSION IS

$$5 + .312w + .104 + .0124w + 2$$

$$= 7.104 + .324w$$

ASSUME A SAFETY FACTOR OF 5, SO THE BREAKING STRENGTH IS

$$5(7.104 + .324w) = 35.52 + 1.62w$$

BY TRIAL AND ERROR FROM TABLE 14.5 USE

1" DIA, 6×19 IMPROVED PLOW STEEL

11 ΔKINETIC ENERGY $= \frac{1}{2}m(v_1^2 - v_2^2) = 1500$ FT-LBF

$$v_1 = \frac{(200)\frac{\text{REV}}{\text{MIN}}(\pi)(30)\frac{\text{IN}}{\text{REV}}}{(60)\frac{\text{SEC}}{\text{MIN}}(12)\frac{\text{IN}}{\text{FT}}} = 26.18 \frac{\text{FT}}{\text{SEC}}$$

$$v_2 = \frac{(175)(\pi)(30)}{(60)(12)} = 22.91 \frac{\text{FT}}{\text{SEC}}$$

$$m = \frac{(1.1)(\rho)(\text{volume})}{g} = \frac{(1.1)(.26)\frac{\text{LBM}}{\text{IN}^3}(\text{volume})\text{ IN}^3}{32.2 \frac{\text{FT}}{\text{SEC}^2}}$$

$$= 8.88\text{EE-}3\,(\text{volume in IN}^3)$$

$$\text{volume} = (\pi)(30)\text{ IN }(12)\text{ IN }(t)\text{ IN} = 1131\,t \text{ IN}^3$$

$$1500 = (.5)(8.88\text{EE-}3)(1131\,t)\frac{\text{LBM-SEC}^2}{\text{FT}}\left[(26.18)^2 - (22.91)^2\right]$$

$$\boxed{t = 1.86 \text{ IN}}$$

12 ASSUME A COEFFICIENT OF FRICTION OF $\mu = .12$

WITH N FRICTION PLANES, THE CONTACT SURFACE AREA IS

$$A = N\frac{\pi}{4}\left((4.5)^2 - (2.5)^2\right)\text{ IN}^2 = 11N \text{ IN}^2$$

WITH A CONTACT PRESSURE OF P = 100 PSI, THE NORMAL FORCE IS

$$F_N = AP = (11N)\text{ IN}^2\,(100)\frac{\text{LBF}}{\text{IN}^2} = 1100N \text{ LBF}$$

THE FRICTIONAL FORCE IS

$$F_f = \mu F_N = (.12)(1100N)\text{ LBF} = 132N \text{ LBF}$$

ASSUME THE FRICTIONAL FORCE IS APPLIED AT THE MEAN RADIUS

$$r_m = \frac{1}{2}\left(\frac{2.5 + 4.5}{2}\right) = 1.75 \text{ IN}$$

THE RESISTING TORQUE IS

$$T_R = r_m F_f = (1.75)\text{ IN }(132N)\text{ LBF} = 231N$$

SETTING THE SLIPPING TORQUE EQUAL TO THE RESISTING TORQUE,

$$3(300)\text{ IN-LBF} = (231N)\text{ IN-LBF}$$

$$N = 3.9$$

USE 4 CONTACT SURFACES - 3 PLATES AND 2 DISCS

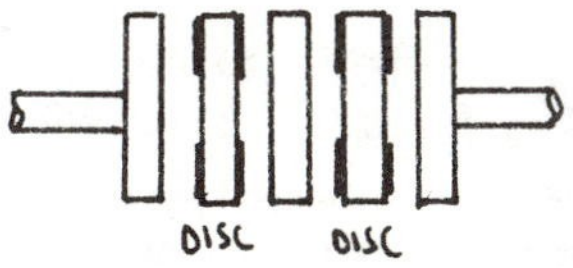

CONCENTRATES

1 DESIGN FOR STATIC LOADING. CHOOSE A SPRING INDEX OF 9 {FROM PAGE 15-11}. FROM EQUATION 15.58

$$d = \frac{8FC^3N_a}{G\delta} = \frac{(8)(50)(9)^3(12)}{(11.5\text{ EE6})(.5)} = .609$$

SO $D = 9d = 9(.609) = 5.481$

2

ASSUME ASTM A230 WIRE. TO GET STARTED, ASSUME $d = .15''$ WIRE.

FROM TABLE 15.4

$$S_{ut} = 205{,}000 \text{ psi}$$

BASED ON EQUATION 15.74 THE ALLOWABLE WORKING STRESS IS

$$\tau = (.3)(S_{ut}) = (.3)(205{,}000) = 61500 \text{ psi}$$

FROM EQUATION 15.62

$$W = \frac{4(10)-1}{4(10)-4} + \frac{.615}{10} = 1.145$$

FROM EQUATION 15.61

$$d = \sqrt{\frac{8CFW}{\pi\tau}} = \sqrt{\frac{8(10)(30)(1.145)}{\pi(61500)}} = .119$$

FROM TABLE 15.2, USE W+M WIRE #11 WITH $d = .1205''$

$$D = 10d = 1.205''$$

FROM EQUATION 15.60, THE SPRING CONSTANT IS

$$K = \frac{\Delta F}{\delta} = \frac{30-20}{.3} = 33.33 \text{ LBF/IN}$$

FROM EQN 15.60

$$N_a = \frac{Gd}{8KC^3} = \frac{(11.5\text{ EE}6)(.1205)}{(8)(33.33)(10)^3} = 5.2$$

SPECIFY A SQUARED AND GROUND SPRING. USE 7.2 COILS WITH 5.2 ACTIVE.

THE SOLID HEIGHT IS

$$7.2(.1205) = .8676''$$

AT SOLID THE MAXIMUM STRESS SHOULD NOT EXCEED S'_{us}. FROM TABLE 15.4

$$S_{ut} \approx 208{,}000 \text{ psi}$$

$$S_{us} = .577(S_{ut}) = 120{,}000 \text{ psi}$$

FROM EQN 15.61, THE FORCE AT SOLID IS

$$F = \frac{\tau\pi d^3}{8DW} = \frac{(120{,}000)(\pi)(.1205)^3}{(8)(1.205)(1.145)} = 59.76$$

SO, THE DEFLECTION AT SOLID IS

$$\frac{F}{K} = \frac{59.76}{33.33} = 1.793''$$

THE MAXIMUM FREE HEIGHT IS

$$.8676 + 1.793 = 2.661 \text{ IN}$$

3 THE POTENTIAL ENERGY ABSORBED IS

$$E_p = mg\Delta h = \frac{(700)(32.2)(10+46)}{(32.2)} = 39200 \text{ IN-LBF}$$

THE WORK DONE BY THE SPRING IS

$$W = \tfrac{1}{2}KX^2$$

$$39200 = \tfrac{1}{2}(K)(10)^2$$

$$K = 784 \text{ LBF/IN}$$

THE EQUIVALENT FORCE IS

$$F = KX = (784)(10) = 7840 \text{ LBF}$$

FROM EQUATION 15.62

$$W = \frac{(4)(7)-1}{(4)(7)-4} + \frac{.615}{7} = 1.213$$

FROM EQUATION 15.61

$$d = \sqrt{\frac{8CFW}{\pi\tau}} = \sqrt{\frac{8(7)(7840)(1.213)}{(\pi)(90{,}000)}} = 1.841$$

$$D = 7d = 12.89$$

FROM EQUATION 15.60

$$N_a = \frac{Gd}{8KC^3} = \frac{(1.2\text{ EE}7)(1.841)}{(8)(784)(7)^3} = 10.27$$

{SAY 10.3}

4 FOR THE STEEL

$$E_s = 2.9\text{ EE}7 \text{ psi}$$
$$\mu_s = .3$$

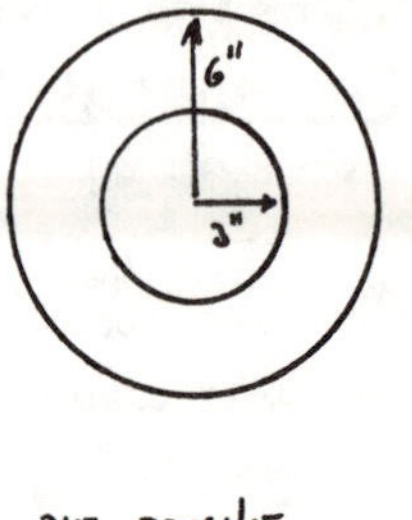

FOR THE CAST IRON

$$E_c = 1.45\text{ EE}7 \text{ psi}$$
$$\mu_c = .27$$
$$S_{ut} = 30{,}000 \text{ psi}$$

ASSUME A SAFETY FACTOR OF 3. THE TENSILE STRENGTH OF THE CAST IRON GOVERNS, SO KEEP

$$\sigma < \frac{30{,}000}{3} = 10{,}000 \text{ psi}$$

FROM PAGE 14-22 FOR THE CAST IRON HUB UNDER INTERNAL PRESSURE,

$$\sigma_{max} = \sigma_{ci} = 10{,}000 \text{ psi}$$

NOW,

$$\sigma_{ci} = 10{,}000 = \frac{[(6)^2+(3)^2]\,p}{[(6)^2-(3)^2]}$$

OR $p = 6000$ psi

$$\sigma_{ri} = -p = -6000 \text{ AT INNER SURFACE}$$

FROM EQUATION 14.88, THE RADIAL INTERFERENCE IN THE CAST IRON IS

$$\Delta r = \frac{r}{E}\left[\sigma_c - \mu(\sigma_r + \sigma_L)\right]$$

$$= \frac{3}{1.45\text{ EE}7}\left[10{,}000 - .27(-6000)\right] = 2.40\text{ EE-}3 \text{ IN}$$

FROM TABLE 14.6 FOR A CYLINDER UNDER EXTERNAL PRESSURE,

σ_{co} IS MAX

$$\sigma_{co} = 6000\frac{-[(3)^2+(0)^2]}{[(3)^2-(0)^2]} = -6000$$

$$\sigma_{ro} = -p = -6000$$

FROM EQUATION 14.88 THE RADIAL INTERFERENCE IN THE STEEL IS

$$\Delta r = \frac{3}{3\text{ EE}7}\left[-6000 - (.3)(-6000)\right] = -4.2\text{ EE-}4$$

THE TOTAL RADIAL INTERFERENCE IS

$$2.40\text{ EE-}3 + 4.2\text{ EE-}4 = 2.82\text{ EE-}3 \text{ IN}$$

5 PROCEED AS IN WARM-UP #1. FIRST FIND THE PITCH DIAMETERS

$$\frac{d_1}{2} + \frac{d_2}{2} = 15$$

$$\frac{d_1}{d_2} = \frac{250}{83.33}$$

SOLVING THESE SIMULTANEOUSLY,

$$d_1 = 22.5 \qquad d_2 = 7.5$$

FROM EQUATION 15.75, THE PITCH LINE VELOCITY IS

$$V = \frac{\pi(\text{RPM})d}{12} = \frac{\pi(7.5)(250)}{12} = 490.9 \text{ FPM}$$

FOR THE PINION, ASSUME $Y = .3$, $w = 1.5''$, AND $\sigma_{max} = 30{,}000$ PSI

THE BARTH SPEED FACTOR IS

$$K_d = \frac{600}{600 + 490.9} = .55$$

FROM EQUATIONS 15.98 AND 15.104

$$\frac{(HP)(33000)}{V} = \frac{\sigma w Y}{K_d P}$$

$$P = \frac{(30000)(1.5)(.3)(490.9)}{(40)(33000)(.55)} = 9.12$$

SO CHOOSE $P = 9$. THE NUMBER OF TEETH ARE

$$N_{PINION} = Pd = 9(7.5) = 67.5 \text{ (SAY 68)}$$

$$N_{GEAR} = 9(22.5) = 202.5 \text{ (SAY 203)}$$

FROM TABLE 15.6, ACTUAL VALUES OF Y ARE

$$Y_{PINION} = .358$$

$$Y_{GEAR} = .377$$

FROM EQUATIONS 15.98 AND 15.104, THE WIDTHS ARE

$$w = \frac{(HP)(33000)PK_d}{(\sigma)Y}$$

$$w_{PINION} = \frac{(40)(33000)(9)(.55)}{(490.9)(30{,}000)(.358)} = .41''$$

ASSUME $\sigma_{MAXIMUM} = 50{,}000$ FOR THE CAST STEEL GEAR

$$w_{GEAR} = \frac{(40)(33000)(9)(.55)}{(490.9)\left(\frac{50{,}000}{3}\right)(.377)} = .706''$$

---

6 THE CENTROID OF THE RIVET GROUP IS

$\left(\frac{2}{3}\right)(9) = 6$ FROM EACH RIVET

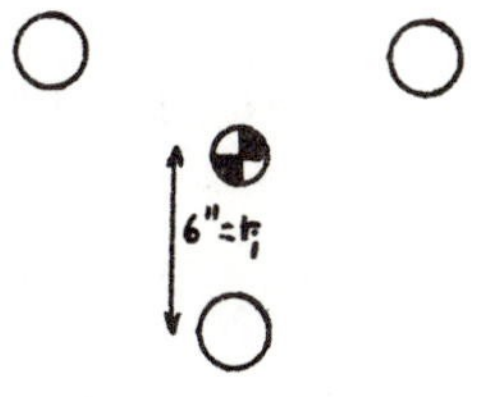

THE VERTICAL SHEAR LOAD CARRIED BY EACH RIVET IS $F_v = P/3$

FOR EACH RIVET, THE RESISTING MOMENT OF INERTIA IS

$$J = A r_i^2$$

$$= (\pi r^2) r_i^2$$

$$= \pi\left(\frac{.75}{2}\right)^2 (6)^2 = 15.9 \text{ IN}^4$$

THE APPLIED MOMENT IS 20P

THE AREA OF EACH RIVET IS

$$A = \frac{\pi}{4} d^2 = \frac{\pi}{4}(.75)^2 = .442$$

SO, THE VERTICAL SHEAR STRESS IS

$$\tau_v = \frac{F_v}{A} = \frac{P}{3(.442)} = .754P$$

THE TWISTING SHEARING STRESS IN THE RIGHT-MOST RIVET (WHICH BY INSPECTION IS THE HIGHEST STRESSED) IS

$$\tau = \frac{Mr}{J} = \frac{\left(\frac{20P}{3}\right)6}{15.94} = 2.509P$$

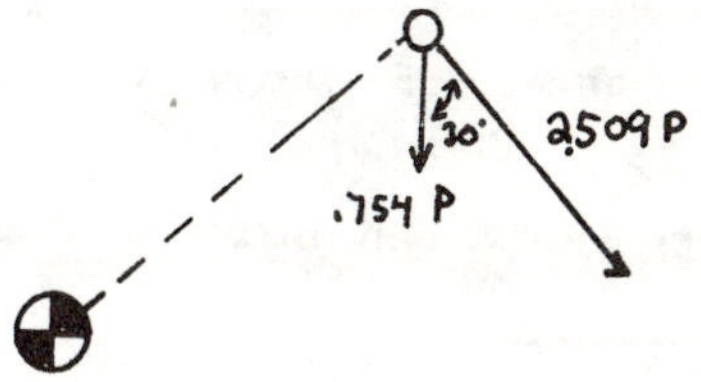

ADDING THESE HEAD-TO-TAIL GIVES A RESULTANT OF 3.185 P.

KEEPING $\tau < 15000$ REQUIRES

$P < 4710$ LBF

---

7 ASSUME THE WELD HAS A THICKNESS w

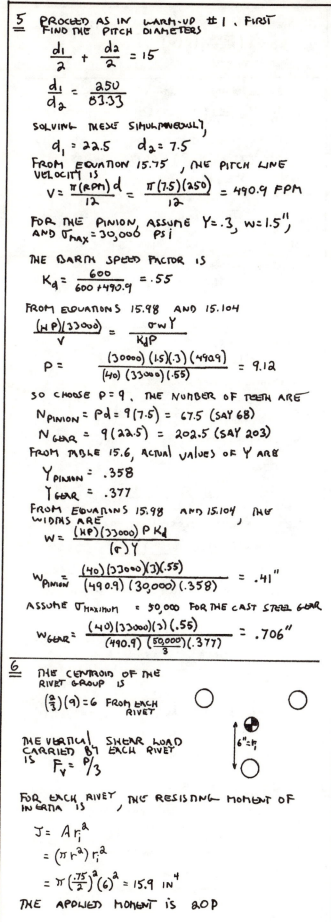

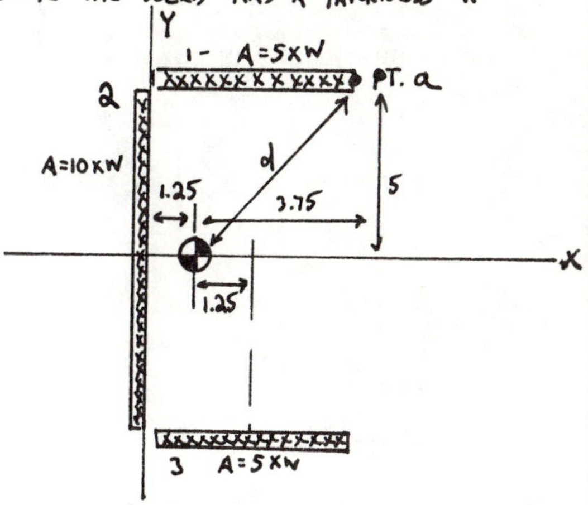

BY INSPECTION, $\bar{y} = 0$

$A_1 = 5w \qquad A_2 = 10w \qquad A_3 = 5w$

$\bar{x}_1 = 2.5 \qquad \bar{x}_2 = 0 \qquad \bar{x}_3 = 2.5$

$$\bar{x}_c = \frac{5w(2.5) + 10w(0) + 5w(2.5)}{5w + 10w + 5w} = 1.25$$

THE CENTROIDAL MOMENT OF INERTIA IN THE X DIRECTION IS

$$I_x = \frac{w(10)^3}{12} + 2\left[\frac{5(w)^3}{12} + (5w)(5)^2\right]$$

$$= 333.33w + .833w^3$$

SINCE W WILL BE SMALL (PROBABLY LESS THAN .5") THE W^3 TERM MAY BE OMITTED. SO

$$I_x = 333.33\,W$$

SIMILARLY THE CENTROIDAL MOMENT OF INERTIA IN THE Y DIRECTION IS

$$I_y = \frac{(10)w^3}{12} + (10w)(1.25)^2 + 2\left[\frac{w(5)^3}{12} + (5w)(1.25)^2\right]$$

$$= .833w^3 + 15.625\,w + 20.833\,w + 15.625\,w$$

$$= .833w^3 + 52.08w$$

$$\approx 52.08W$$

THE POLAR MOMENT OF INERTIA IS

$$J = I_x + I_y = 385.4\,W$$

THE MAXIMUM SHEAR WILL OCCUR AT POINT a SINCE IT IS FARTHEST AWAY.

$$d = \sqrt{(3.75)^2 + (5)^2} = 6.25$$

THE APPLIED MOMENT IS

$$Fx = (10000)(12 + 3.75) = 157{,}500 \text{ IN-LBF}$$

SO THE TORSIONAL SHEAR STRESS IS

$$\tau = \frac{Mc}{J} = \frac{(157{,}500)(6.25)}{385.4} = \frac{2554.2}{w} \text{ psi}$$

THIS SHEAR STRESS IS AT RIGHT ANGLES TO THE LINE d. THE STRESS CAN BE DIVIDED INTO X AND Y COMPONENTS:

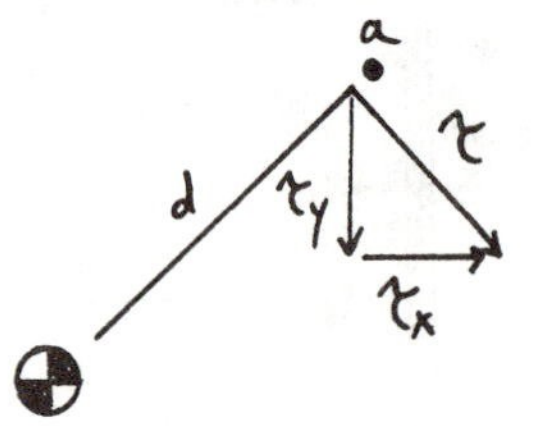

$$\tau_y = \frac{3.75}{6.25}\left(\frac{2554.2}{w}\right) = \frac{1532.5}{w}$$

$$\tau_x = \frac{5}{6.25}(\tau) = \frac{2043.4}{w}$$

IN ADDITION THE VERTICAL LOAD OF 10,000 LBF MUST BE SUPPORTED. THE VERTICAL SHEAR STRESS IS

$$A = 10w + 5w + 5w = 20w$$

$$\tau_y = \frac{10{,}000}{20w} = \frac{500}{w}$$

THE RESULTANT SHEAR STRESS IS

$$\tau = \sqrt{\left(\frac{2043.4}{w}\right)^2 + \left(\frac{1532.5 + 500}{w}\right)^2} = \frac{2882.1}{w}$$

ASSUMING AN ALLOWABLE LOAD OF 8000 PSI FOR AN UNSHIELDED WELD IN SHEAR, THE REQUIRED THROAT THICKNESS IS

$$\frac{2882.1}{w} = 8000$$

OR $W = .360''$ (SAY 3/8")

8 THE ALUMINUM STRENGTH PER INCH OF BOND IS

$$S = (15000)(1)(.020)$$
$$= 300 \text{ LBF}$$

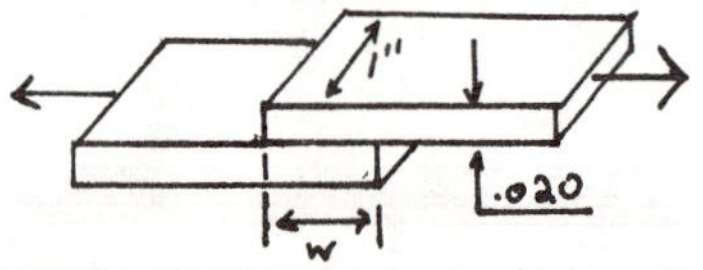

THE WIDTH OF BOND REQUIRED IS

$$300 = \left(\frac{1500}{2}\right)(1)(w)$$

$$w = .4''$$

9 THE CENTROIDAL MOMENT OF INERTIA IS

$$\frac{\pi}{4}r^4 = \frac{\pi}{4}(1)^4 = .7854$$

AT THE STEP, THE MOMENT IS

$$M = (2500)\left(4 - \frac{5}{16}\right) = 9218.8 \text{ IN-LBF}$$

$$\frac{D}{d} = \frac{3}{2} = 1.5$$

$$\frac{r}{d} = \frac{\frac{5}{16}}{2} = .156$$

FROM PAGE 14-33 $K \approx 1.51$

THE BENDING STRESS IS

$$\sigma = K\frac{Mc}{I} = 1.51\left(\frac{(9218.8)(1)}{.7854}\right) = 17{,}724 \text{ psi}$$

10 THE CONTACT AREA IS

$$A = \pi(2)(2) = 12.566 \text{ IN}^2$$

ASSUMING A DENSITY OF $.283\ \text{LBM/IN}^3$ OR $489\ \text{LBM/FT}^3$

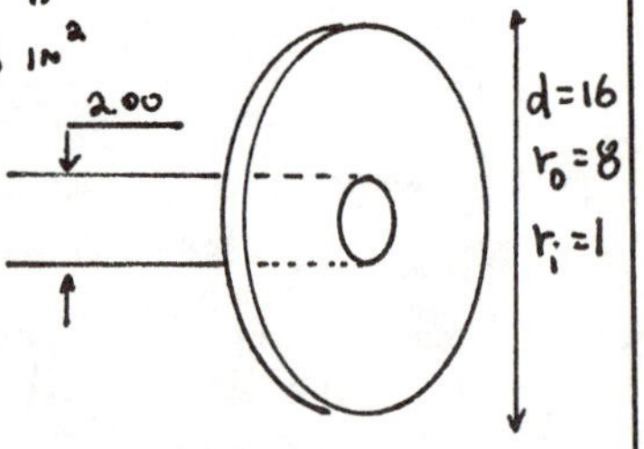

THE FLYWHEEL MASS IS

$$m = \frac{w}{g} = \frac{e(\text{VOL})}{g}$$

$$= \frac{(489)\left(\frac{2}{12}\right)\text{IN}(\pi)\left(\frac{r}{12}\right)^2}{32.2} = .0552\,r^2$$

THE TANGENTIAL VELOCITY AT RADIUS r IS

$$v = \frac{\text{RPM}(2\pi r)}{60} = \frac{(3500)(2\pi)r}{(60)\frac{\text{SEC}}{\text{MIN}}(12)\frac{\text{IN}}{\text{FT}}} = 30.54r \text{ FPS}$$

THE CENTRIFUGAL FORCE FOR A LUMPED MASS IS

$$F_c = \frac{mv^2}{r}$$

INTEGRATING THIS OVER THE DISK'S RADIUS,

$$F_c = \int \frac{mv^2}{r} = \int_{r=1}^{r=8} \frac{(.0552)r^2(30.54)^2 r^2}{r}\,dr$$

$$= \frac{51.485}{4}\Big[r^4\Big]_1^8 = 52707 \text{ LBF}$$

THE PRESSURE DECREASE AT THE CONTACT SURFACE IS

$$\frac{52707}{12.566} = 4194.4$$

SO, THE REQUIRED CONTACT PRESSURE IS

$$p = 4194.4 + 1250 = 5444 \text{ psi}$$

FROM TABLE 14.6, THE MAXIMUM STRESS IS

$$\sigma_{ci} = \frac{[(8)^2 + (1)^2]\,5444}{[(8)^2 - (1)^2]} = 5616.8 \text{ psi}$$

$\sigma_{r_i} = -p = -5444$

FROM EQN 14.88, FOR THE DISK

$$\Delta r = \frac{r}{E}\left[\sigma_c - \mu\sigma_r\right]$$

$$= \frac{1}{2.9\text{EE}7}\left(5616.8 - .3(-5444)\right) = 2.5\text{ EE-4}$$

FOR THE SHAFT,

$$\sigma_{co} = \frac{-\left[(1)^2+(0)^2\right]5444}{\left[(1)^2+(0)^2\right]} = -5444$$

$\sigma_{ro} = -p = -5444$

FROM EQUATION 14.88

$$\Delta r = \frac{1}{2.9\text{EE}7}\left(-5444 - .3(-5444)\right) = -1.3\text{ EE-4}$$

THE TOTAL RADIAL INTERFERENCE IS

2.5 EE-4 + 1.3 EE-4 = 3.8 EE-4 IN

11

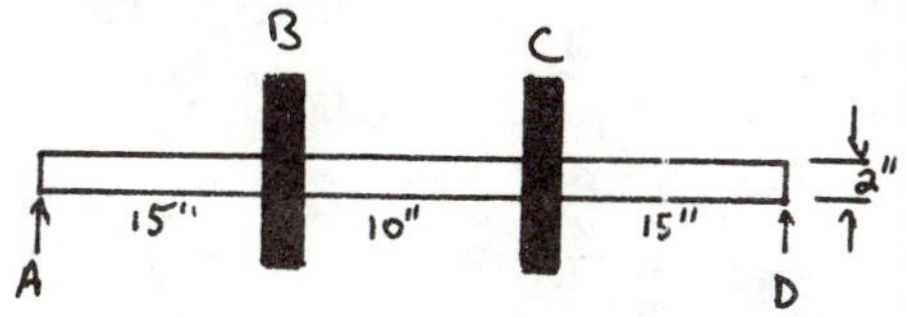

$I_{SHAFT} = \frac{1}{4}\pi r^4 = (.25)(\pi)(1)^4\text{ IN}^4 = .7854\text{ IN}^4$

CALCULATE DEFLECTIONS DUE TO 100# LOAD

$$\delta_{B,100} = \frac{Fa^2b^2}{3EIL} = \frac{(100)(15)^2(25)^2}{(3)(3\text{ EE}7)(.7854)(40)}$$

$= 4.97$ EE-3 IN

$$\delta_{C,100} = \left(\frac{Fbx}{6EIL}\right)\left(L^2-b^2-x^2\right) =$$

$$= \frac{(100)(15)(15)}{(6)(3\text{ EE}7)(.7854)(40)}\left((40)^2-(15)^2-(15)^2\right)$$

$= 4.58$ EE-3 IN

CALCULATE DEFLECTIONS DUE TO 75 # LOAD

$$\delta_{C,75} = \frac{(75)(25)^2(15)^2}{(3)(3\text{ EE}7)(.7854)(40)} = 3.73\text{ EE-3 IN}$$

$$\delta_{B,75} = \frac{(75)(15)(15)}{(6)(3\text{ EE}7)(.7854)(40)}\left((40)^2-(15)^2-(15)^2\right)$$

$= 3.43$ EE-3 IN

THE TOTAL DEFLECTIONS ARE

$\delta_B = 4.97$ EE-3 + 3.43 EE-3 IN = 8.4 EE-3 IN

$\delta_C = 8.31$ EE-3 IN

THE CRITICAL SHAFT SPEED IS

$$f = \frac{1}{2\pi}\sqrt{\frac{g\sum w_i\delta_i}{\sum w_i\delta_i^2}} =$$

$$= \frac{1}{2\pi}\sqrt{\frac{386\left[(100)(8.4\text{ EE-3}) + (75)(8.3\text{ EE-3})\right]}{(100)(8.4\text{ EE-3})^2 + (75)(8.3\text{ EE-3})^2}}$$

= **34.2 HZ = 2052 RPM**

12

$N_S = 24$ $\omega_S = +50$ RPM

$N_P = 40$

$N_R = 104$ $\omega_R = 0$

$$TV = -\frac{N_R}{N_S} = -\frac{104}{24} = -4.333$$

$$\omega_S = (TV)(\omega_R) + \omega_C(1-TV)$$

$$\omega_C = \frac{+50}{(1-(-4.333))} = \boxed{+9.376\text{ RPM CLOCKWISE}}$$

13

$$\omega_R = \omega_A\left(-\frac{N_A}{N_R}\right) = (-100)\left(\frac{-30}{100}\right) = 30\text{ RPM (clockwise)}$$

$$TV = -\frac{N_R}{N_S} = -\frac{80}{40} = -2$$

$$\omega_S = (TV)(\omega_R) + \omega_C(1-TV)$$

$$\omega_S = (-2)(+30) + 60(1-(-2)) = \boxed{+120\text{ RPM (clockwise)}}$$

14

$N_S = 10(5) = 50$ $\omega_S = 0$

$N_P = 10(2.5) = 25$

$N_R = 10(10) = 100$ $\omega_R = +1500$ RPM

$$TV = -\frac{N_R}{N_S} = \frac{-100}{50} = -2$$

$$\omega_S = (TV)(\omega_R) + \omega_C(1-TV)$$

$$0 = (-2)(1500) + \omega_C(1-(-2))$$

$\omega_C = +1000$ RPM

$$T_{INPUT} = \frac{(33000)(HP)}{(RPM)(2\pi)} = \frac{(33000)(15)}{(1500)(2\pi)} = \boxed{52.52\text{ FT-LBF}}$$

$$T_{OUTPUT} = \frac{(33000)(15)}{(1000)(2\pi)} = \boxed{78.78\text{ FT-LBF}}$$

TIMED

1

$\mu = 1.184$ EE-6 REYNS

$N = 20$ RPS

$W = 880$ LBF

$$\frac{c}{r} = \frac{\text{RADIAL CLEARANCE}}{\text{JOURNAL RADIUS}} = \frac{1}{1000}$$

$r = 1.5$

$\ell = 3.5$

THE UNIT LOAD IS

$$p = \frac{W}{2(r)\ell} = \frac{880}{(2)(1.5)(3.5)} = 83.81 \text{ psi}$$

THE BEARING CHARACTERISTIC NUMBER IS

$$S = \left(\frac{r}{c}\right)^2 \frac{\mu N}{p} = (1000)^2 \frac{(1.184\text{EE-}6)(20)}{83.81} = .283$$

$$\frac{\ell}{d} = \frac{\ell}{2r} = \frac{3.5}{3} = 1.17$$

FROM FIGURE 15.45 THE FRICTION VARIABLE IS 6 , SO THE COEFFICIENT OF FRICTION IS

$$f = 6\left(\frac{c}{r}\right) = \frac{6}{1000} = .006$$

THE FRICTION TORQUE IS

$$T = fWr = (.006)(880)(1.5) = 7.92 \text{ IN-LBF}$$

THE FRICTION HORSEPOWER IS

$$HP = \frac{(7.92)\text{ IN-LBF}(20)\frac{1}{\text{sec}}(2\pi)}{(12)\frac{\text{IN}}{\text{FT}}(550)\frac{\text{FT-LBF}}{\text{HP-SEC}}} = .151 \text{ HP}$$

FROM FIGURE 15.46 THE FILM THICKNESS VARIABLE IS .64, SO THE FILM THICKNESS IS

$$h_o = c(.64) = r\left(\frac{1}{1000}\right)(.64) = .00096$$

BECAUSE THIS IS OUT OF THE RECOMMENDED REGION THE LOAD IS PROBABLY TOO MUCH FOR THE BEARING

---

2 BECAUSE OF THEIR HARDNESS, ASSUME

E = 3.0 EE7 FOR THE BOLTS

E = 2.9 EE7 FOR THE VESSEL

THE BOLT STIFFNESS IS

$$K_b = \frac{AE}{L} = \frac{6\left(\frac{\pi}{4}\right)\left(\frac{3}{8}\right)^2 3\text{ EE7}}{\frac{3}{8} + \frac{3}{4}} = 1.767 \text{ EE }7 \ \frac{\text{LBF}}{\text{IN}}$$

THE PLATE/VESSEL CONTACT AREA IS AN ANNULUS WITH INSIDE AND OUTSIDE DIAMETERS OF 8.5" AND 11.5" RESPECTIVELY

$$\text{AREA} = \frac{\pi}{4}\left[(11.5)^2 - (8.5)^2\right] = 47.12 \text{ IN}^2$$

THE VESSEL/PLATE STIFFNESS IS

$$K_m = \frac{AE}{L} = \frac{(47.12)(2.9\text{EE7})}{\frac{3}{8} + \frac{3}{4}} = 1.215 \text{ EE}9$$

LET X BE THE DECIMAL PORTION OF THE PRESSURE LOAD (P) TAKEN BY THE BOLTS. THE INCREASE IN THE BOLT LENGTH IS

$$\delta_b = \frac{xP}{K_b}$$

THE PLATE/VESSEL DEFORMATION IS

$$\delta_m = \frac{(1-x)P}{K_m}$$

BUT $\delta_b = \delta_m$

SO $$\frac{x}{1.767 \text{ EE7}} = \frac{1-x}{1.215 \text{ EE9}} \quad \text{OR } x = .0143$$

THE END PLATE AREA EXPOSED TO PRESSURE IS

$$\frac{\pi}{4}(8.5)^2 = 56.75 \text{ IN}^2$$

THE FORCES ON THE END PLATE ARE

$$F_{max} = pA = 350(56.75) = 19860 \text{ LBF}$$

$$F_{min} = 50(56.75) = 2838 \text{ LBF}$$

THE STRESS AREA IN THE BOLT (PAGE 14-34) IS .0876 IN^2

THE STRESSES IN EACH BOLT ARE

$$\sigma_{max} = \frac{3700 + \frac{(.0143)(19860)}{6}}{.0876} = 42778 \text{ psi}$$

$$\sigma_{min} = \frac{3700 + \frac{(.0143)(2838)}{6}}{.0876} = 42244 \text{ psi}$$

THE MEAN STRESS IS

$$\sigma_{mean} = \tfrac{1}{2}(42778 + 42244) = 42511$$

THE ALTERNATING STRESS IS (USING THE THREAD CONCENTRATION FACTOR OF 2)

$$\sigma_{alt} = \left(\tfrac{1}{2}\right)(2)(42778 - 42244) = 534 \text{ psi}$$

FOR THE BOLT MATERIAL, USE EQN 15.20

$$S'_{END} = .5 S_{ut} = .5(110{,}000) = 55{,}000$$

DERATING FACTORS FROM OTHER SOURCES ARE

$K_a = .55$ (SURFACE FINISH)

$K_b = .85$ (SIZE)

$K_c = .90$ (RELIABILITY)

$$S_e = (.55)(.85)(.90)(55000) = 23141 \text{ psi}$$

PLOTTING THE POINT ON THE MODIFIED GOODMAN DIAGRAM,

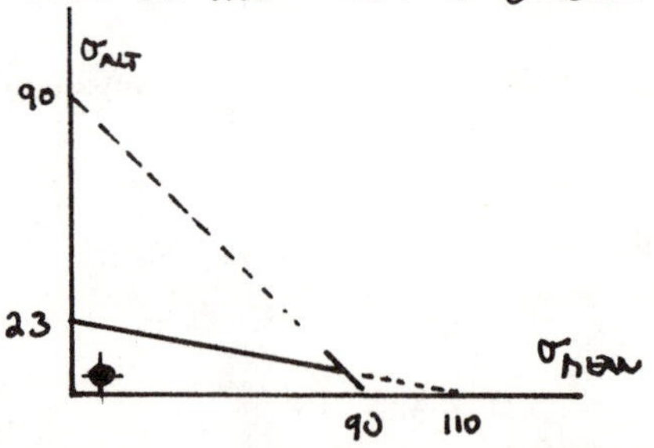

IT IS APPARENT THAT THE FACTOR OF SAFETY IS QUITE HIGH AND THE BOLTS ARE IN NO DANGER OF FAILING. HOWEVER, THE PLATE SHOULD ALSO BE CHECKED.

THE STRESS IN THE FLANGE/VESSEL IS

$$\sigma_{min} = \frac{(6)\,3700 - (1-.0143)(19860)}{47.12} = 55.69 \text{ PSI}$$

$$\sigma_{max} = \frac{6(3700) - (1-.0143)(2838)}{47.12} = 411.8$$

THESE STRESSES ARE VERY LOW

---

3 ASSUME THE DIAMETERS ARE MEAN DIAMETERS.

THE SPRING INDEXES ARE

$$C_{INNER} = \frac{D}{d} = \frac{1.5}{.177} = 8.475$$

$$C_{outer} = \frac{2.0}{.2253} = 8.877$$

FOR oil-HARDENED STEEL, G = 11.5 EE6 PSI.

ASSUME THE NUMBER OF COILS IS THE NUMBER OF TOTAL COILS. THE NUMBER OF ACTIVE COILS IS 2 LESS THAN THE TOTAL SINCE THE SPRINGS HAVE SQUARED AND GROUND ENDS.

$$N_{a,inner} = 12.75 - 2 = 10.75$$

$$N_{a,outer} = 10.25 - 2 = 8.25$$

FROM EQUATION 15.60, THE SPRING RATES (STIFFNESSES) ARE

$$k_{INNER} = \frac{(11.5\text{ EE6})(.177)}{(8)(8.475)^3(10.75)} = 38.88 \text{ LB/IN}$$

$$k_{outer} = \frac{(11.5\text{ EE6})(.2253)}{(8)(8.877)^3(8.25)} = 56.12 \text{ LB/IN}$$

SINCE THE INNER SPRING IS LONGER BY (4.5 − 3.75) = .75", THE INNER SPRING WILL ABSORB THE FIRST

$$(38.88)(.75) = 29.16 \text{ LB}$$

THE REMAINING (150 − 29.16) = 120.84 WILL BE SHARED BY BOTH SPRINGS.

THE COMPOSITE SPRING CONSTANT IS

$$(38.88) + 56.12 = 95.0 \text{ LB/IN}$$

THE DEFLECTION OF THE OUTER SPRING IS

$$\delta_{outer} = \frac{120.84}{95} = 1.272 \text{ IN}$$

THE TOTAL DEFLECTION OF THE INNER SPRING IS

$$\delta_{INNER} = .75 + 1.272 = \boxed{2.022"}$$

THE TOTAL FORCE EXERTED BY THE INNER SPRING IS

$$F = k\delta = (38.88)(2.022) = 78.62 \text{ LB}$$

THE WAHL FACTOR IS

$$W = \frac{(4)(8.475)-1}{(4)(8.475)-4} + \frac{.615}{8.475} = 1.173$$

FROM EQUATION 15.61, THE SHEAR STRESS IS

$$\tau_{INNER} = \frac{(8)(8.475)(78.62)(1.173)}{\pi\,(.177)^2} = 63{,}528 \text{ PSI}$$

ACCORDING TO THE MAXIMUM SHEAR STRESS THEORY,

$$\tau_{max} = .5\,S_{yt}$$

UNFORTUNATELY, THE YIELD STRENGTH IS NOT GIVEN. FROM TABLE 15.4 FOR OIL TEMPERED STEEL,

$$S_{UT} = 215 - \left(\frac{.177 - .15}{.20 - .15}\right)(215 - 195)$$

$$= 204 \text{ ksi}$$

FROM FOOTNOTE 17 ON PAGE 15-14,

$$S_{yt} = .75\,S_{UT} = .75(204) = 153 \text{ ksi}$$

THEN, $\tau_{max} = .5(153) = 76.5$ KSI

THE .577 MULTIPLIER FACTOR IS NOT USED BECAUSE IT WAS DERIVED FROM DISTORTION ENERGY THEORY NOT FROM THE MAXIMUM SHEAR STRESS THEORY.

THE FACTOR OF SAFETY IS

$$FS = \frac{76.5}{63.5} = \boxed{1.2}$$

ANSWER WILL VARY DEPENDING ON YOUR METHOD OF CALCULATING S_{yt}

THE INNER AND OUTER SPRINGS SHOULD BE WOUND WITH OPPOSITE DIRECTION HELIXES. THIS WILL MINIMIZE RESONANCE AND PREVENT COILS FROM ONE SPRING ENTERING THE OTHER SPRING'S GAPS.

---

4 THIS IS CLEARLY AN AUTOMOBILE DIFFERENTIAL SO BOTH OUTPUT SHAFTS MUST TURN IN THE SAME DIRECTION

THIS IS IDENTICAL TO THE FOLLOWING GEAR SET

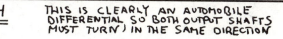

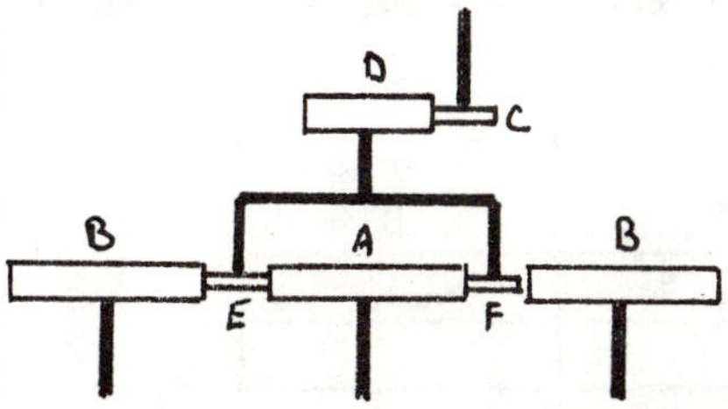

THE DIFFERENCE BETWEEN THIS AND A CONVENTIONAL EPICYCLIC GEAR TRAIN IS THAT THE RING GEAR IS REPLACED WITH TWO EXTERNAL GEARS, B.

$$\omega_C = +600 \text{ RPM}$$

$$\omega_D = \frac{-18}{54}(600) = -200 \text{ RPM}$$

$$\omega_A = -50 \text{ RPM}$$

FROM EQN 12.89

$$TV = \frac{N_B}{N_A} = \frac{30}{30} = -1$$

TV IS CLEARLY NEGATIVE. LOOK AT THE ACTUAL GEAR SET (NOT THE EQUIVALENT GEAR SET ABOVE) HOLD GEAR D STATIONARY, TURN GEAR A. THEN, GEAR B MOVES IN THE OPPOSITE DIRECTION OF GEAR A.

FROM EQUATION 15.92

$$\omega_A = (TV)\omega_B + \omega_D(1-TV)$$

$$-50 = (-1)\omega_B - 200(1-(-1))$$

$$\omega_B = -350$$

5 $\omega_r = 0$

$\omega_s = 1000$ RPM

$\omega_c = \frac{\omega_s}{3} = 333.3$ RPM

FROM EQUATION 15.92

$$1000 = (TV)\,0 + 333.3(1-TV)$$

$$TV = -2$$

FROM EQUATION 15.91

$$2 = \frac{N_{RING}}{N_{SUN}}$$

ASSUME $N_S = 40$ AND $N_R = 80$. SINCE $P = 10$, THEN $D_S = 4$, $D_R = 8$, $D_P = 2$

CHECK: $8 = 4 + (2\times 2)$ ✓

FROM EQUATION 15.93

$$\frac{20}{40} = -\frac{1000-333.3}{\omega_P - 333.3}$$

$$\boxed{\omega_P = -1000 \text{ RPM}}$$

6 FIRST, SIMPLIFY THE PROBLEM. INPUT #2 CAUSES GEAR D-E TO TURN AT

$$(-75)\left(\frac{-28}{32}\right) = 65.625 \text{ RPM (CW)}$$

SO, $\omega_D = 65.625$

NOW, FOLLOW THE PROCEDURE ON PAGE 15-17

STEP 1: GEARS A AND D HAVE THE SAME CENTER.

A	D

STEP 2:

	A	D
1	ω_{ARM}	ω_{ARM}
2		
3		

STEP 3: SINCE ALL SPEEDS ARE KNOWN CHOOSE GEAR D ARBITRARILY AS THE "UNKNOWN".

	A	D
1	ω_{ARM}	ω_{ARM}
2		
3		ω_D

STEP 4:

	A	D
1	ω_{ARM}	ω_{ARM}
2		$\omega_D - \omega_{ARM}$
3		ω_D

STEP 5: WE WANT THE RATIO $\left(\frac{\omega_A}{\omega_D}\right)$.

THE TRANSMISSION PATH D→A IS A COMPOUND MESH.

$$\omega_A = \omega_D\left(\frac{N_B N_D}{N_A N_C}\right) = \omega_D\left(\frac{(63)(56)}{(68)\,N_C}\right)$$

$$= \frac{51.8823\,\omega_D}{N_C}$$

SO, PUT $\frac{51.8823}{N_C}(\omega_D - \omega_{ARM})$ INTO COLUMN A

	A	D
1	ω_{ARM}	ω_{ARM}
2	$\frac{51.8823}{N_C}(\omega_D - \omega_{ARM})$	$\omega_D - \omega_{ARM}$
3		ω_D

STEP 6: INSERT THE KNOWN VALUES INTO ω

	A	D
1	150	150
2	$\frac{51.8823}{N_C}(65.625-150)$	$65.625-150$
3	-250	65.625

STEP 7:

ROW 1 + ROW 2 = ROW 3

$$150 + \frac{51.8823}{N_C}(65.625-150) = -250$$

SOLVING FOR N_C GIVES

$$N_C = 10.94$$

SAY, $\boxed{N_C = 11}$

RESERVED FOR FUTURE USE

Dynamics

1 $\omega = \left(\frac{\#REV}{SEC}\right)(2\pi)$

$$= \frac{(28)\frac{MI}{HR}(5280)\frac{FT}{MI}(12)\frac{IN}{FT}(2\pi)}{\pi(24)\,IN\,(3600)\frac{SEC}{HR}}$$

$$= 41.07\ \frac{RAD}{SEC}$$

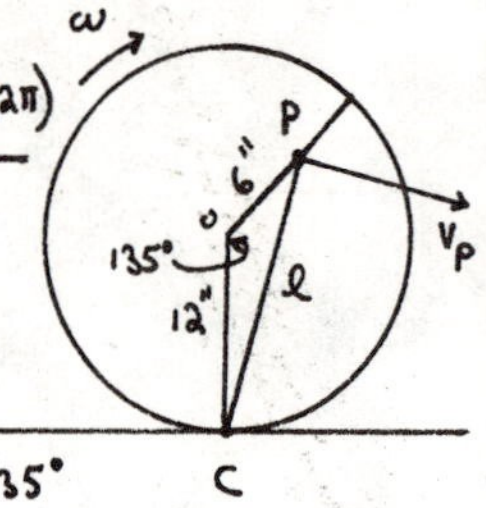

FIND DISTANCE ℓ

$$\ell^2 = (12)^2 + (6)^2 - 2(12)(6)\cos 135°$$

$$\ell = 16.79\ IN$$

$$V_p = \omega \ell = (41.07)\frac{RAD}{SEC}(16.79)\,IN = 689.6\ \frac{IN}{SEC}$$

$$= 57.46\ FPS$$

2 $x = (V_o \cos\phi)t$

$$= (900\cos\phi)t$$

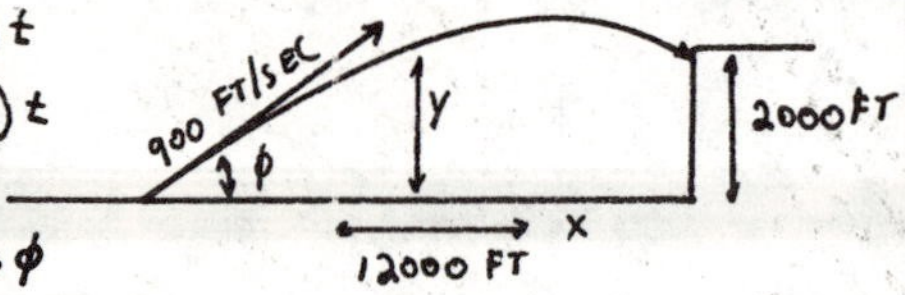

OR $t = \frac{12000}{900\cos\phi}$

$$= \frac{13.33}{\cos\phi}$$

$$y = (V_o \sin\phi)t - \tfrac{1}{2}gt^2 = (900\sin\phi)t - \left(\tfrac{1}{2}\right)(32.2)t^2$$

SUBSTITUTING $y = 2000$ AND $t = 13.33/\cos\phi$

$$2000 = \frac{900\sin\phi\,(13.33)}{\cos\phi} - (16.1)\left(\frac{13.33}{\cos\phi}\right)^2$$

$$1 = 6\tan\phi - \frac{1.43}{\cos^2\phi}$$

BUT $\frac{1}{\cos^2\phi} = 1 + \tan^2\phi$

SO $\tan^2\phi - 4.2\tan\phi + 1.7 = 0$

$$\tan\phi = \begin{Bmatrix} 3.747 \\ -.453 \end{Bmatrix} \qquad \boxed{\phi = \begin{Bmatrix} 75.06° \\ -24.37° \end{Bmatrix}}$$

CONCENTRATES

1 THE STATIC DEFLECTION IS THAT CAUSED BY THE 300 LBM MAGNET

$$\delta_{ST} = \frac{300\ LBM}{1000\ \frac{LBF}{IN}} = .3\ IN$$

$$f_{NAT} = \frac{1}{2\pi}\sqrt{\frac{g}{\delta_{ST}}} = \frac{1}{2\pi}\sqrt{\frac{386}{.3}} = \boxed{5.71\ HZ}$$

MINIMUM TENSION OCCURS AT THE UPPER LIMIT OF TRAVEL. THE DECREASE IN TENSION AT THAT POINT IS THE SAME AS THE INCREASE AT THE LOWER LIMIT CAUSED BY THE 200# OF SCRAP

$$F_{MIN} = 300 - 200 = \boxed{100\ LBF}$$

2 IF THE SPRINGS REDUCE THE FORCE FROM 25 LBF TO 3 LBF, THE TRANSMISSIVITY IS

$$TR = \frac{-3}{25} = -.12$$

{TR AND β ARE NEGATIVE ANYTIME THEY ARE <1}

THE ANGULAR FORCING FREQUENCY IS

$$\omega_f = 2\pi f = 2\pi\ \frac{RAD}{SEC}\left(\frac{1200\ \frac{REV}{MIN}}{60\ \frac{SEC}{MIN}}\right) = 125.66\ \frac{RAD}{SEC}$$

BUT $TR = \frac{1}{1 - \left(\frac{\omega_f}{\omega}\right)^2}$ SO $\omega = 41.13\ \frac{RAD}{SEC}$

FOR THE SPRING-MOUNTED MASS,

$$\omega = \sqrt{K/M} \quad SO \quad 41.13 = \sqrt{\frac{K(386)}{800}}$$

$$K = 3506.1\ \frac{LBF}{IN}$$

$$K_{SPRING} = \frac{K}{4} = 876.5\ \frac{LBF}{IN}$$

THE STATIC DEFLECTION IS

$$\delta_{ST} = \frac{W}{K} = \frac{800\ LBM}{3506.1\ \frac{LBF}{IN}} = .228\ IN$$

THE ADDED DEFLECTION DUE TO THE OSCILLATORY FORCE IS

$$\frac{3\ LBF}{(3506.1)\frac{LBF}{IN}} = \boxed{.000856\ IN}$$

3 THE MOMENT OF INERTIA OF THE ARM ABOUT POINT A IS

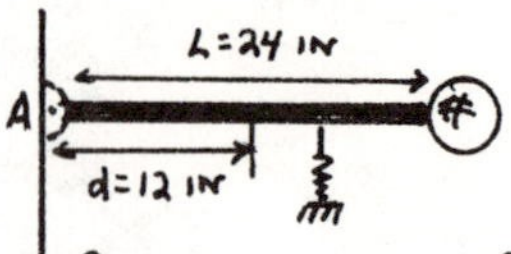

$$J_{A,ARM} = \tfrac{1}{12}ML^2 + Md^2$$

$$= \tfrac{1}{12}\left(\tfrac{5}{386}\right)(24)^2 + \left(\tfrac{5}{386}\right)(12)^2 = 2.487\ LB\text{-}IN\text{-}SEC^2$$

THE MOMENT OF INERTIA OF THE CONCENTRATED LOAD IS

$$J_{A,LOAD} = ML^2 = \left(\tfrac{3}{386}\right)(24)^2 = 4.477\ LB\text{-}IN\text{-}SEC^2$$

$$J_{total} = 2.487 + 4.477 = 6.964\ LBM\text{-}IN\text{-}SEC^2$$

THE MOMENT CAUSING DEFLECTION {WHICH IS RESISTED IN TOTAL BY THE SPRING MOMENT} IS FOUND BY SUMMING MOMENTS ABOUT POINT A.

$$\Sigma M_A\ (\circlearrowright): 3(24) + 5(12) - M_{SPRING} = 0$$

SO $M_{SPRING} = 132\ IN\text{-}LBF$

THE DEFLECTION AT THE TIP IS

$$(.55)\left(\tfrac{24}{16}\right) = .825''$$

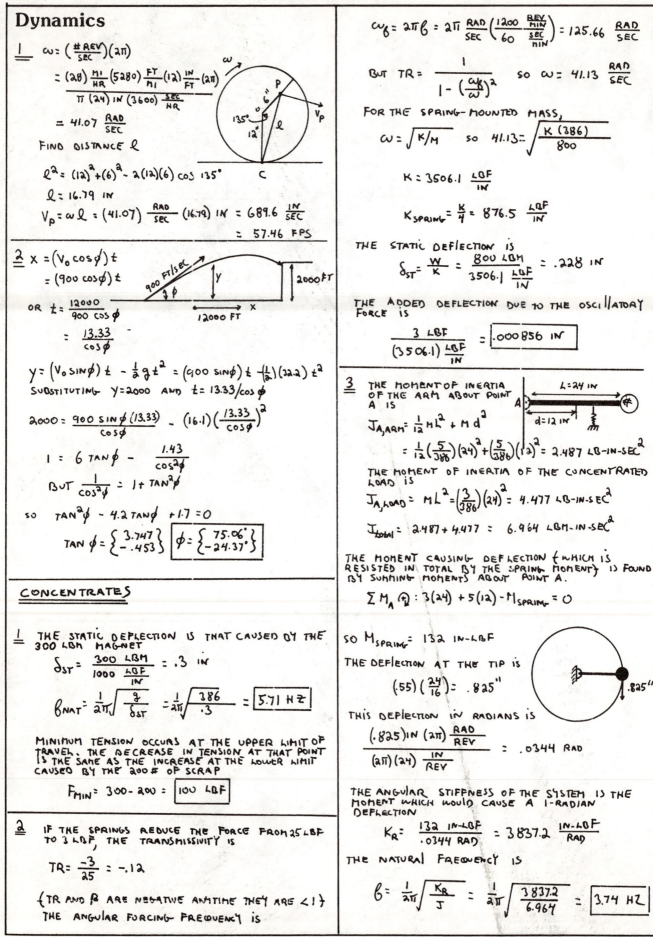

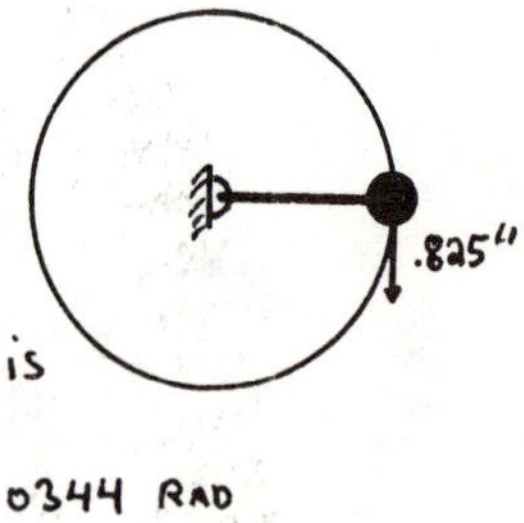

THIS DEFLECTION IN RADIANS IS

$$\frac{(.825)\,IN\,(2\pi)\frac{RAD}{REV}}{(2\pi)(24)\frac{IN}{REV}} = .0344\ RAD$$

THE ANGULAR STIFFNESS OF THE SYSTEM IS THE MOMENT WHICH WOULD CAUSE A 1-RADIAN DEFLECTION

$$K_R = \frac{132\ IN\text{-}LBF}{.0344\ RAD} = 3837.2\ \frac{IN\text{-}LBF}{RAD}$$

THE NATURAL FREQUENCY IS

$$f = \frac{1}{2\pi}\sqrt{\frac{K_R}{J}} = \frac{1}{2\pi}\sqrt{\frac{3837.2}{6.964}} = \boxed{3.74\ HZ}$$

4

$$M = \frac{(1)\,oz}{(16)\frac{oz}{lbm}(386)\frac{in}{sec^2}} = .000162\ \frac{lbm\text{-}sec^2}{in}$$

$$\omega_f = \frac{(2\pi)\frac{rad}{rev}(800)\frac{rev}{min}}{(60)\frac{sec}{min}} = 83.78\ \frac{rad}{sec}$$

$r = 5$ in

$$V_t = \omega r = (83.78)\frac{rad}{sec}(5)\,in = 418.9\ in/sec$$

$$F_c = \text{CENTRIFUGAL FORCE} = \frac{M V_t^2}{r} = \frac{(.00162)(418.9)^2}{5}$$

$$= 5.69\ lbf$$

THE NATURAL FREQUENCY IS

$$\omega = \sqrt{\frac{Kg}{W}} = \sqrt{\frac{(4)(1000)(386)}{50}} = 175.7\ \frac{rad}{sec}$$

$$\frac{C}{C_{crit}} = \frac{1}{8} \text{ AND } \beta = \frac{1}{\sqrt{\left[1-\left(\frac{\omega_f}{\omega}\right)^2\right]^2 + \left[2\left(\frac{C}{C_{crit}}\right)\left(\frac{\omega_f}{\omega}\right)^2\right]^2}}$$

so $\beta = 1.29$

THE MAGNIFIED EXCURSION IS

$$\delta = \frac{\beta F_c}{K} = \frac{(1.29)(5.69)\,lbf}{(4)(1000)\frac{lbf}{in}} = \boxed{.00184\ in}$$

TIMED

1 ASSUME "DYNAMIC FORCE" MEANS "DYNAMIC UNBALANCE"

THE ANGULAR FORCING FREQUENCY IS

$$\omega_f = \frac{(1200)\,2\pi}{60} = 125.7\ rad/sec$$

THE CENTRIFUGAL FORCE DUE TO ROTATING THIS UNBALANCE IS GIVEN BY EQUATIONS 16.20 AND 16.21

$$F_c = m a_N = m r \omega^2$$

$$= \left(\frac{3.6}{32.2}\right)\left(\frac{3}{12}\right)(125.7)^2 = 441.6\ lbf$$

THE TRANSMISSIVITY IS GIVEN AS $-.05$. (NEGATIVE BY DEFINITION BECAUSE IT IS LESS THAN ONE.)

THE NATURAL FREQUENCY CAN BE FOUND FROM EQUATION 16.110

$$TR = \frac{1}{1-\left(\frac{\omega_f}{\omega}\right)^2}$$

$$-.05 = \frac{1}{1-\left(\frac{125.7}{\omega}\right)^2}$$

$\omega = 27.43$ rad/sec

FOR A SPRING-MOUNTED MASS, THE SPRING CONSTANT CAN BE FOUND FROM

$$\omega = \sqrt{K/m}$$

$$27.43 = \sqrt{\frac{K}{\left(\frac{175}{386}\right)}}$$

{386 is g in in/sec²}

$K = 341.1$ lb/in (TOTAL)

SINCE THE 4 CORNER SPRINGS ARE IN PARALLEL,

$$K_{total} = K_1 + K_2 + K_3 + K_4$$

SO,

$$K_{each} = \frac{341.1}{4} = \boxed{85.28\ lb/in}$$

THE AMPLITUDE OF VIBRATION (PEAK TO PEAK) IS

$$x = 2\left(\frac{F}{K}\right) = \frac{2(441.6)}{341.1} = \boxed{2.59\ \text{inches}}$$

2 THE 1ST PLATE UNDER THE LOAD IS A SIMPLE BEAM

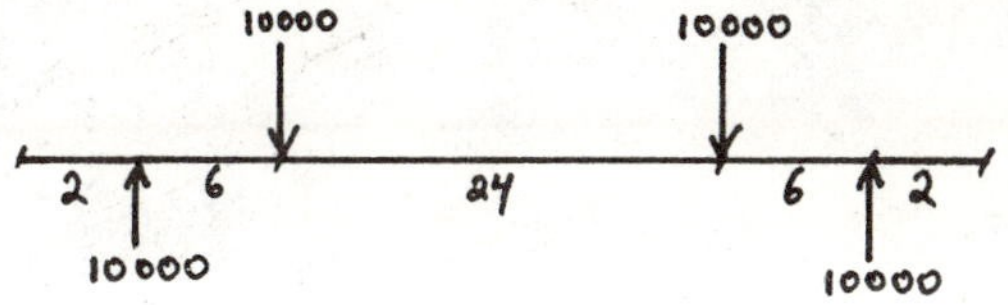

ASSUME $E = 2.9\ EE7$ psi

$$I = \frac{bh^3}{12} = \frac{(30)(1/2)^3}{12} = .3125\ in^4$$

THE DEFLECTION AT THE POINT OF LOADING IS GIVEN BY CASE 10 ON PAGE . {THE OVERHANG CONTRIBUTES NOTHING TO THE RIGIDITY.}

$$y = \frac{Fx}{6EI}\left[(3a)(L-a) - x^2\right]$$

$$= \frac{(10{,}000)(6)}{(6)(2.9\ EE7)(.3125)}\left[(3)(6)(36-6) - (6)^2\right]$$

OR, $y = .556"$

THE 2ND PLATE IS LOADED EXACTLY THE SAME AS THE TOP PLATE, ONLY UPSIDE DOWN. THEREFORE, ITS DEFLECTION IS ALSO .556".

THE TOTAL DEFLECTION OF 8 PLATES IS

$$y_{total} = (8)(.556) = \boxed{4.448"}$$

OF COURSE THIS RESULT ASSUMES THE YIELD POINT IS NOT EXCEEDED.

FOR CASE 10 AGAIN,

$$M_{max} = Fa = (10{,}000)(6) = 60{,}000\ in\text{-}lb$$

$$\sigma_{max} = \frac{M_{max}\,c}{I} = \frac{(60{,}000)(.25)}{.3125} = \boxed{48{,}000\ psi}$$

THE TOTAL SPRING CONSTANT IS

$$K = \frac{F}{x} = \frac{20{,}000}{4.448} = 4496\ lb/in$$

FROM EQUATION 16.91

$$f = \frac{1}{2\pi}\sqrt{\frac{Kg}{W}}$$

$$= \frac{1}{2\pi}\sqrt{\frac{(4496)(386)}{20{,}000}} = \boxed{1.48\ Hz}$$

RESERVED FOR FUTURE USE

Noise Control

WARM-UPS

1 METHOD 1 USE FIGURE 17.1

$\Delta L = 40 - 35 = 5$

FROM FIGURE 17.1, INCREMENT = 1.2

$L_{total} = 40 + 1.2 = 41.2$ dB

METHOD 2 USE EQUATION 17.6

$$L_{total} = 10 \log_{10}\left[10^{\left(\frac{40}{10}\right)} + 10^{\left(\frac{35}{10}\right)}\right]$$

$= 41.19$ dB

2 IF THE BACKGROUND SOUND LEVEL IS 43 dB, USE EQN 17.6

$$45 = 10 \log_{10}\left[10^{\left(\frac{43}{10}\right)} + 10^{\left(\frac{L}{10}\right)}\right]$$

$$\text{ANTILOG}\left(\frac{45}{10}\right) = 10^{4.3} + 10^{L/10}$$

$$11670.2 = 10^{L/10}$$

$$\log_{10}(11670.2) = \frac{L}{10}$$

$L = 40.67$ dB

3 BEFORE BEING ENCLOSED, THE OBSERVED SOUND LEVEL IS 100 dBA. AFTER BEING ENCLOSED, THE OBSERVED SOUND LEVEL IS $110 - 30 = 80$.

THE DIFFERENCE IS $100 - 80 = 20$ dB

4 FROM TABLE 17.2, 4 HOURS

5

6 THE GENERAL SOUND PRESSURE LEVEL IS

$$L_{total} = 10 \log_{10}\left[10^{(85/10)} + 10^{(90/10)} + 10^{(92/10)} + 10^{(87/10)} + 10^{(82/10)} + 10^{(78/10)} + 10^{(65/10)} + 10^{(54/10)}\right]$$

$= 95.6$ dB

IF THE A-WEIGHTED SOUND LEVEL IS WANTED (THE PROBLEM IS NOT SPECIFIC) THEN CORRECTIONS FROM TABLE 17.3 MUST BE ADDED TO THE MEASUREMENTS, IGNORING THE 31.5 HZ FREQUENCY,

$$L_{total} = 10 \log_{10}\left[10^{\frac{85-26.2}{10}} + 10^{\frac{90-16.1}{10}} + 10^{\frac{92-8.6}{10}} + 10^{\frac{87-3.2}{10}} + 10^{\frac{82-0}{10}} + 10^{\frac{78+1.2}{10}} + 10^{\frac{65+1}{10}} + 10^{\frac{54-1}{10}}\right]$$

$= 88.6$ dBA

7 $L_{500} = 87$

$L_{1000} = 82$

$L_{2000} = 78$

$\quad\quad\;\; 247$

$\frac{247}{3} = 82.3$

8 $\frac{S}{A} = .5$ MEANS 50% OF THE SOUND ENERGY IS REMOVED. FROM EQUATION 17.14,

$\Delta L = 10 \log\left(\frac{.5}{1}\right) = -3.01$ (DECREASE)

CONCENTRATES

1 A_W = WALL AREA $= 20(100 + 100 + 400 + 400) = 20{,}000$ FT^2

A_{CF} = CEILING + FLOOR AREA $= 2(100 + 400) = 80{,}000$ FT^2

FROM PAGE 17-9, CHOOSE NRC $= \bar{\alpha} = .02$ FOR POURED CONCRETE.

$S_1 = (20{,}000 + 80{,}000)(.02) = 2000$

AFTER TREATMENT,

$$S_2 = (.4)(.8)(20{,}000) + (.6)(.02)(20{,}000) + (.02)(80{,}000)$$

$= 8240$

FROM EQN 17.14

$\Delta L = 10 \log\left(\frac{8240}{2000}\right) = 6.15$

2 REFER TO THE PROCEDURE ON PAGE 17-8

AS FOUND

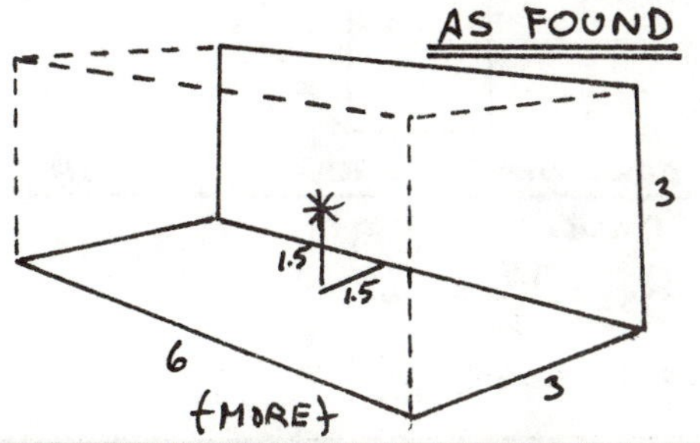

(MORE)

CONCENTRATES # 2 CONTINUED

AREA NAME	A (AREA)	B	AB
FRONT	18	1	18
TOP	18	1/3	6
SIDES	18	1/3	6
	54		30

THE EXISTING SABIN RATIO IS

$$R_{S1} = \frac{30}{54} = .556$$

AFTER BEING ENCLOSED,

AS PROPOSED

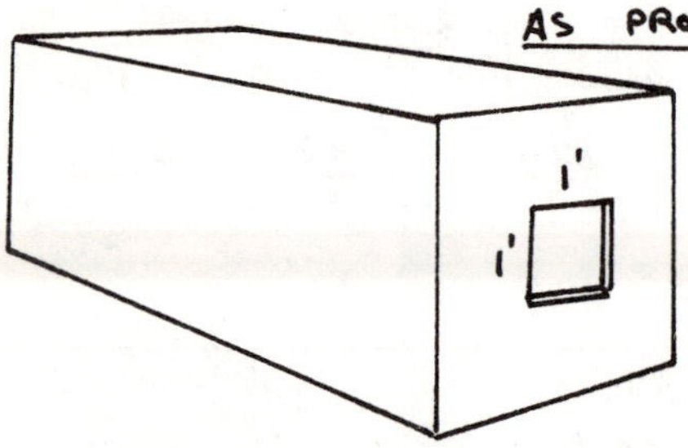

AREA NAME	AREA	B	AB
SIDE OPENINGS	2	$\frac{1}{3}$	2/3

THE NEW SABIN RATIO IS

$$R_{S2} = \frac{2/3}{2} = .333$$

SO, $\Delta L = 10 \text{ LOG}\left(\frac{.556}{.333}\right) = 2.23$

3

AS FOUND

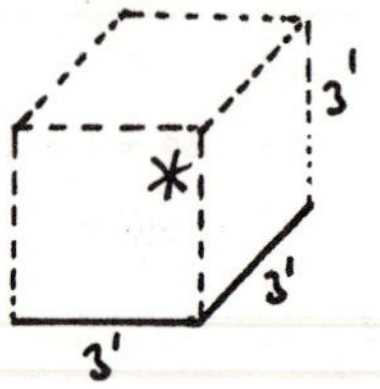

AREA NAME	AREA	B	AB
FRONT	9	1	9
BACK	9	1/6	1.5
TOP	9	1/3	3
SIDES	18	1/3	6
	45		19.5

SO $R_{S1} = \frac{19.5}{45} = .433$

AS PROPOSED

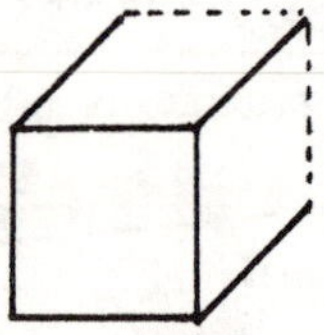

AREA NAME	AREA	B	AB
BACK	9	1/6	1.5

$$R_{S2} = \frac{1.5}{9} = .167$$

$$\Delta L = 10 \text{ LOG}\left(\frac{.433}{.167}\right) = 4.14$$

4 THE SABINS FOR EACH ABSORBING SURFACE ARE:

FLOOR: $S_1 = (1000)(.03) = 30$

CEILING (ASSUME SMOOTH) $\alpha = .03$

$S_2 = (1000)(.03) = 30$

WALLS AND GLASS (BOTH HAVE SAME $\alpha = .03$)

$A = 10(20+20+50+50) = 1400$

$S_3 = (1400)(.03) = 42$

OCCUPANTS (FROM PAGE 17-9)

ABOUT 5 SABINS EACH

$S_4 = (15)(5) = 75$

DESKS (ABOUT $1\frac{1}{2}$ SABIN EACH)

$S_5 = (15)(1\frac{1}{2}) = 22.5$

MISCELLANEOUS

$S_6 = 5$ (GIVEN)

$S_{total} = 30+30+42+75+22.5+5$
$= 204.5$

AFTER TREATMENT,

$S_2 = (.7)(1000) = 700$

SO

$S_{total} = 30+700+42+75+22.5+5$
$= 874.5$

SO $\Delta L = 10 \text{ LOG}\left(\frac{874.5}{204.5}\right) = 6.31$

5 ORIGINALLY

$A_{CEILING} = (20)(30) = 600$

$S_1 = (.5)(600) = 300$

AFTER PARTITIONING

$S_x = \frac{1}{2}(300) = 150$

$S_z = 150$

$A_o = (20)(2) = 40$

USING THE PROCEDURE ON PAGE 17-8

$$PR = \left[\frac{\frac{40}{150} + \frac{150}{150}}{1 + \frac{150}{150}}\right]\left(\frac{150}{40}\right) = 2.375$$

$$L_{IL} = 10 \text{ LOG}(2.375) = 3.76$$

6

$A_{walls} =$
$10(15+15+20+20)$
$= 700$

$A_{CEILING} = 20(15)$
$= 300$

$A_{FLOOR} = 300$

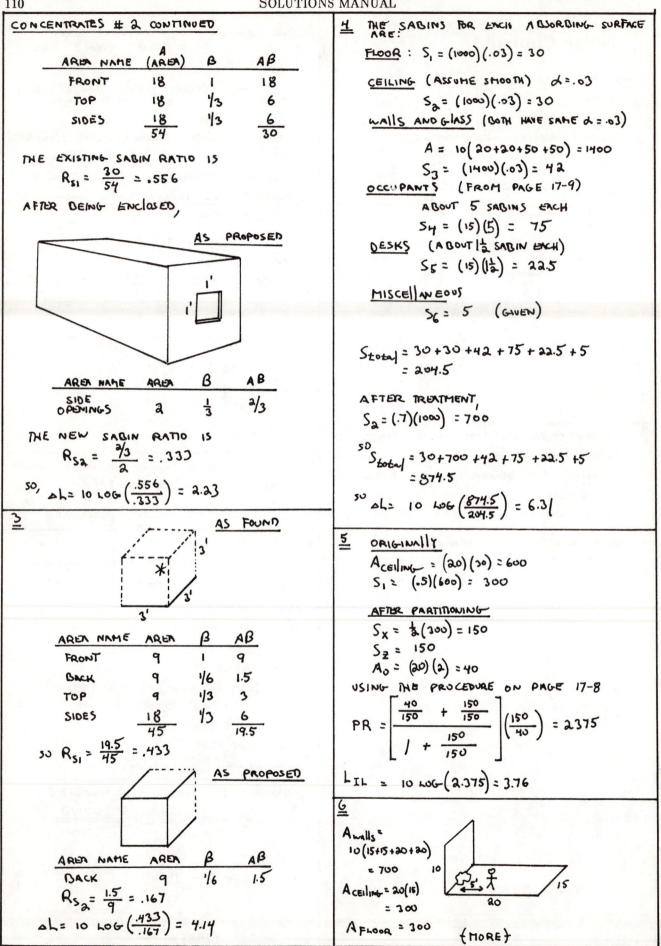

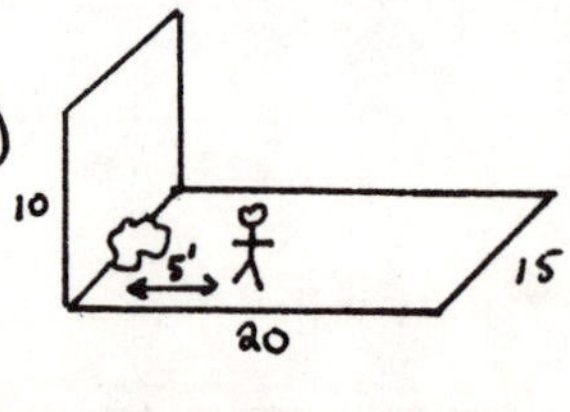

{MORE}

CONCENTRATES # 6 CONTINUED

FROM EQN 17.9

$$\bar{\alpha} = \frac{700(.06)+300(.03)+300(.5)}{700+300+300} = .155$$

FROM EQN 17.7

$$R = \frac{.155(700+300+300)}{1-.155} = 238.5$$

FROM EQN 17.10,

$$L_p = L_W + 10 \log_{10}\left[\frac{Q}{4\pi r^2} + \frac{4}{R}\right] + 10.5$$

Q = 4 BECAUSE THE SOURCE IS AT THE INTERSECTION OF 2 WALLS

$$L_p = 65 + 10 \log\left[\frac{4}{4\pi(5)^2} + \frac{4}{238.5}\right] + 10.5$$

$$= 60.19$$

ADDING IN THE BACKGROUND NOISE,

$$L_{total} = 10 \log\left[10^{\frac{60.19}{10}} + 10^{\frac{50}{10}}\right]$$

$$= 60.58$$

7 FROM THE FAN

ROTATION: $\frac{600 \text{ RPM}}{60}$ = 10 HZ

DRIVING BLADES: $\frac{(600)(8)}{60}$ = 80 HZ

FAN BLADES: $\frac{(600)(64)}{60}$ = 640 HZ

FROM THE MOTOR

ROTATION: $\frac{1725}{60} \approx$ 29 HZ

POLES: $\frac{(1725)(4)}{60}$ = 115 HZ

ELECTRICAL HUM 60 HZ

PULLEYS

MOTOR PULLEY {SAME AS MOTOR} 29 HZ

FAN PULLEY {SAME AS FAN} 10 HZ

BELT

$$\text{BELT SPEED} = \pi D (\text{RPS})$$

$$= \pi(4)\left(\frac{1725}{60}\right) = 361.3 \text{ IN/SEC}$$

$$\text{FREQUENCY} = \frac{361.3}{72} = 5 \text{ HZ}$$

8 $f_{FORCED} = \frac{1725}{60} = 28.75$ HZ

FROM EQN 16.92,

$$f_{NATURAL} = \frac{1}{2\pi}\sqrt{g/\delta} = \frac{1}{2\pi}\sqrt{\frac{386 \text{ IN/SEC}^2}{.02}} = 22.11 \text{ HZ}$$

FROM EQUATION 16.110

$$TR = \frac{1}{\left(\frac{28.75}{22.11}\right)^2 - 1} = 1.448$$

THIS IS AN INCREASE IN FORCE.

RESERVED FOR FUTURE USE

Nuclear Engineering

WARM-UPS

1

USING CARBON-BASED AMU VALUES, THE MASS INCREASE IS

$(1.007825 + 1.008665 - 2.0141) = .00239$ AMU

OR $(.00239)(931.481) = 2.226$ MeV INCREASE

THUS $2.75 - 2.226 = .5238$ MeV IS SHARED BY THE NEUTRON + HYDROGEN ATOM. SINCE BOTH HAVE THE SAME MASS THE NEUTRON WILL RECEIVE HALF $= .262$ MeV

2

$\lambda = .693/6.47 = .1071$

$.05 = EXP(-.1071\,t)$

$t = 27.97$ DAYS

3

ASSUME THE SHIELD WILL BE 10 CM THICK, FOR A 2 MeV SOURCE,

$\mu_e/\rho = .0457,$

OR $\mu_e = (.0457)(11.34) = .5182$

THEN $\mu_e X = 5.182$ AND $B = 2.78$ BY INTERPOLATION, THEN

$.01 = 2.78\ EXP[-(.5182)X]$

OR $X = 10.86$

SINCE OUR INITIAL ESTIMATE WAS CLOSE, A SECOND ITERATION IS NOT NEEDED.

$$\phi_U = (EE6)\,e^{-\mu_e X} = (EE6)\,e^{-5.182} = 5.63\ EE3\ \frac{\gamma}{cm^2\text{-}s}$$

WHICH IS UNCOLLIDED FLUX.

THE BUILD-UP FLUX IS

$\phi_B = 2.78\,\phi_U = 1.56\ EE4$

THE DOSE IS GIVEN BY $.0659\ E_0(\phi_B)\left(\frac{\mu_m}{\rho}\right)_{AIR}$

AND SINCE $\left(\frac{\mu_m}{\rho}\right)$ FOR AIR AND 2 MeV GAMMAS IS $(.0238)$

$\text{DOSE} = .0659(2)(1.56\ EE4)(.0238)$

$= 48.9$ mR/HR

5

FOR 20°C GOLD, $\sigma_a = 98$ BARNES AND $\rho = 19.32$

$$\bar{\sigma}_a = \frac{98}{1.128}\sqrt{293/273} = 82.75\ b$$

$$N = \frac{(.4909)(19.32)(6.023\ EE23)}{197} = 2.9\ EE22$$

THE ACTIVATED GOLD (Au^{198}) HAS $t_{1/2} = 2.7\ d$, SO

$\lambda = .693/2.7 = .2567\ d$

$$A = \frac{(EE8)(82.75)(EE\text{-}24)(2.9\ EE22)\left[1 - e^{(-.2567)(1)}\right]}{3.7\ EE10}$$

$= 1.468\ EE\text{-}3$ CURIE

6

FROM AN ISOTROPIC POINT SOURCE

$$\phi = \frac{S_0\,e^{-r/L}}{4\pi \bar{D} r}$$

FROM EQN 18.77

$\bar{D} = L^2 \Sigma_a$, SO

$$\bar{D} = (2.85)^2 \frac{(.66\ EE\text{-}24)(1)(6.023\ EE23)}{18} = .18$$

(ACTUALLY, $\bar{D} = .16$)

$$\phi = \frac{(EE7)\,e^{-20/2.85}}{(4\pi)(.16)20} = 2.23\ EE2\ \frac{\text{NEUTRONS}}{cm^2\text{-}s}$$

7

$\sigma_f = 4.18,\ \sigma_a = 7.68$

$P\{\text{FISSION}\} = 4.18/7.68 = .544$

8

ASSUME A SHIELD 20 CM THICK. FOR A 1 MeV GAMMA IN IRON, $\mu_e/\rho = .0595$. FOR IRON, $\rho = 7.87$ SO $\mu_e = .4683$ AND $\mu_e X = 9.366$. $B = 14.93$ BY INTERPOLATION.

$$\phi_B = \frac{(14.93)(EE8)\,e^{-.4683X}}{4\pi X^2} = \frac{1.19\ EE8\ e^{-.4683X}}{X^2}$$

THEN, THE EXPOSURE RATE IS

$$1 = \frac{.0659\,(1)(1.19\ EE8)\,e^{-.4683X}(.0280)}{X^2}$$

OR $X = 14.8$ BY TRIAL + ERROR.

CONCENTRATES

1

IF THE CONVERSION RATIO IS NOT ASSUMED, CALCULATE FOR FAST FISSION

U-238: $\sigma_a = .59 \quad \sigma_f = .5$

Pu-239: $\sigma_a = 1.95 \quad \sigma_f = 1.8$

$$\eta = .8(2.45)\left(\frac{.5}{.59}\right) + (.2)(2.95)\left(\frac{1.8}{1.95}\right) = 2.21$$

ASSUME $\epsilon = 1.05$, SO

$$CR = (2.21)(1.05) - 1 = 1.305$$

$$\text{LINEAR } t_d = \frac{(2000)(1000)}{(1.305-1)(1.23)(1000)} = 5331 \text{ DAYS}$$

$$\text{EXPONENTIAL} = (.693)(5331) = 3695 \text{ DAYS}$$

2

$$P = \frac{\bar{\phi}\Sigma_f V}{3.1\,EE10} = \frac{\left(\frac{4.5\,EE15}{3.29}\right)(.005)\left(\frac{4}{3}\pi(40)^3\right)}{3.1\,EE10}$$

$$= 5.914\,EE7 \text{ WATTS}$$

3

$$r_1 = \sqrt{(25.4)^2/\pi} = 14.33 \text{ CM}$$

AND FROM EQNS 18.119 AND 18.120

$$E = 1 + \frac{(14.33)^2}{2(59)^2}\left[\frac{\ln\left(\frac{14.33}{1.02}\right)}{1-\left(\frac{1.02}{14.33}\right)^2} - .75 + \left(\frac{1.02}{28.66}\right)^2\right] = 1.0563$$

$$F \approx 1 + \frac{1}{2}\left(\frac{r_0}{2L}\right)^2 - \frac{1}{12}\left(\frac{r_0}{2L}\right)^4 + \frac{1}{48}\left(\frac{r_0}{2L}\right)^6$$

AND $L = 1.55$ FOR NATURAL URANIUM

$$F \approx 1.0551$$

$$\Sigma_{aF} = \frac{(18.7)(6.023\,EE23)(7.68)EE{-24}(.984)}{(237.98)(1.128)} = .3171$$

$$\Sigma_{aM} = \frac{(1.6)(6.023\,EE23)(4.65\,EE{-27})}{(12)(1.128)} = 3.311\,EE{-4}$$

$$V_F \propto \pi(1.02)^2 = 3.2685$$

$$V_M = (25.4)^2 - 3.2685 = 641.8915$$

THEN $$\frac{\phi_M}{\phi_F} = 1.0551 + \left(\frac{3.2685}{641.8915}\right)\left(\frac{.3171}{3.311\,EE{-4}}\right)(.0563) = 1.33$$

AND $$1/f = 1 + (1.33)\left(\frac{3.311\,EE{-4}}{.3171}\right)\left(\frac{641.8915}{3.2685}\right) = 1.2727$$

$$f = .7857$$

RESERVED FOR FUTURE USE

Modeling of Engineering Systems

1

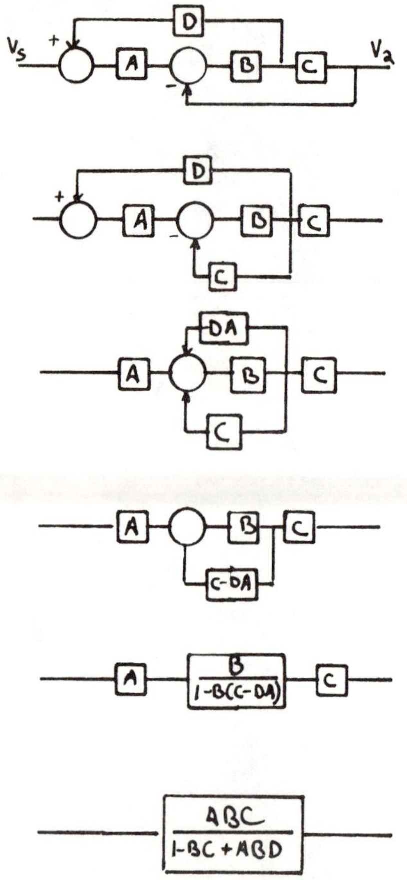

2

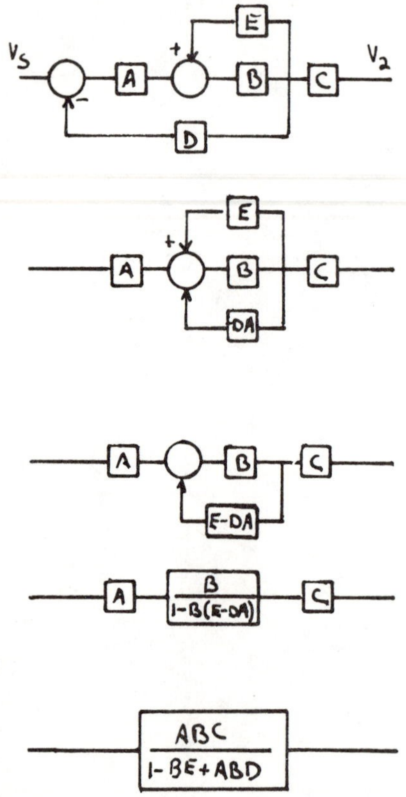

3.1

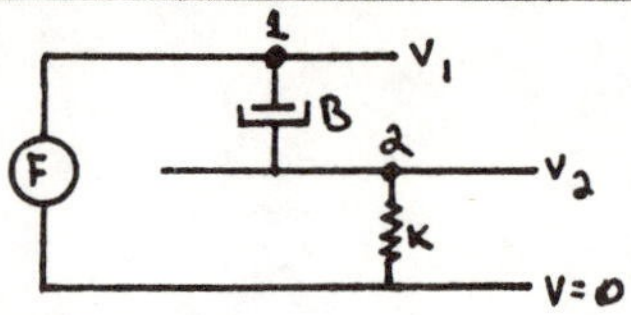

AT NODE 1

$F = B(v_1 - v_2)$ OR $F = B(x_1' - x_2')$

AT NODE 2

$F = K(x_2 - 0)$

3.2

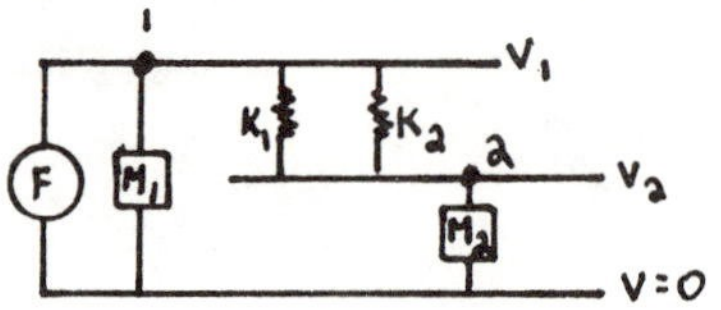

AT NODE 1

$F = M_1 a_1 + (K_1 + K_2)(x_1 - x_2)$

$= M_1 x_1'' + (K_1 + K_2)(x_1 - x_2)$

AT NODE 2

$0 = M_2 a_2 + (K_1 + K_2)(x_2 - x_1)$

$= M_2 x_2'' + (K_1 + K_2)(x_2 - x_1)$

3.3 DESPITE ITS APPEARANCES, THIS IS A ROTATIONAL SYSTEM.

τ = APPLIED ROTATIONAL TORQUE = FL

θ = ROTATED ANGLE

x_1 = ARC DISTANCE

x_2 = ARC DISTANCE

I = MOMENT OF INERTIA OF BEAM ABOUT AN END (FROM P. 9-20) $= \left(\frac{1}{3}\right) ML^2$

M = RESISTING MOMENT $= Kx_2 = K\ell\theta$

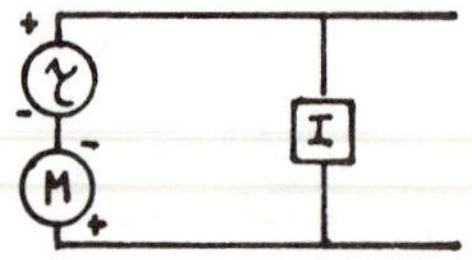

IN GENERAL, $\tau = I\alpha$, SO

$FL - K\ell\theta = \frac{1}{3} ML^2 \alpha$

BUT $\alpha = \theta''$

$\theta'' + \left(\frac{3K\ell}{ML^2}\right)\theta = FL$

3.4

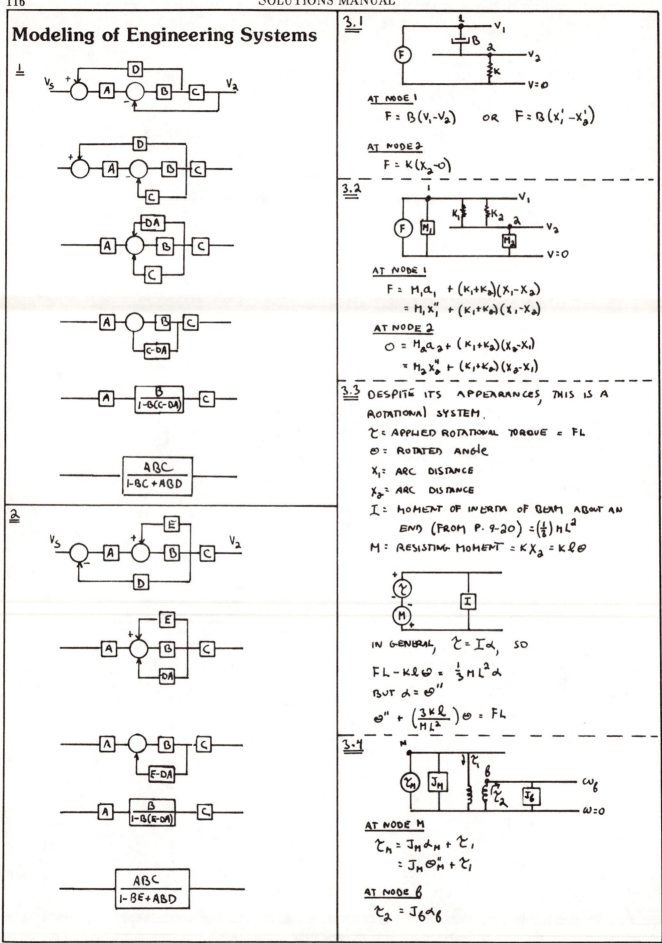

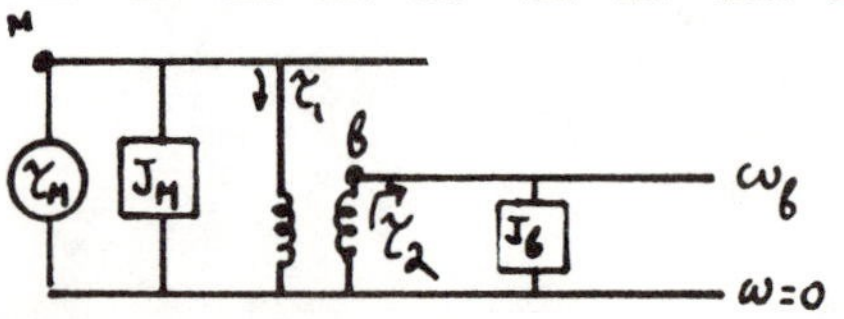

AT NODE M

$\tau_M = J_M \alpha_M + \tau_1$

$= J_M \theta_M'' + \tau_1$

AT NODE b

$\tau_2 = J_b \alpha_b$

BUT $\tau_2 = \left(\frac{N_2}{N_1}\right)\tau_1$

SO $\left(\frac{N_2}{N_1}\right)\tau_1 = J_6\,\theta_6''$

THE THIRD EQUATION REQUIRED IS

$\theta_M = \left(\frac{N_2}{N_1}\right)\theta_6$

3.7

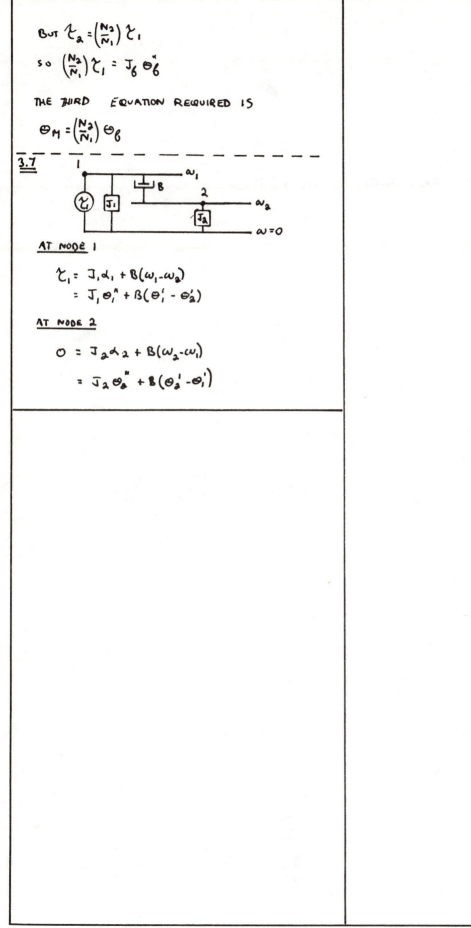

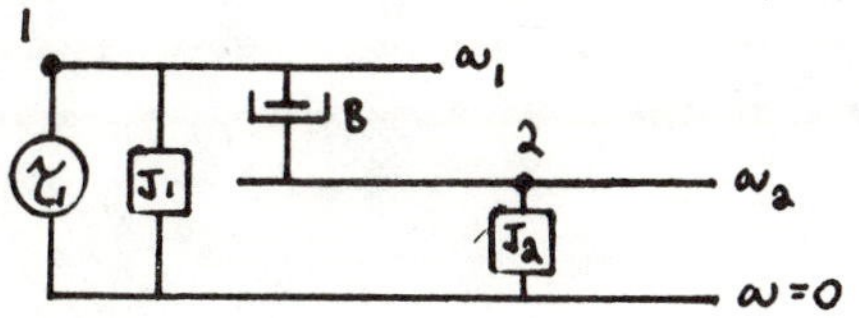

AT NODE 1

$$\tau_1 = J_1\alpha_1 + B(\omega_1 - \omega_2)$$
$$= J_1\theta_1'' + B(\theta_1' - \theta_2')$$

AT NODE 2

$$0 = J_2\alpha_2 + B(\omega_2 - \omega_1)$$
$$= J_2\theta_2'' + B(\theta_2' - \theta_1')$$

RESERVED FOR FUTURE USE

SAMPLE EXAMINATION

THIS IS AN OPEN BOOK EXAMINATION

You may use textbooks, handbooks, *bound* reference materials, any battery-operated silent calculators or slide rule. However, no (writing) tablets, unbound tables or unbound notes are permitted in the examination room. Sufficient paper for scratchwork is provided in each solution pamphlet. You are not permitted to exchange reference materials or aids during the examination.

You will have four hours in which to work each part of the test. Your score will be determined by the number of problems through four (4) you solve correctly. Each correct solution counts ten points. The maximum possible score for this part of the examination is 40 points. Partial credit for partially correct solutions will be given.

Work only four of the problems according to your proctor's instructions. Do not submit solutions or partial solutions for more than four problems.

State and justify all assumptions.

After checking your solutions and making sure that you have followed the instructions printed above, turn in all examination materials to your proctor.

You are advised to use your time effectively.

Take four hours to complete any four problems from PART 1. Do not work any problems from PART 2 during the first four hours. Take an additional four hours to complete any four problems from PART 2. Do not go back and work problems from PART 1 during the second four hours.

PART 1

1-1: A city engineer must determine the capitalized cost of perpetual service associated with constructing and maintaining a new water storage facility. The facility is to be constructed in two stages. The first water tank will have an initial cost of $40,000. The second tank will be built ten years later and will cost $60,000. When the second tank is built, a $7000 pumping station also must be provided. Pump life is 8 years, and, when removed, it is possible to replace pumps with identical units costing $7000. The annual cost of operating the pump is $800 the first year, and this cost increases $100 each year until the end of the service life for any given pump. The city has specified an interest rate of 10% for all analyses.

1-2: The output of a 500 hp turbine is transmitted at 5000 rpm through a wet clutch. The clutch consists of 40 steel plates .2″ thick, 9″ outside diameter, coefficient of friction of .1, and specific heat of .1 BTU/lbm-°F. The load is accelerated from a standstill to full speed in 30 seconds. The system must work if the cooling and lubrication system fails, as high reliability is required. Find the required shaft size, the required engagement pressure of the clutch, and the expected temperature rise during engagement.

1-3: A 2′ by 2′ square sandwich is constructed of $\frac{1}{8}''$ copper between two $\frac{1}{2}''$ layers of asbestos. The copper is dissipating 800 watts. The sandwich is placed in

a large room with air and wall temperatures of 70°F. Neglecting the end effects of the plate, find the copper and asbestos surface temperatures. The plate is placed vertically in the room. Use the following emissivities and conductivities.

copper: $k = 200$ BTU-ft/hr-ft^2-°F $e = .5$
asbestos: $k = 0.05$ BTU-ft/hr-ft^2-°F $e = .9$

1-4: An orbiting spacecraft has an argon gas manipulator for deployment of a solar panel. The arm of the manipulator is connected to a 3″ diameter piston with an 8″ stroke requiring a 36 pound thrust in one direction only. The argon gas is compressed to 1500 psig in a 0.1 ft^3 bottle. The bottle temperature is maintained at 380°R. There is a pressure regulator in the gas line. How many cycles will the bottle of argon gas provide? What will be the pressure in the gas bottle at the time of launch if the ambient conditions are 14.7 psia and 540°R?

1-5: A 2″ diameter solid 1020 cold-drawn steel shaft on simple supports carries two pulleys at a noncritical speed. The shaft ultimate strength is 69,000 psi. Yield stress is 48,000 psi. The endurance limit in complete reversal is 35,000 psi. Poisson's ratio is .283. It is required that the deflection not exceed .04 inches at any point and that the maximum angle of twist not exceed .3°. Does the shaft meet the specifications?

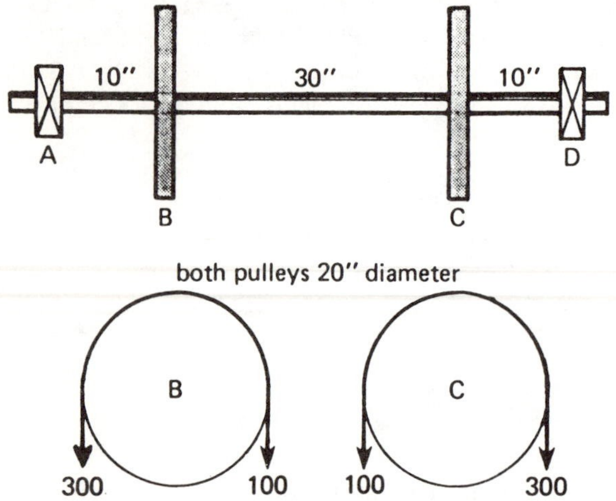

1-6: An aircraft power generator and an air conditioning system running on an open cycle use compressed air supplied by one of the main engines. The turbine performs a polytropic expansion according to $p(v)^{1.2} =$ constant. The heat exchanger uses ram air from outside the aircraft. The turbine mechanical efficiency is 85%. Find the cooling provided in tons and the turbine work output in kilowatts, and compare these quantities to those that would have been derived assuming an ideal gas and an isentropic turbine.

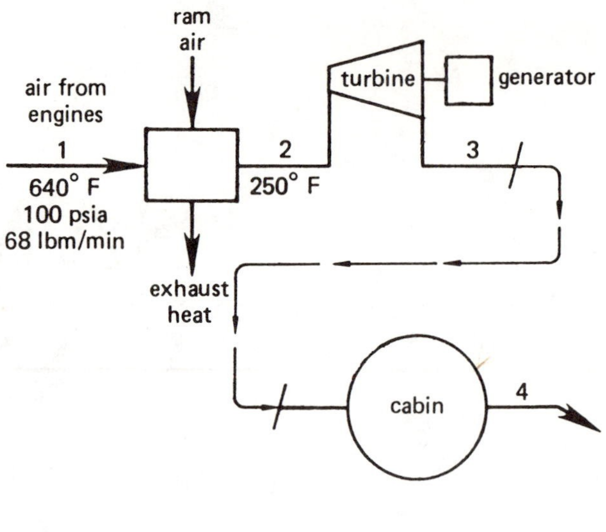

1-7: A power plant is connected to a supply of 150 psi steam at 500°F. The load is constant and requires throttling between points 1 and 2 to 100 psi in order to hold the load. The turbine exhausts to 2 psia. In order to conserve energy, it is desired to bleed off a fraction of the steam at point 3 (which is at 30 psia) for feed water heating so that the throttle can be opened full without changing the load. The isentropic efficiency in both turbine stages is 86%. What is the maximum percentage of steam which can be bled off without reducing the work output? What is the maximum temperature to which returned water at 60 psia and 60°F can be heated?

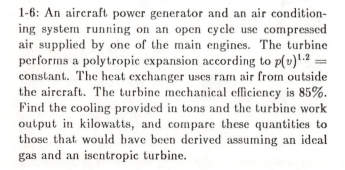

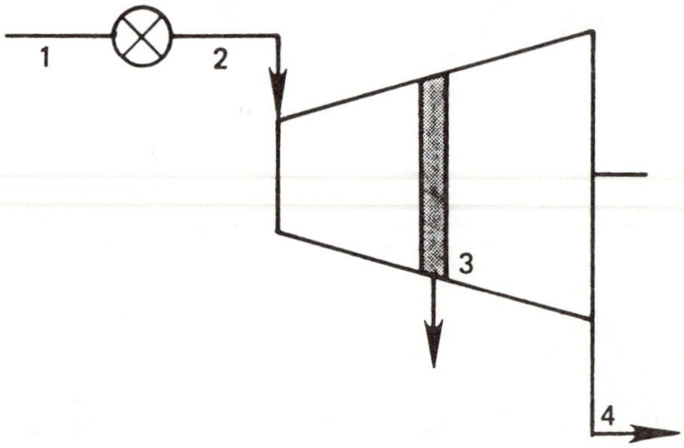

1-8: A 5′ diameter tank is partially filled with 70°F water. The atmospheric pressure is 29″ Hg. The suction line of 2″ schedule 40 commercial steel pipe with square-cut ends is inserted in the tank 4 feet below the initial surface. The suction line is 5′ long. The NPSHR of the pump attached to the suction line is given by

$$\text{NPSHR} = 0.046\,(\text{gpm})^{1.5} \text{ in feet}$$

What is the maximum flow for the pump?

1-9: A road grader with 40″ diameter wheels moves at 20 mph. The driving wheel is driven by a planetary gear train as shown. All gears have 20 teeth except for the fixed ring gear which has 60 teeth. Find the rpm of the input shaft and the overall input-to-output ratio.

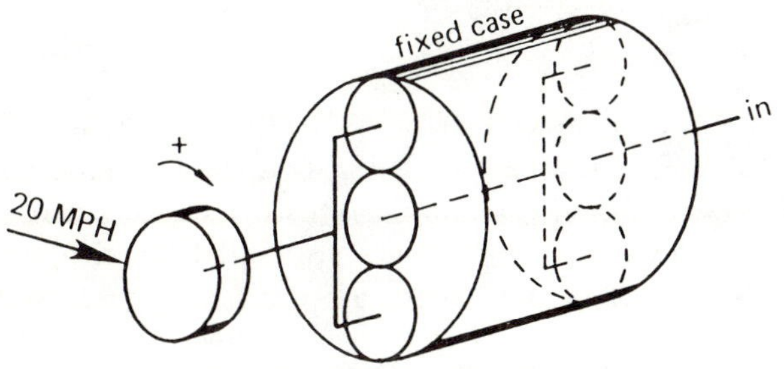

1-10: A stainless steel pressure transducer diaphragm is clamped rigidly around its circumference. The diaphragm is .5″ in diameter and has the following mechanical properties.

yield strength: 95,000 psi

elastic modulus: 2.76 EE7 psi

Poisson's ratio: .305

If a minimum deflection of .00001″ is required for electrical conductivity with the stop, and if the maximum overpressure is 2, based on the normal stress, what is the useful pressure range? What is the natural frequency of the diaphragm?

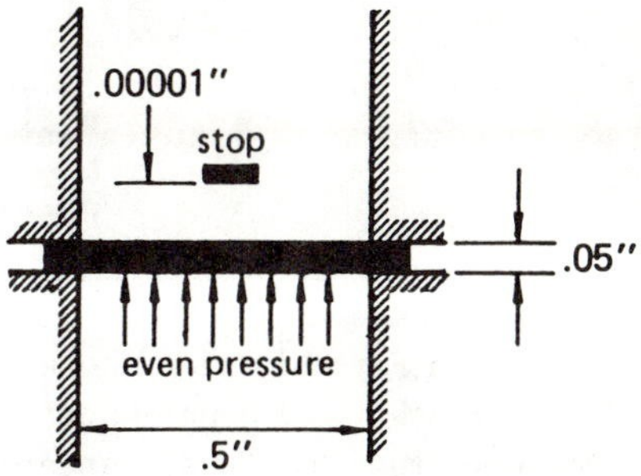

PART 2

Do not work any of the following problems during the first four hours. The following problems are to be attempted during the second four hours. You may not go back and work problems from PART 1 during the second four hours.

2-1: Previous years' records show that the cost of operating and owning a bulldozer are as given in the table below. The alternatives available are to salvage and replace at the end of the third year, salvage and replace at the end of the sixth year, or salvage and replace at the end of the eighth year. Which alternative would you specify? Use 10% interest rate.

year	item	amount
0	initial cost	$24,000
1	operation and maintenance	6,000
2	operation and maintenance	7,500
3	operation and maintenance	9,000
	SALVAGE VALUE	8,400
	overhaul	7,600
4	operation and maintenance	8,000
5	operation and maintenance	8,900
6	operation and maintenance	9,800
	SALVAGE VALUE	4,000
	overhaul	10,000
7	operation and maintenance	9,900
8	operation and maintenance	9,900
	SALVAGE VALUE	0

2-2: It is desired to use 100 gpm of saturated 210°F steam from a turbine to heat 100 gpm of cold water. The waste turbine steam temperature remains constant at 210°F in the heat exchanger to be installed. The cold water enters the heat exchanger at 70°F and is to be heated to 140°F. Copper tubes $\frac{7}{8}''$ O.D. and .75″ I.D. are to be used on a staggered 1.25″ pitch. Velocity in the tubes is to be 5 to $5\frac{1}{2}$ fps. The diameter of the shell is to be 10″. The heat exchanger is to be baffled with 25% (area) baffles spaced 6″ apart. The waste steam from the turbine is to make four passes across the tubes. After six months of operation, the fouling factor outside the tubes is .0005 and inside the tubes is .001. The shell is single pass, and the overall coefficient of heat transfer is 250 BTU/hr-ft^2-°F initially.

(a) How many tubes are required?

(b) How long is the heat exchanger allowing an extra 20% for contingency?

(c) What is the final overall coefficient of heat transfer?

(d) Verify that $U = 250$ BTU/hr-ft^2-°F initially.

2-3: Two concrete storage tanks 100′ apart are connected by a section of 8″ schedule 40 steel pipe with flanged ends. At assembly it is discovered that the pipe is .5″ too short. The flanges are bolted up with extra long bolts, and the flanges are drawn together as the nuts are tightened. The ambient temperature is 80°F. The range of temperatures seen by the pipe is 40°F to 110°F. At which temperature should proof testing be done? Will 1000 psig be a safe proof pressure? Use the following mechanical properties:

$E = 3$ EE7 psi

Poisson's ratio = .3

coefficient of linear thermal expansion = 6.5 EE−6 1/°F

yield strength = 30,000 psi

2-4: A counterflow heat exchanger has 100 ft^2 of a surface area. It is tested in clean condition, and it is found that a gas flow of 5000 lbm/hr ($c_p = 0.25$ BTU/lbm-°F) will cool from 500°F to 300°F in heating a water flow from 80°F to 200°F. After some time in service, a fouling factor of .05 hr-°F-ft^2/BTU is expected. Find the outlet temperature of the gas and water streams in the fouled condition. Assume the same flow rates as in the clean test.

2-5: A heat engine using ideal carbon dioxide goes through the following processes.

	pt	T	p
a to b: isothermal compression	a	580°R	15psia
b to c: isentropic heating	b	580	42
c to d: isothermal expansion	c	1400	42
d to a: isentropic cooling	d	1400	15

Draw the p-V and the T-s diagrams. Calculate the thermal efficiency of this process, assuming that the specific heats are constant. Calculate the thermal efficiency of this process, assuming that the specific heats vary.

2-6: A large compressor feeds a tank as shown below. The tank supplies compressed air to three outlets which run continually. 300 cfm of 90°F, 14.7 psia air enter the compressor. How long can the system run with the conditions shown if the tank is charged initially with 300 psig, 90°F air?

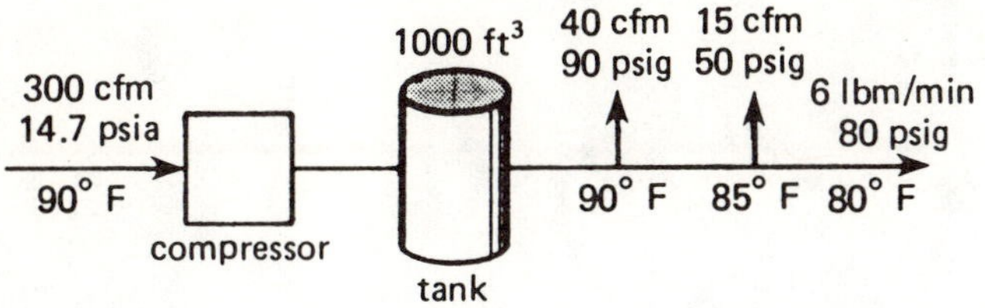

2-7: 140 gallons of fuel oil (high heating value = 20,000 BTU/lbm, specific gravity = .86, ultimate analysis: carbon −.86, hydrogen −.13, ash −.01) are burned per hour. Stack gas enters a heat exchanger at 500°F and drops to 300°F while heating air from 70°F to 140°F. If the excess air is 20% by weight, how many cfm of air are heated? What is the heat transfer surface area if $U = 4$ BTU/hr-ft²-°F?

2-8: An air conditioned building is to be kept at 76°F and 50% relative humidity when the ambient conditions are 96°F db and 76°F wb. The total heat load is 150,000 BTUH with the sensible load being 80% of the total. 800 cfm of outside air are conditioned to 58°F before entering the building. Find

—the required flow rate.
—the humidity in grains per pound of supply air.
—the ton rating of the air conditioner.

2-9: A plate clutch transmits 25 horsepower. The steel plates have a coefficient of friction of .1. The clutch and shaft start from zero speed and reach 1500 rpm in 30 seconds. The maximum diameter of the clutch is 6″. Each plate is .2″ thick. The clutch material has a density of 491 lbm/ft³ and a specific heat of .107 BTU/lbm-°F.

—What is the required shaft diameter?
—How many plates are required for the clutch?
—What will be the temperature of the clutch after one cycle of braking with a braking time equal to the acceleration time?
—Would it be practical to use the outside of the clutch as a disk brake?

Use a factor of safety of 2 in all appropriate calculations.

2-10: A tank is pressurized to 4 psig and is 5′ in diameter. The water level in the tank drops from 10 feet to 7 feet by discharging to the atmosphere through a 1″ diameter short pipe at the bottom of the tank. The pipe has square-cut ends. Ignoring pipe exit losses, how long does it take for the 3-foot change in level if the 4 psig pressurization remains constant during discharge?

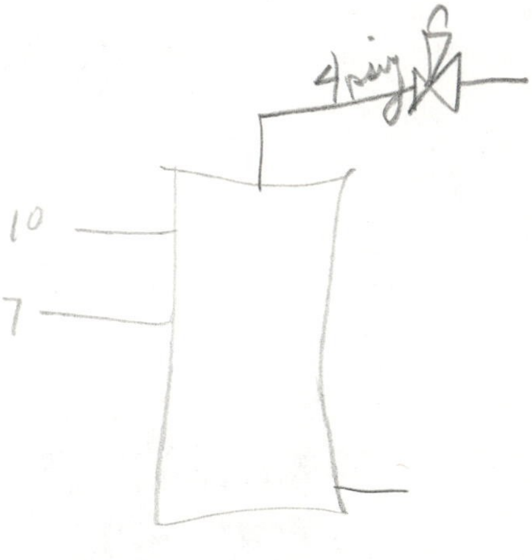

RESERVED FOR FUTURE USE

Solutions to
MECHANICAL ENGINEERING
Sample Examination

1-1 CAPITALIZED COST IS THE PRESENT WORTH OF A PERPETUAL PROJECT

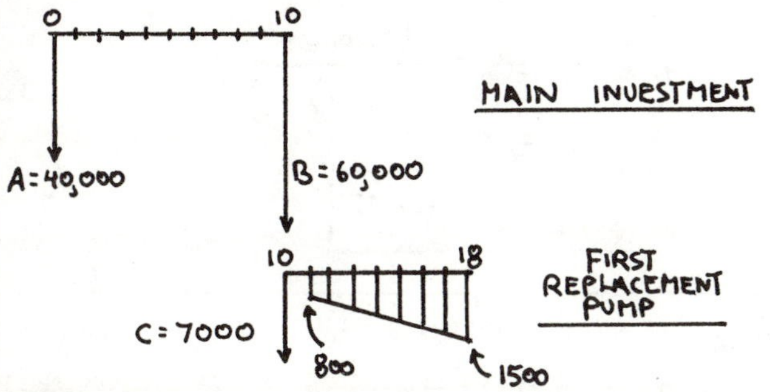

THE PRESENT WORTH OF CASH FLOWS A AND B IS

$$PW = -40{,}000 - 60{,}000(P/F,10,10)$$
$$= -40{,}000 - 60{,}000(.3855)$$
$$= -63{,}130$$

NOW WORK WITH THE FIRST REPLACEMENT PUMP, IT'S ANNUALIZED COST IS

$$-7000(A/P,10\%,8) - 800 - 100(A/G,10\%,8)$$
$$= -7000(.1874) - 800 - 100(3.0045)$$
$$= -2412.25$$

THE PRESENT WORTH AT $t=10$ OF -2412.25 EVERY YEAR FOREVER IS

$$PW(t=10) = -2412.25(P/A,10\%,\infty)$$
$$= -2412.25/.10 = -24122.5$$

THE PRESENT WORTH AT $t=0$ OF -241225 AT $t=10$ IS

$$PW(t=0) = -24122.5(P/F,10\%,10)$$
$$= -24122.5(.3855)$$
$$= -9299.22$$

THE CAPITALIZED COST IS

$$-63{,}130 - 9299.22 = \boxed{72429.22}$$

1-2 ASSUME THE SHAFT IS HARDENED AISI 1330 STEEL WITH A TENSILE YIELD STRENGTH OF 100,000 PSI, USING THE DISTORTION ENERGY THEORY, THE YIELD STRENGTH IN SHEAR IS

$$S_{ys} = (.577)(100{,}000) = 57{,}700 \text{ PSI}$$

USING A FACTOR OF SAFETY OF 4 BECAUSE OF THE HIGH RELIABILITY REQUIREMENT, THE ALLOWABLE SHEAR STRESS IS

$$\frac{57{,}700}{4} = 14{,}425 \text{ PSI}$$

FROM EQUATION 14.72, THE MAXIMUM TORSIONAL SHEAR STRESS IS

$$\tau_{MAX} = \frac{16T}{\pi d^3}$$

FROM EQUATION 14.78, THE HORSEPOWER IS

$$HP = \frac{T(2\pi)\,RPM}{33000} \quad \{T \text{ IS IN ft-lbf}\}$$

COMBIN THESE TWO EQUATIONS GIVES

$$d^3 = \frac{(16)(33000)(HP)}{(2\pi)(RPM)\,\pi\,\tau_{MAX}}$$

SO, THE REQUIRED SHAFT SIZE IS

$$d = \sqrt[3]{\frac{(16)(33000)\frac{FT\text{-}LB}{HP\text{-}MIN}(500)\,HP\,(12)\frac{IN}{FT}}{(2\pi)\frac{RAD}{REV}(5000)\frac{REV}{MIN}(\pi)(14425)\frac{LB}{IN^2}}}$$

$$d = \boxed{1.306''}$$

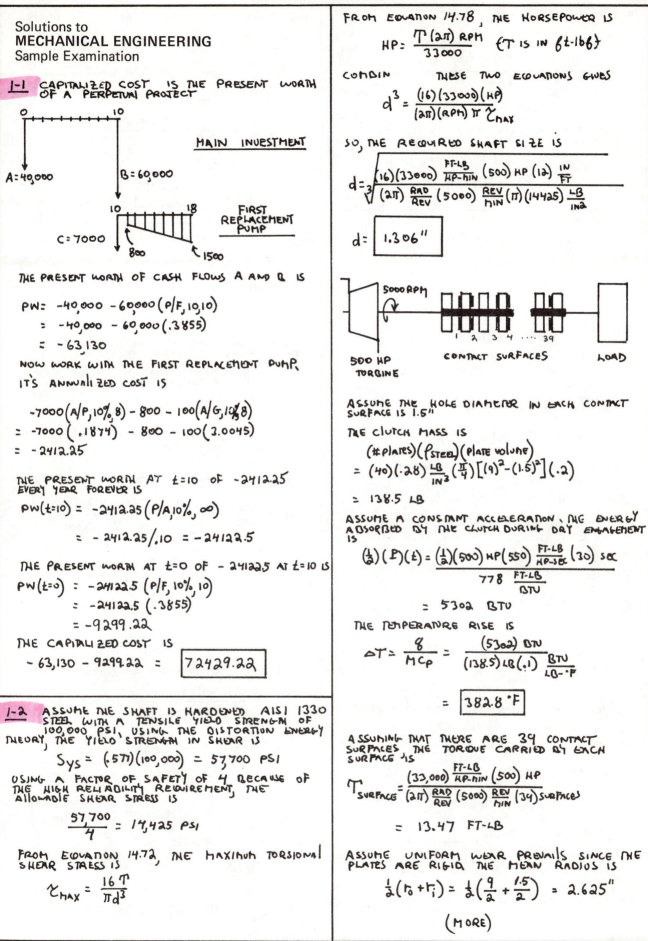

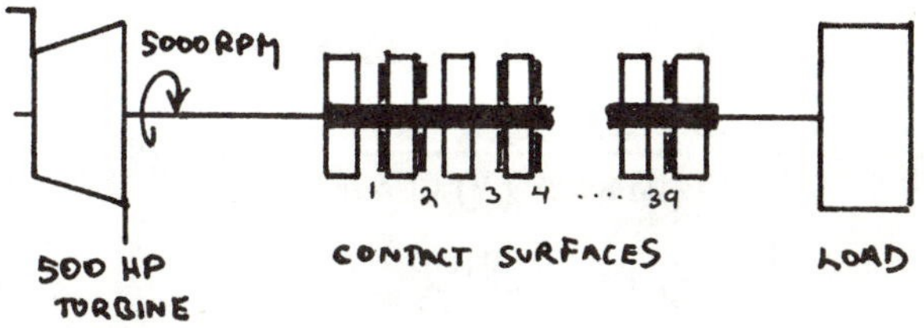

ASSUME THE HOLE DIAMETER IN EACH CONTACT SURFACE IS 1.5"

THE CLUTCH MASS IS

$$(\#\text{plates})(\rho_{STEEL})(\text{plate volume})$$
$$= (40)(.28)\frac{LB}{IN^3}\left(\frac{\pi}{4}\right)\left[(9)^2-(1.5)^2\right](.2)$$
$$= 138.5 \text{ LB}$$

ASSUME A CONSTANT ACCELERATION, THE ENERGY ABSORBED BY THE CLUTCH DURING DRY ENGAGEMENT IS

$$\left(\tfrac{1}{2}\right)(P)(t) = \frac{\left(\tfrac{1}{2}\right)(500)\,HP\,(550)\frac{FT\text{-}LB}{HP\text{-}SEC}(30)\,SEC}{778\ \frac{FT\text{-}LB}{BTU}}$$
$$= 5302 \text{ BTU}$$

THE TEMPERATURE RISE IS

$$\Delta T = \frac{q}{MC_p} = \frac{(5302)\,BTU}{(138.5)\,LB\,(.1)\frac{BTU}{LB\text{-}°F}}$$
$$= \boxed{382.8\ °F}$$

ASSUMING THAT THERE ARE 39 CONTACT SURFACES, THE TORQUE CARRIED BY EACH SURFACE IS

$$T_{SURFACE} = \frac{(33{,}000)\frac{FT\text{-}LB}{HP\text{-}MIN}(500)\,HP}{(2\pi)\frac{RAD}{REV}(5000)\frac{REV}{MIN}(39)\,\text{SURFACES}}$$
$$= 13.47 \text{ FT-LB}$$

ASSUME UNIFORM WEAR PREVAILS SINCE THE PLATES ARE RIGID THE MEAN RADIUS IS

$$\tfrac{1}{2}(r_o + r_i) = \tfrac{1}{2}\left(\frac{9}{2} + \frac{1.5}{2}\right) = 2.625''$$

(MORE)

PROBLEM 1-2, CONTINUED

THE AVERAGE FRICTIONAL FORCE IS

$$F_f = \frac{T}{r} = \frac{(13.47)\text{ FT-LB}\,(12)\text{ IN/FT}}{2.625 \text{ IN}}$$

$$= 61.6 \text{ LB}$$

SINCE THIS FRICTIONAL FORCE IS ALSO EQUAL TO (P)(A),

$$P = \frac{F}{A} = \frac{61.6 \text{ LB}}{\left(\frac{\pi}{4}\right)\left[(4)^2-(1.5)^2\right]} = \boxed{.996 \text{ PSI}}$$

THE THERMAL RESISTANCE OF THE ASBESTOS IS

$$R_t = \frac{K}{L} = \frac{.05}{\left(\frac{.5}{12}\right)} = 1.2 \frac{\text{BTU}}{\text{HR-FT}^2\text{-°F}}$$

FROM $q = \frac{\Delta T}{R_t}$

$$(341.2)\frac{\text{BTU}}{\text{HR-FT}^2} = \frac{(T_{COPPER} - 217)\text{ °F}}{(1.2)\frac{\text{BTU}}{\text{HR-FT}^2}}$$

$$\boxed{T_{COPPER} = 501.3 \text{ °F}}$$

1-3 $q_t = 800$ WATTS

$= (800)\text{ WATTS}\,(3.412)\frac{\text{BTU}}{\text{HR-WATT}}$

$= 2729.6$ BTU/HR

q PER SQUARE FOOT OF SURFACE IS

$$q = \frac{q_t}{A} = \frac{(2729.6)\text{ BTU/HR}}{(2)\text{ SIDES }(4)\text{ FT}^2\text{/SIDE}}$$

$$= 341.2 \frac{\text{BTU}}{\text{HR-FT}^2}$$

THIS HEAT TRANSFER IS THE RESULT OF CONVECTION AND RADIATION

$$q_{CONV} = h\Delta T = h(T_s - 70)$$

WHERE T_s = SURFACE TEMPERATURE

$$q_{RAD} = \epsilon\sigma\left[(T_s)^4 - (T_w)^4\right]$$

$$= (.1713 \text{ EE-8})(.9)\left[(T_s+460)^4 - (530)^4\right]$$

SO,

$$341.2 = h(T_s - 70) + (.1542 \text{ EE-8})\left[(T_s+460)^4 - (530)^4\right]$$

UNFORTUNATELY h WILL DEPEND ON T_s ALSO, SO, THIS IS AN ITERATIVE PROBLEM, ASSUME $T_s = 200$. THEN

$$T_{FILM} = \tfrac{1}{2}(200+70) = 135 \text{ °F}$$

FOR 135°F AIR,

$$N_{Pr} = .72, \quad N_{Gr} = (2)^3(200-70)(1.44 \text{ EE6})$$

$$= 1.5 \text{ EE9}$$

SO $N_{Pr}N_{Gr} = (.72)(1.5 \text{ EE9}) = 1.08 \text{ EE9}$

FROM TABLE 10.7

$$h = .19(\Delta T)^{.33} = .19(200-70)^{.33} = .947$$

BY TRIAL AND ERROR,

$$T_s = T_{ASBESTOS} = \boxed{217 \text{ °F}}$$

THIS IS CLOSE ENOUGH TO THE ORIGINAL ESTIMATE SO THAT h DOES NOT HAVE TO BE RECALCULATED

1-4 THE SWEPT PISTON VOLUME IS

$$V = AL = \left(\frac{\pi}{4}\right)(3)^2(8) = 56.55 \text{ IN}^3$$

THE REQUIRED PRESSURE AT THE PISTON FACE IS

$$P = \frac{F}{A} = \frac{36}{\left(\frac{\pi}{4}\right)(3)^2} = 5.09 \text{ PSIG}$$

ASSUME THE PRESSURE GIVEN IS WITH RESPECT TO A VACUUM, THEREFORE, PSIG = PSIA.

FOR ARGON (FROM PAGE 6-15) THE REDUCED PRESSURE AND TEMPERATURE ARE

$$\frac{P}{P_c} = \frac{1500}{705.0} = 2.13$$

$$\frac{T}{T_c} = \frac{380}{272.2} = 1.4$$

THE CORRESPONDING COMPRESSIBILITY FACTOR IS .77. THE MASS OF THE BOTTLED ARGON IS

$$M = \frac{PV}{ZRT} = \frac{(1500)\text{ LB/IN}^2(144)\text{ IN}^2\text{/FT}^2(.1)\text{ FT}^3}{(.77)(38.70)\text{ FT/°R}\,(380)\text{ °R}}$$

$$= 1.908 \text{ LB}$$

SIMILARLY, THE MASS OF ARGON WHEN THE PRESSURE HAS BEEN REDUCED TO 5.09 PSIG IS

$$M = \frac{PV}{RT} = \frac{(5.09)(144)(.1)}{(38.70)(380)}$$

$$= .005$$

THE MASS OF ARGON USED PER CYCLE IS

$$M = \frac{(5.09)(144)(56.55)}{(12)^3(38.70)(380)} = 1.63 \text{ EE-3 LB}$$

THE MAXIMUM # OF CYCLES BEFORE THE TANK CONTENTS DROP BELOW PSIG IS

$$N = \frac{1.908 - .005}{1.63 \text{ EE-3}} = \boxed{1167.5 \text{ CYCLES}}$$

A FIRST ESTIMATE OF THE PRESSURE ON THE GROUND IS

$$P = \frac{MRT}{V} = \frac{(1.908)(38.70)(540)}{(.1)(144)} = 2769 \text{ PSIA}$$

THE REDUCED PROPERTIES ARE

$$\frac{P}{P_c} = \frac{2769}{705.0} = 3.93 \qquad \frac{T}{T_c} = \frac{540}{272.2} = 1.98$$

FROM FIGURE 6.11, Z = .96 SO

$$P_{GROUND} = (.96)(2769) = 2658 \text{ PSIA}$$

1-5

$I = \frac{1}{4}\pi r^4 = \frac{1}{4}(\pi)(1)^4 = .7854 \text{ IN}^4$

ASSUME E = 3 EE7 PSI

THIS IS CASE 10 ON PAGE 14-28

$$Y_{MAX} = \frac{(400)(10)}{(3\text{ EE}7)(.7854)}\left[\frac{(50)^2}{8} - \frac{(10)^2}{6}\right]$$

$= .05''$

DOES NOT MEET DEFLECTION SPECS

THE TWIST IS

$$\Theta = \frac{TL}{GJ} \quad \text{(IN RADIANS)}$$

$T = (300-100)\left(\frac{20}{2}\right) = 2000 \text{ IN-LBF}$

$L = 30$

FROM EQUATION 14.77

$$G = \frac{E}{2(1+\mu)} = \frac{3\text{ EE}7}{2(1+.283)} = 1.17\text{ EE}7 \text{ PSI}$$

$$J = \frac{\pi d^4}{32} = \frac{\pi(2)^4}{32} = 1.571 \text{ IN}^4$$

$$\Theta = \frac{(2000)\text{ IN-LB }(30)\text{ IN}}{(1.17\text{ EE}7)\text{ PSI }(1.571)\text{ IN}^4} = 3.26\text{ EE-3 RAD}$$

$$\phi = \frac{\Theta(360)}{2\pi} = \frac{(3.26\text{ EE-3})(360)}{2\pi} = .187^\circ$$

DOES MEET TWIST SPECS

1-6 REFER TO THE PROBLEM DIAGRAM.

AT POINT 2

$\dot{M} = 68 \text{ LBM/MIN}$

$T_2 = 250^\circ F = 710^\circ R$

NEGLECT ANY PRESSURE DROP ACROSS THE HEAT EXCHANGER, OR ARBITRARILY SPECIFY 1 OR 2 PSI DIFFERENCE

$P_2 = 100 \text{ PSIA}$

$$\dot{V}_2 = \frac{WRT}{P} = \frac{(68)(53.3)(710)}{(100)(144)}$$

$= 178.7 \text{ CFM}$

$h_2 = 169.98 \text{ BTU/LBM}$ {FROM PAGE 6-35}

AT POINT 3

$$P_3(V_3)^{1.2} = P_2(V_2)^{1.2}$$

$$\dot{V}_3 = 178.7\left[\frac{100}{12}\right]^{1/1.2} = 1045.9 \text{ CFM}$$

$$T_3 = (710)\left[\frac{12}{100}\right]^{\frac{1.2-1}{1.2}} = 498.6^\circ R$$

$= 38.6^\circ F$

$h_3 = 119.1 \text{ BTU/LBM}$

THE TURBINE WORK IS

$$W = \frac{(.85)(68)\frac{\text{LBM}}{\text{MIN}}(169.98-119.1)\frac{\text{BTU}}{\text{LBM}}(17.57)\frac{\text{WATTS-MIN}}{\text{BTU}}}{(1000)\frac{\text{WATTS}}{\text{KW}}}$$

$= $ **51.67 KW**

THE COOLING EFFECT IS

$$\frac{(68)\frac{\text{LBM}}{\text{MIN}}(.24)\frac{\text{BTU}}{\text{LBM-}^\circ\text{F}}(80^\circ F - 38.6^\circ F)}{(200)\frac{\text{BTU}}{\text{MIN-TON}}}$$

$= $ **3.38 TONS**

FOR AN ISENTROPIC TURBINE,

$$W = \dot{M}C_pT_2\left[1-\left(\frac{P_3}{P_2}\right)^{\frac{K-1}{K}}\right]$$

$$= (68)(.24)(710)\left[1-\left(\frac{12}{100}\right)^{\frac{1.4-1}{1.4}}\right](.85)(.01757)$$

CONVERSION FROM BTU/MIN TO KW

$= $ **78.6 KW (ISENTROPIC)**

IF ISENTROPIC,

$$T_3 = 710\left(\frac{12}{100}\right)^{\frac{1.4-1}{1.4}} = 387.4^\circ R$$

$= -72.6^\circ F$

THE COOLING EFFECT IS

$$\frac{(68)(.24)(80-(-72.6))}{200}$$

$= $ **12.45 TONS (ISENTROPIC)**

1-7

BEFORE

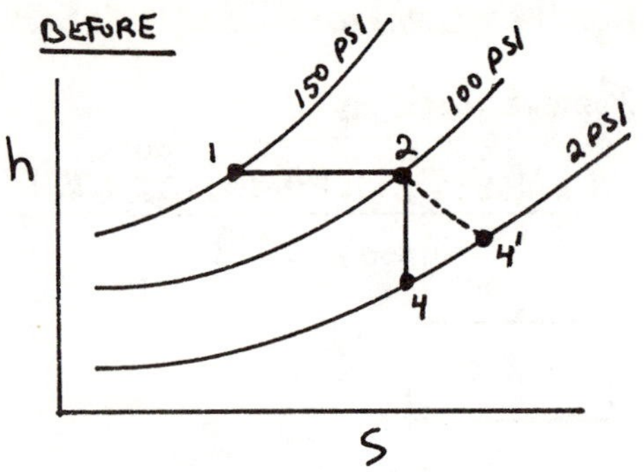

$h_1 = 1274.3$ BTU/LBM

$h_2 = 1274.3$

$S_4 = S_2$

$h_4 = 990$ BTU/LBM

$W_{IDEAL} = 1274.3 - 990 = 284.3$ BTU/LBM

$W_{ACTUAL} = .86(284.3) = 244.5$ BTU/LBM

AFTER

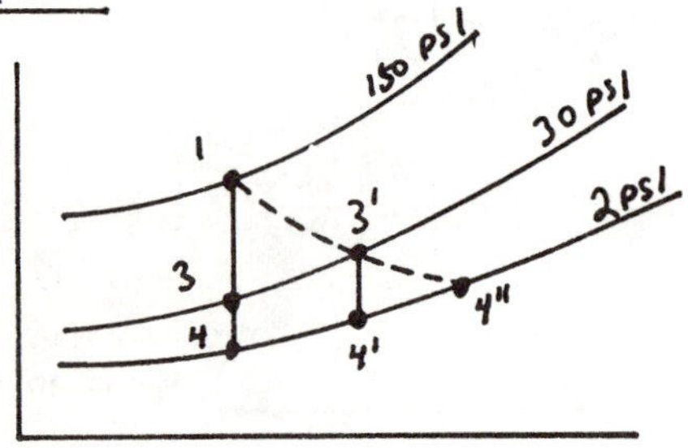

THE TURBINE WORK IS

$$W = (h_1 - h_3') + (1-x)(h_3' - h_4'')$$

$h_1 = 1274.3$

$h_3 = 1137$

$h_4 = 965$

$h_3' = h_1 - .86(h_1 - h_3) = 1156.2$

$h_4'' = h_1 - .86(h_1 - h_4) = 1008.3$

THEN, SOLVING FOR X

$$244.5 = 1274.3 - 1156.2 + (1-x)(1156.2 - 1008.3)$$

$x = .145$

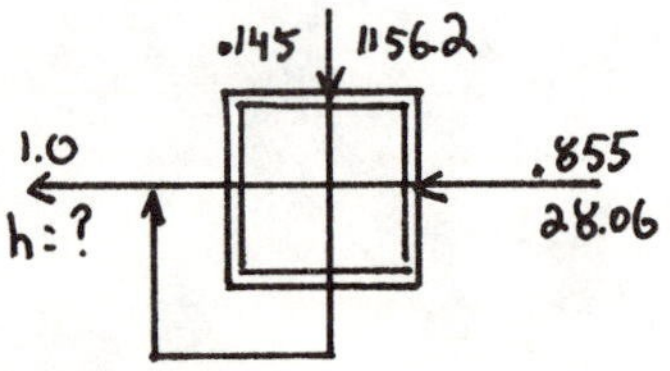

$h = .855(28.06) + .145(1156.2) = 191.64$ BTU/LB

$T \approx 223$ °F NEGLECTING DRIP PUMP WORK

1-8

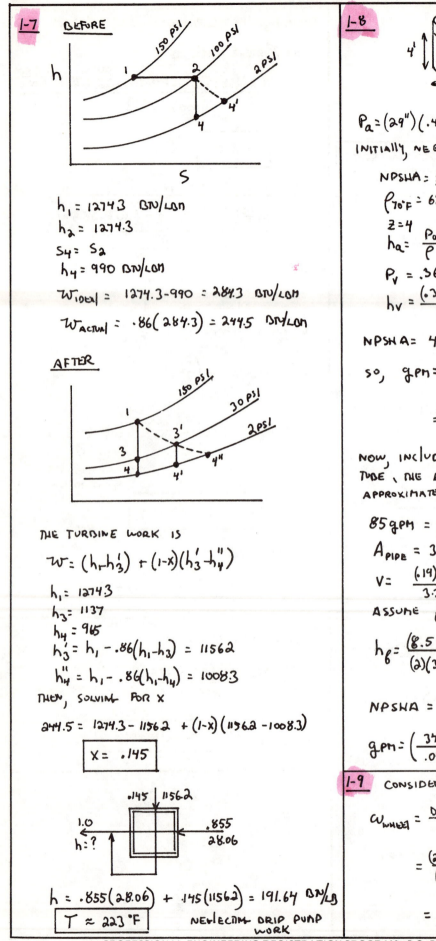

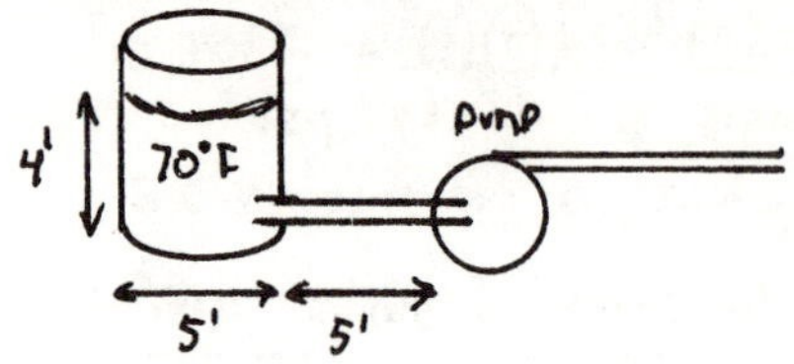

$P_a = (29'')(.491) = 14.24$ PSIA

INITIALLY, NEGLECT FRICTION AND INLET LOSSES.

$NPSHA = z + h_a - h_{vapor}$

$\rho_{70°F} = 62.3$ LBM/FT³

$z = 4$

$$h_a = \frac{P_a}{\rho} = \frac{(14.24)(144)}{62.3} = 32.91 \text{ FT}$$

$P_v = .3629$ PSIA AT 70°F

$$h_v = \frac{(.3629)(144)}{62.3} = .84$$

$NPSHA = 4 + 32.91 - .84 = 36.07$

$$\text{SO, gpm} = \left(\frac{NPSHR}{.046}\right)^{1/1.5}$$

$$= \left(\frac{36.07}{.046}\right)^{1/1.5} = 85 \text{ gpm}$$

NOW, INCLUDE FRICTION. ASSUME A REENTRANT TUBE, THE EQUIVALENT ENTRANCE LOSS IS APPROXIMATELY 8.5 FT.

$85 \text{ gpm} = (85)(.00223) = .19$ CFS

$A_{PIPE} = 3.356$ IN²

$$V = \frac{(.19)(144)}{3.356} = 8.15 \text{ FPS}$$

ASSUME $f = .02$

$$h_f = \frac{(8.5+5)(8.15)^2(.02)}{(2)(32.2)\left(\frac{2.067}{12}\right)} = 1.62 \text{ FT}$$

$NPSHA = 36.07 - 1.62 = 34.45$ FT

$$\text{gpm} = \left(\frac{34.45}{.046}\right)^{1/1.5} = \boxed{82.5 \text{ gpm}}$$

1-9 CONSIDER THE WHEEL, ITS ROTATION IS

$$\omega_{WHEEL} = \frac{\text{DISTANCE TRAVELED}}{\text{CIRCUMFERENCE}}$$

$$= \frac{(20)\frac{MI}{HR}(5280)\frac{FT}{MI}(12)\frac{IN}{FT}}{(3600)\frac{SEC}{HR}(2\pi)\frac{RAD}{REV}(20)\text{ IN}}$$

$= +2.8$ REV/SEC CLOCKWISE

(MORE)

PROBLEM 1-9 CONTINUED

THE WHEEL IS ATTACHED TO THE PLANET CARRIER OF THE FIRST STAGE, SO

$$\omega_{c,1} = 2.8$$

SINCE THE RIM GEAR IS STATIONARY,

$$\omega_{R,1} = 0$$

$$TV = -\frac{N_R}{N_S} = -\frac{60}{20} = -3$$

FROM EQUATION 15.92

$$\omega_{S,1} = TV(\omega_{R,1}) + (\omega_{c,1})(1-TV)$$

$$= (-3)(0) + (2.8)(1-(-3)) = 11.2 \ \frac{REV}{SEC}$$

BUT FOR THE SECOND STAGE,

$$\omega_{c,2} = \omega_{S,1}$$

AND

$$\omega_{S,2} = (-3)(0) + (11.2)(1-(-3)) = 44.8 \ \frac{REV}{SEC}$$

(SINCE THIS IS TWO SIMPLE PLANETARY TRAINS IN SERIES, EACH WITH THE SAME TRAIN VALUE)

$$RPM = (60)(44.8) = 2688$$

$$\text{OVERALL RATIO} = \frac{44.8}{2.8} = \boxed{16}$$

1-10 CONSIDER THIS AS A THIN PLATE WITH BUILT-IN EDGES. USE PAGE 14-24. IF $\delta = .00001$ IN, THEN

$$\delta = \frac{(3/16)\,p\,(r)^4(1-\mu^2)}{E t^3}$$

$$.00001 = \frac{(3/16)(p)(.25)^4(1-(.305)^2)}{(2.76\ EE7)(.05)^3}$$

OR

$$\boxed{p = 51.94 \ \text{PSIG}}$$

IF THE '2' MEANS A SAFETY FACTOR OF 2, THE STRESS IS

$$\frac{95{,}000}{2} = \frac{(3/4)(p)(.25)^2}{(.05)^2}$$

$$\boxed{p = 2533 \ \text{PSIG}}$$

CONSIDER THE PLATE UNDER THE ACTION OF ITS OWN DISTRIBUTED WEIGHT,

$$\rho = .295 \ \text{LB/IN}^3$$

$$t = .05''$$

SO, THE UNIFORM LOAD PER SQUARE INCH IS

$$p = (.295)(.05) = .01475 \ \text{LB/IN}^2$$

FROM PAGE 14-24, THE STATIC DEFLECTION IS

$$\delta_{ST} = \frac{\left(\frac{3}{16}\right)(.01475)(.25)^4(1-(.305)^2)}{(2.76\ EE7)(.05)^3}$$

$$= 2.84\ EE\text{-}9 \ \text{IN}$$

THEN

$$f = \frac{1}{2\pi}\sqrt{g/\delta_{ST}} = \frac{1}{2\pi}\sqrt{\frac{386}{2.84\ EE\text{-}9}}$$

$$= \boxed{58674 \ \text{HZ}}$$

2-1 THE BEST ALTERNATIVE WILL MINIMIZE THE ANNUAL COST

ALTERNATIVE #1

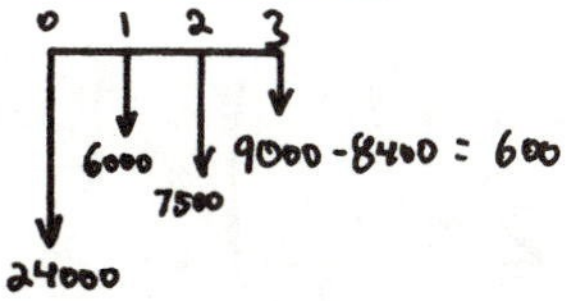

$$PW(1) = -24000 - 6000(P/F,10\%,1) - 7500(P/F,10\%,2) - 600(P/F,10\%,3)$$

$$= -24000 - 6000(.9091) - 7500(.8264) - 600(.7513)$$

$$= -36103$$

$$EUAC(1) = -36103(A/P,10\%,3)$$

$$= -36103(.4021) = \underline{-14517}$$

ALTERNATIVE #2

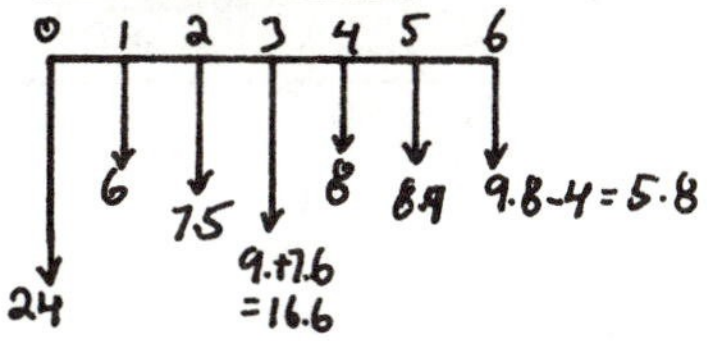

$$PW(2) = -24000 - 6000(.9091) - 7500(.8264) - 16600(.7513) - 8000(.6830) - 8900(.6209) - 5800(.5645) = -62388$$

$$EUAC(2) = -62388(A/P,10\%,6)$$

$$= -62388(.2296) = \underline{-14324}$$

ALTERNATIVE #3

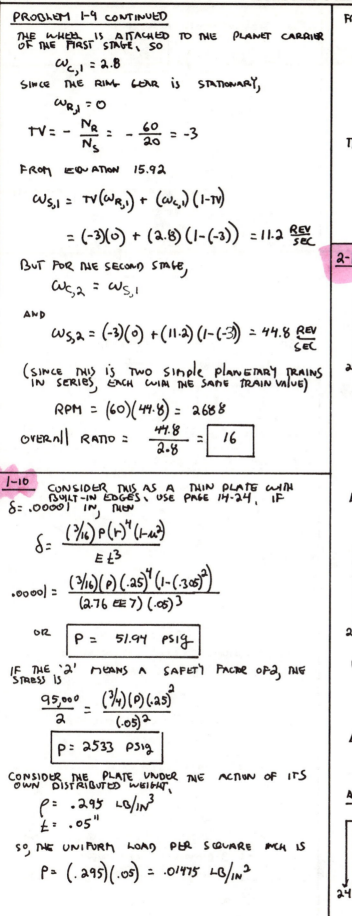

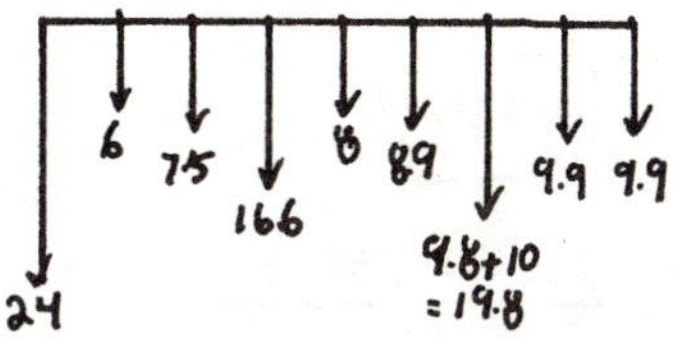

PROBLEM 2-1 CONTINUED

$$PW(3) = -24000 - 6000(.9091) - 7500(.8264) - 16600(.7513) - 8000(.6830) - 8900(.6209) - 19800(.5645) - 9900(.5132) - 9900(.4665)$$

$$= -79990$$

$$EUAC(3) = -79990(A/P, 10\%, 8) = -79990(.1874)$$

$$= \underline{-14990}$$

ALTERNATIVE #2 HAS THE LOWEST EUAC

2-2

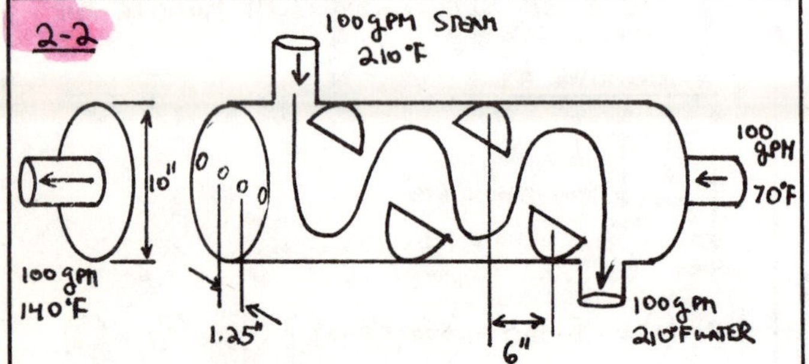

$$\Delta T_M = \frac{70-140}{\ln\frac{70}{140}} = 101\ °F$$

ASSUME THE VALUE OF U GIVEN CORRESPONDS TO A_o, THEN

$$A_o = \frac{q}{U_o \Delta T_M}$$

$$= \frac{(100)\frac{gal}{MIN}(60)\frac{MIN}{HR}(.1337)\frac{FT^3}{GAL}(.998)\frac{BTU}{LBM\text{-}°F}(62)\frac{LBM}{FT^3}(140-70)\ °F}{(250)\frac{BTU}{HR\text{-}FT^2\text{-}°F}\ (101)\ °F}$$

$$= 137.6\ FT^2$$

SINCE $V = \frac{Q}{A}$,

$$A_{PIPES} = \frac{(100)\frac{gal}{MIN}(.1337)\frac{FT^3}{GAL}(144)\frac{IN^2}{FT^2}}{(60)\frac{SEC}{MIN}(5.5)\frac{FT}{SEC}}$$

$$= 5.8\ IN^2$$

$$A_{SINGLE\ PIPE} = \left(\frac{\pi}{4}\right)(d^2) = \left(\frac{\pi}{4}\right)(.75)^2 = .442$$

$$\#\ TUBES = \frac{5.8}{.442} = 13.1$$

SAY **14 TUBES**

THEN,

$$A_{SURFACE} = 137.6\ FT^2 = \frac{(\pi)(.875)(14)\ L}{(12)}$$

$$L = 42.9'$$

ALLOW 20% EXCESS LENGTH, SO

$$L = (1.2)(42.9) = \boxed{51.5'}$$

$$\frac{1}{U_{FINAL}} = \frac{1}{250} + .001 + .0005$$

$$U_{FINAL} = \boxed{181.8}$$

NOW, CALCULATE THE VALUE OF $U_{INITIAL}$ ORIGINALLY.

ASSUME $h_{STEAM} = 2000$

$$V_{PIPE} = \left(\frac{13.1}{14}\right)5.5 = 5.15\ FPS$$

FOR THE INSIDE FILM, USE EQN 10.45

$$h_i = \frac{150(1+.011(100))(5.15)^{.8}}{(.75)^{.2}} = 1238$$

FOR COPPER, K = 220

SO

$$\frac{1}{U_o} = \frac{1}{2000} + \frac{(.875-.75)/12}{220} + \frac{1}{1238}$$

$$\boxed{U_o = 738}$$

THE VALUE OF 250 GIVEN SEEMS LOW

2-3 TEST AT 40°F SINCE THIS WILL PLACE ADDITIONAL LONGITUDINAL STRESS ON THE PIPE

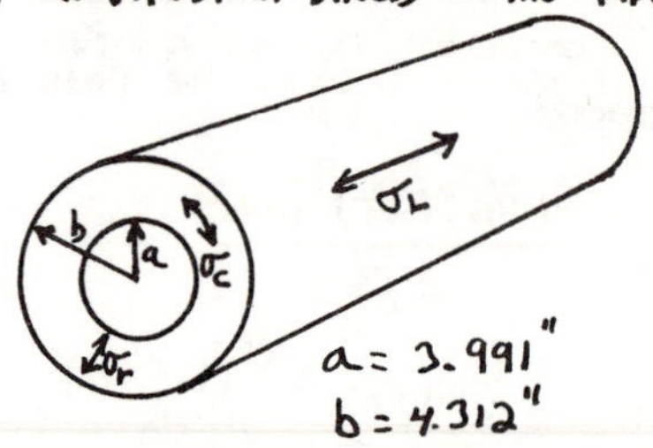

$a = 3.991"$

$b = 4.312"$

$$\epsilon = \frac{.5}{(100)(12)} = .000417$$

THE ADDITIONAL THERMAL STRAIN AT 40°F IS

$$\epsilon_{th} = \alpha \Delta T = (6.5\ EE\text{-}6)(80-40)$$

$$= 2.6\ EE\text{-}4$$

$$\sigma_{LONG} = E\epsilon = (3\ EE7)(.000417 + .00026)$$

$$= 20310\ PSI$$

$$\sigma_{ri} = -p = -1000\ PSI$$

$$\sigma_{ci} = \frac{(a^2+b^2)\,p}{b^2-a^2} = 12{,}952\ PSI$$

SINCE THERE IS NO TORSIONAL STRESS, σ_{LONG}, σ_r, AND σ_c ARE PRINCIPAL STRESSES. USE VON MISES THEORY FOR 3 DIMENSIONAL LOADING:

$$\sigma' = \sqrt{\tfrac{1}{2}\left[(20310-(-1000))^2 + (-1000-12952)^2 + (12952-20310)^2\right]}$$

$$= 18{,}747\ PSI$$

SINCE 18,747 < 30,000 THE PIPE WILL NOT YIELD AND 1000 PSI IS SAFE. HOWEVER, THIS IS A POOR DESIGN.

2-4 $A_o = 100\ FT^2$

300 ← 500

80 → 200

$$\Delta T_M = \frac{220-300}{\ln\left(\frac{220}{300}\right)}$$

$= 258°F$

$$q = UA_o \Delta T_M = \dot{M} C_p \Delta T$$

SO

$$U = \frac{(5000)\frac{LBM}{HR}(.25)\frac{BTU}{LBM\text{-}°F}(500-300)°F}{(100)\ FT^2\ (258)\ °F}$$

$= 9.69 \frac{BTU}{HR\text{-}°F\text{-}FT^2}$

AT THE BULK TEMPERATURE OF $\frac{1}{2}(80+200) = 140$, $C_p \approx 1.00$

$$\dot{M}_1 C_{p1} \Delta T_1 = \dot{M}_2 C_{p2} \Delta T_2$$

$$\dot{M}_{WATER} = \frac{(5000)(.25)(500-300)}{(1)(200-80)} = 2083\ \frac{LBM}{HR}$$

$$\frac{1}{U_{NEW}} = \frac{1}{9.69} + .05$$

OR $\boxed{U_{NEW} = 6.53}$

THE REMAINDER OF THIS PROBLEM IS AN NTU (NUMBER OF TRANSFER UNITS) PROBLEM.

$C = \dot{M} c_p$

$C_{WATER} = (2083)(1) = 2083$

$C_{GAS} = (5000)(.25) = 1250$

SO $C_{GAS} = C_{MIN}$

$$\frac{C_{MIN}}{C_{MAX}} = \frac{1250}{2083} = .6$$

$$NTU = \frac{UA}{C_{MIN}} = \frac{(6.53)(100)}{1250} = .52$$

FROM AN EFFECTIVENESS (ϵ) CHART, $\epsilon = .367$, FROM THE DEFINITION,

$$\epsilon = \frac{q}{C_{MIN}(T_{HOT,IN} - T_{COLD,IN})}$$

SO

$$q = (.367)(1250)(500-80) = 192{,}675\ BTU/HR$$

$$T_{FINAL,WATER} = 80 + \frac{192{,}675}{(2083)(1)} = 172.5\ °F$$

$$T_{FINAL,GAS} = 500 - \frac{192{,}675}{(5000)(.25)} = 345.9\ °F$$

2-5 THIS IS A CARNOT CYCLE - SEE FIGURE 7.29

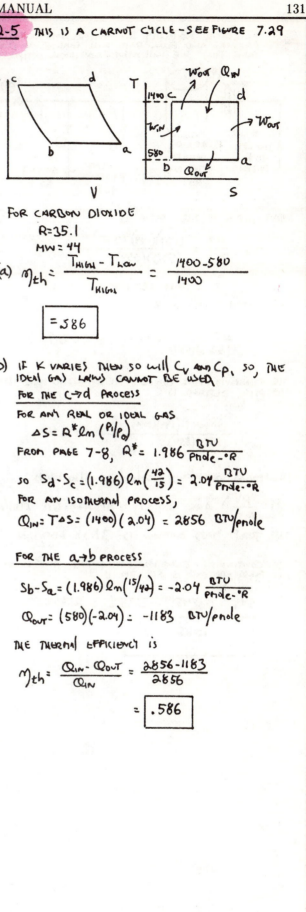

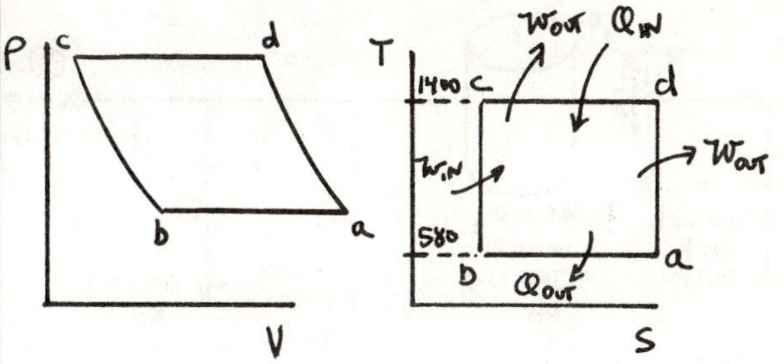

FOR CARBON DIOXIDE

R = 35.1

MW = 44

(a) $$\eta_{th} = \frac{T_{HIGH} - T_{LOW}}{T_{HIGH}} = \frac{1400-580}{1400}$$

$\boxed{= .586}$

(b) IF K VARIES THEN SO WILL C_v AND C_p, SO, THE IDEAL GAS LAWS CANNOT BE USED.

FOR THE C→d PROCESS

FOR ANY REAL OR IDEAL GAS

$$\Delta S = R^* \ln(P_1/P_2)$$

FROM PAGE 7-8, $R^* = 1.986 \frac{BTU}{PMole\text{-}°R}$

SO $S_d - S_c = (1.986)\ln\left(\frac{42}{15}\right) = 2.04 \frac{BTU}{PMole\text{-}°R}$

FOR AN ISOTHERMAL PROCESS,

$Q_{IN} = T\Delta S = (1400)(2.04) = 2856\ BTU/Pmole$

FOR THE a→b PROCESS

$S_b - S_a = (1.986)\ln(15/42) = -2.04 \frac{BTU}{PMole\text{-}°R}$

$Q_{OUT} = (580)(-2.04) = -1183\ BTU/Pmole$

THE THERMAL EFFICIENCY IS

$$\eta_{th} = \frac{Q_{IN} - Q_{OUT}}{Q_{IN}} = \frac{2856-1183}{2856}$$

$= \boxed{.586}$

2-6 ALTHOUGH NORMALLY THE COMPRESSOR OUTPUT PRESSURE AND FLOW RATE WILL CHANGE AS THE TANK FILLS, WE WILL ASSUME AN IDEAL COMPRESSOR.

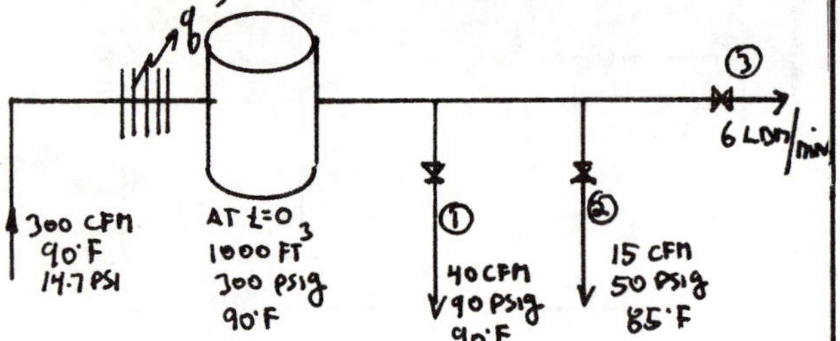

THE INPUT TO THE TANK IS

$$\dot{m} = \frac{p\dot{V}}{RT} = \frac{(14.7)(144)(300)}{(53.3)(550)}$$

$$= 21.66 \text{ LBM/MIN}$$

THE INITIAL TANK CONTENT IS

$$\frac{(300+14.7)(144)(1000)}{(53.3)(550)} = 1545.9 \text{ LBM}$$

THE MINIMUM TANK CONTENT REQUIRED TO MEET THE 90 PSIG DEMAND IS

$$\frac{(90+14.7)(144)(1000)}{(53.3)(550)} = 514 \text{ LBM}$$

THE MASS OUTFLOWS (CALCULATED ALSO FROM $\dot{m} = \frac{p\dot{V}}{RT}$) ARE 20.57, 4.81, AND 6.0 LBM/MIN.

THE TOTAL MASS OUTFLOW IS 31.38 LBM/MIN.

THE CONTINUITY EQUATION GIVING THE TIME WHEN THE TANK CONTENT IS 514 LBM IS

$$514 = 1545.9 - 31.38t + 21.66t$$

$$\boxed{t = 106.2 \text{ MIN}}$$

2-7

$$(140)\frac{\text{gal}}{\text{HR}}(.1337)\frac{\text{FT}^3}{\text{GAL}}(.86)(62.4)\frac{\text{LBM}}{\text{FT}^3} = 1004.5 \frac{\text{LBM}}{\text{HR}}$$

THE CARBON CONTENT IS

$$(.86)(1004.5) = 863.9 \frac{\text{LBM}}{\text{HR}}$$

THE HYDROGEN CONTENT IS

$$(.13)(1004.5) = 130.6 \frac{\text{LBM}}{\text{HR}}$$

THE STOICHIOMETRIC AIR REQUIREMENTS FOR THE CARBON ARE

$$(863.9)\frac{\text{LBM}}{\text{HR}}(2.67)\frac{\text{LBM}}{\text{LBM}} = 2306.6 \frac{\text{LBM}}{\text{HR}} \text{ OXYGEN}$$

$$(863.9)(8.78) = 7585 \frac{\text{LBM}}{\text{HR}} \text{ NITROGEN}$$

FOR THE HYDROGEN, THE AIR REQUIREMENTS ARE

$$(130.6)(8.0) = 1044.8 \frac{\text{LBM}}{\text{HR}} \text{ OXYGEN}$$

$$(130.6)(26.3) = 3434.8 \frac{\text{LBM}}{\text{HR}} \text{ NITROGEN}$$

THE TOTAL OXYGEN IS

$$2306.6 + 1044.8 = 3351.4 \frac{\text{LBM}}{\text{HR}} \text{ OXYGEN}$$

SIMILARLY, TOTAL NITROGEN IS

$$7585 + 3434.8 = 11019.8 \frac{\text{LBM}}{\text{HR}} \text{ NITROGEN}$$

THE EXCESS OXYGEN AND NITROGEN ARE

$$(.2)(3351.4) = 670.3$$

$$(.2)(11019.8) = 2204$$

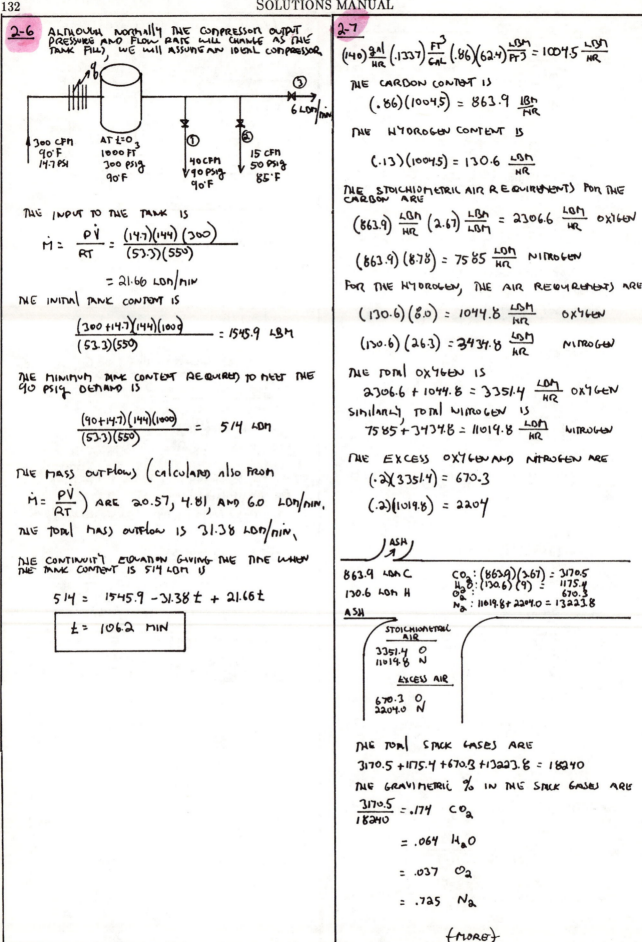

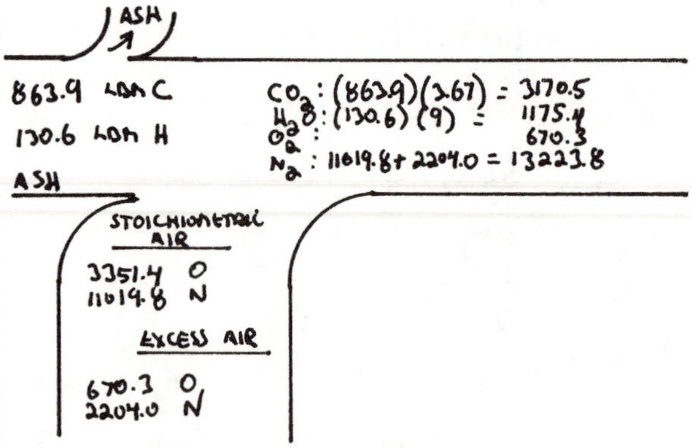

THE TOTAL STACK GASES ARE

$$3170.5 + 1175.4 + 670.3 + 13223.8 = 18240$$

THE GRAVIMETRIC % IN THE STACK GASES ARE

$$\frac{3170.5}{18240} = .174 \quad CO_2$$

$$= .064 \quad H_2O$$

$$= .037 \quad O_2$$

$$= .725 \quad N_2$$

(MORE)

PROBLEM 2-7 CONTINUED

EVALUATE THE C_p VALUES AT 400°F,

GAS	G	C_p AT 400
CO_2	.174	.23
H_2O	.064	.51
O_2	.037	.22
N_2	.725	.25

THE GRAVIMETRICALLY WEIGHTED C_p AVERAGE IS

$(.174)(.23) + (.064)(.51) + (.037)(.22) + (.725)(.25)$

$= .26$

$q = \dot{M} C_p \Delta T = (18240) \frac{LBM}{HR} (.26) \frac{BTU}{LBM\text{-}°F} (500-300)°F$

$= 948,480 \frac{BTU}{HR}$

IF ALL OF THIS HEAT IS TRANSFERRED TO THE AIR,

$q = \dot{M} C_p \Delta T$

$948,480 = (\dot{M})(.24)(140-70)$

$\dot{M}_{AIR} = 56457.1 \ LBM/HR$

ASSUMING $\rho_{AIR} = .075 \ LBM/FT^3$ (14.7 PSIA, 70°F)

THEN $Q = \frac{\dot{M}}{\rho} = \frac{56457.1}{.075} = \boxed{7.53 \ EE5 \ CFH}$

SINCE THIS IS A COUNTER FLOW HEAT EXCHANGER, THE LOGARITHMIC MEAN TEMPERATURE DIFFERENCE IS NEEDED.

$$\Delta T = 360 \begin{cases} 500 \rightarrow 300 \\ 140 \leftarrow 70 \end{cases} \Delta T = 230$$

$\Delta T_M = \dfrac{360-230}{\ln\left(\frac{360}{230}\right)} = 290°$

$q = UA \Delta T_M$, SO

$A = \dfrac{948480}{(4)(290)} = \boxed{817.2 \ FT^2}$

2-8 REFER TO THE FIGURE BELOW:

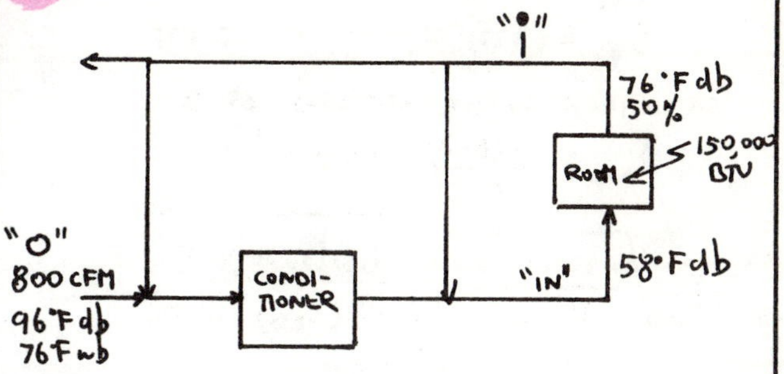

THE TEMPERATURE OF THE AIR ENTERING THE ROOM IS GIVEN AS 58°.

FROM EQUATION 11.66, THE REQUIRED FLOW RATE IS

$CFM = \dfrac{q}{C_p \rho \Delta T} = \dfrac{(.8)(150,000)(55.3)}{(60)(76-58)}$

$= \boxed{6144 \ CFM}$

FROM THE PSYCH CHART,

$h_o = 39.4 \ BTU/LBM$

$h_i = 28.7 \ BTU/LBM$

SINCE THE SENSIBLE HEAT RATIO IS GIVEN, DRAW A LINE THROUGH POINT i WITH A SLOPE OF .8. THEN, LOCATE $T_{db} = 58°F$ ON THAT LINE. THIS IS POINT "IN"

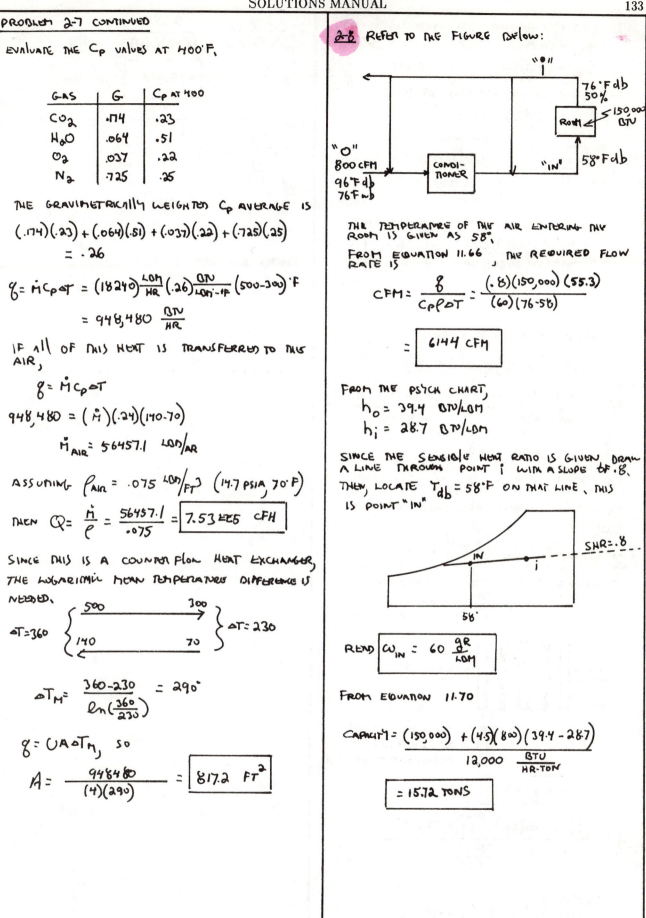

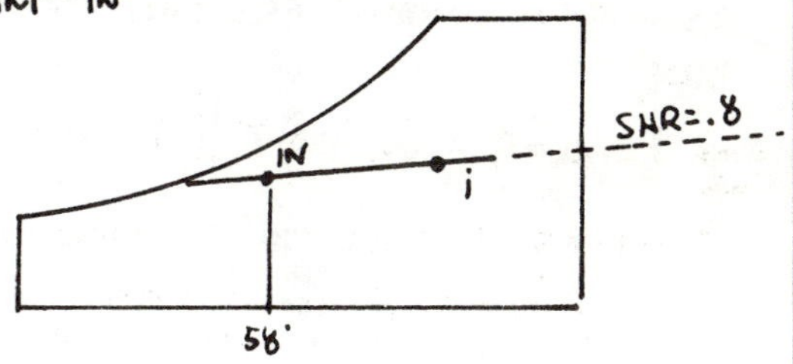

READ $\boxed{\omega_{IN} = 60 \ \dfrac{gr}{LBM}}$

FROM EQUATION 11.70

$CAPACITY = \dfrac{(150,000) + (4.5)(800)(39.4-28.7)}{12,000 \frac{BTU}{HR\text{-}TON}}$

$\boxed{= 15.72 \ TONS}$

2-9 SELECT HARDENED AISI 1040 WITH

$S_{yT} = 86$ KSI

$S_{ys} = (.577)(86{,}000) = 49{,}622$ PSI

SO, WITH A FACTOR OF SAFETY OF 2,

$$\tau_{max} = \frac{49{,}622}{2} = 24{,}811$$

$$d = \sqrt[3]{\frac{16T}{\pi \tau_{max}}} = \sqrt[3]{\frac{(16)(33{,}000)(HP)}{(\pi)(\tau_{max})(2\pi)(RPM)}}$$

$$= \sqrt[3]{\frac{(16)(33000)(12)(25)}{(\pi)^2(2)(24811)(1500)}} = \boxed{.6''}$$

CHOOSE THE INSIDE DIAMETER OF THE FRICTION SURFACE TO BE 1"

THE MEAN TORQUE-CARRYING RADIUS IS

$$r_M = \tfrac{1}{2}(r_o + r_i) = \tfrac{1}{2}\left(3 + \frac{1.0}{2}\right) = 1.75$$

$$T = \frac{(33000)(HP)(12)}{(2\pi)(RPM)} = \frac{(33000)(25)(12)}{(2\pi)(1500)}$$

$= 1050$ IN-LB

THE TANGENTIAL FORCE IS $\frac{1050}{1.75} = 600$ LBF

ASSUME A CONTACT PRESSURE OF 100 PSI, THEN,

$$F = \mu p A = (.1)(100)\,\pi\left[(3)^2 - (.5)^2\right] N = 600$$

WHERE N IS THE # OF CONTACT SURFACES

SO, N = 2.18 (SAY 3 SURFACES MINIMUM ALTHOUGH 2 SURFACES would BE A simpler DESIGN).

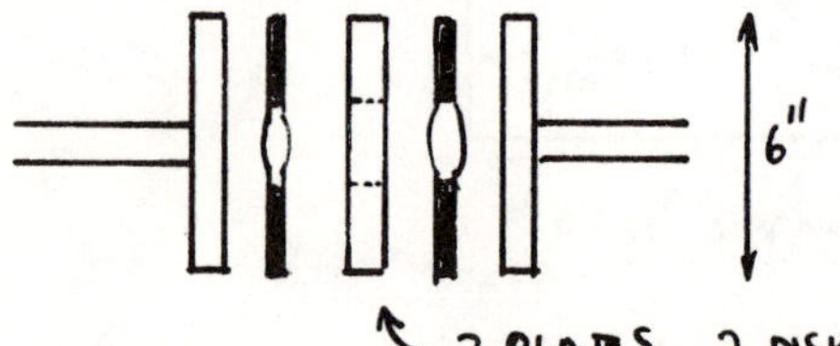

3 PLATES, 2 DISKS GIVES 4 CONTACT SURFACES, OTHER ARRANGEMENTS POSSIBLE,

$\boxed{\text{3 PLATES}}$

$$q = \frac{\left(\frac{1}{2}\right)(25)\,HP\,(550)\,\frac{FT\text{-}LBF}{HP\text{-}SEC}\,(30)\,SEC}{(778)\,\frac{FT\text{-}LBF}{BTU}}$$

$= 265.1$ BTU

$$M = \frac{(3)(\pi)\left[(3)^2 - (.5)^2\right](.2)(491)}{(12)^3} = 4.69 \text{ LBM}$$

$$\Delta T = \frac{q}{M c_p} = \frac{265.1}{(4.69)(.107)}$$

$$= \boxed{528.3°F}$$

2-10

$$h_1 = 10 + \frac{4(144)}{62.4} = 19.23'$$

$h_2 = 16.23$

$A_t = \frac{\pi}{4}(5)^2 = 19.63$ FT^2

$A_o = .00600$ (SCHED 40)

FROM EQUATION 3.97

$$t = \frac{(2)(19.63)\left(\sqrt{19.23} - \sqrt{16.23}\right)}{(.00600)\sqrt{(2)(32.2)}}$$

$$= \boxed{290.7 \text{ SEC}}$$